Solid Base Catalysis: A New Frontier in Industrial Sustainability

Edited by

Ravi Tomar

Department of Chemistry
SRM Institute of Science and Technology
Delhi NCR Campus, Modinagar
Uttar Pradesh, India
SERB-TARE Fellow
Indian Institute of Technology Delhi, India

Vivek Mishra

Amity Institute of Click-Chemistry
Research and Studies (AICCRS)
Amity University Campus, Sector-125, NOIDA-201313
SERB-TARE Fellow
University of Delhi, Delhi, India

Kapil Arya

Innovation Protection Unit (IPU)
CSIR, NISCAIR Building 3rd Floor
14 Satsang Vihar Marg, New Delhi–110067, India

&

Garima Pandey

Department of Chemistry
SRM Institute of Science & Technology
Delhi-NCR Campus, Modinagar Ghaziabad-201204, India

Solid Base Catalysis: A New Frontier in Industrial Sustainability

Editors: Ravi Tomar, Vivek Mishra, Kapil Arya & Garima Pandey

ISBN (Online): 979-8-89881-147-1

ISBN (Print): 979-8-89881-148-8

ISBN (Paperback): 979-8-89881-149-5

First published in 2025.

Bentham Science Publishers Pte. Ltd.
No. 9 Raffles Place
Office No. 26-01
Singapore 048619
Singapore
Email: subscriptions@benthamscience.net

CONTENTS

PREFACE

In recent years, the field of solid base catalysis has emerged as a groundbreaking area of research and application, driven by the growing demand for sustainable industrial processes. "***Solid Base Catalysis: A New Frontier in Industrial Sustainability***" aims to capture this dynamic and rapidly evolving domain, highlighting both recent advancements and the potential future directions of solid base catalysis.

Solid base catalysts have garnered significant attention due to their ability to facilitate a wide range of chemical reactions with high efficiency and selectivity. This versatility makes them pivotal in numerous industrial applications, from organic synthesis to biomass conversion, and even in the development of eco-friendly processes. The advancements in this field not only promise enhanced performance but also align with the broader goal of achieving sustainability in chemical manufacturing.

The book is structured to provide a comprehensive overview of the current state-of-the-art in solid base catalysis. The first chapter delves into the latest methods for synthesizing and characterizing these catalysts, offering insights into various techniques such as precipitation, sol-gel methods, and template-assisted approaches. This foundational knowledge sets the stage for subsequent discussions on more specialized applications.

Following the introduction, each chapter explores different facets of solid base catalysis, reflecting its diverse applications. From supported solid-phase catalysis in organic synthesis to innovative approaches for biomass utilization and chemical conversions, the book covers a broad spectrum of topics. Chapters dedicated to biomass oxidation, multicomponent reactions, and drug synthesis illustrate the practical implications of solid base catalysis in addressing real-world challenges.

Significant emphasis is placed on recent advancements in the field. Detailed discussions on the catalytic mechanisms, optimization strategies, and the development of novel materials offer readers a deep understanding of current trends. Chapters on C-C, C-N, and C-S bond formations, as well as reduction reactions, showcase the versatility and potential of solid base catalysts in a variety of chemical transformations.

As we move towards a more sustainable future, the role of solid base catalysis becomes increasingly critical. The final chapters reflect on the historical context and recent progress, providing a comprehensive perspective on how these catalysts contribute to greener and more efficient chemical processes.

We extend our gratitude to the contributors of this volume, whose expertise and dedication have made this book possible. Their insights and research are a testament to the vibrant and continually advancing field of solid base catalysis. We also hope that this book serves as a valuable resource for researchers, practitioners, and students alike, fostering further innovation and application in the realm of industrial sustainability.

It is our belief that the knowledge presented in these pages will inspire new ideas and drive continued progress in the field of solid base catalysis, ultimately contributing to a more sustainable and efficient chemical industry.

Ravi Tomar
Department of Chemistry
SRM Institute of Science and Technology
Delhi NCR Campus, Modinagar
Uttar Pradesh, India
SERB-TARE Fellow
Indian Institute of Technology Delhi
India

Vivek Mishra
Amity Institute of Click-Chemistry
Research and Studies (AICCRS)
Amity University Campus, Sector-125, NOIDA-201313
SERB-TARE Fellow
University of Delhi, Delhi
India

Kapil Arya
Innovation Protection Unit (IPU)
CSIR, NISCAIR Building 3rd Floor
14 Satsang Vihar Marg, New Delhi–110067
India

&

Garima Pandey
Department of Chemistry
SRM Institute of Science & Technology
Delhi-NCR Campus, Modinagar Ghaziabad-201204
India

List of Contributors

Abha R. Vyas	Department of Pharmaceutical Chemistry, L. J. Institute of Pharmacy, L. J. University, Ahmedabad, 382210, Gujarat, India
Abubakar Suraja Ibrahim	Department of Chemistry & Biochemistry, Sharda School of Engineering & Science, Sharda University, Greater Noida, UP-201310, India
Anil Kumar Teotia	Microbiology Division, Indian Pharmacopoeia Commission, Ghaziabad, Uttar Pradesh, India
Ajam Chand Shaikh	Department of Science, Alliance University, Central Campus, Chikkahadage Cross, Chandapura-Anekal, Main Road, Bengaluru, Karnataka, 562106, India
Divyani P. Patel	Department of Chemistry, Government College Daman (Affiliated to Veer Narmad South Gujarat University, Surat), Daman, U. T. of Dadra & Nagar Haveli and Daman & Diu, India
Garima Pandey	Department of Chemistry, SRM Institute of Science and Technology, Delhi NCR Campus, Modinagar Ghaziabad, U.P., India
Jaya Tuteja	School of Basic Science, Galgotias University, India
Kajal Pawar	Department of Chemistry, Institute of Chemical Technology, Matunga, Mumbai, 400019, India
Monika Kamboj	Department of Applied Chemistry, Amity School of Applied Sciences, Amity University Uttar Pradesh, Lucknow Campus, Lucknow, 226028, U.P., India
Mohan Varkolu	Department of Chemistry, Koneru Lakshmaiah Education Foundation, RVS Nagar, Azziz Nagar (PO), Moinabad Road, Hyderabad, Telangana, 500075, India
Noopur Srivastava	Department of Chemistry & Biochemistry, Sharda School of Engineering & Science, Sharda University, Greater Noida, UP, 201310, India
Nisha Saxena	PG Department of Chemistry, M.R.M. College (Lalit Narayan Mithila University), Darbhanga, Bihar, 846004, India
Neha Yadav	Legal Affairs, IP, SABIC, Bengaluru, Karnataka, India
Pavan More	Department of Chemistry, Institute of Chemical Technology, Matunga, Mumbai, 400019, India
Prasad Thota	Microbiology Division, Indian Pharmacopoeia Commission, Ghaziabad, Uttar Pradesh, India
Ruchi S. Pathak	Department of Pharmaceutical Chemistry, L. J. Institute of Pharmacy, L. J. University, Ahmedabad, 382210, Gujarat, India
Radhika N. Kachhadiya	Department of Pharmaceutical Chemistry, L. J. Institute of Pharmacy, L. J. University, Ahmedabad, 382210, Gujarat, India
Sangeeta Bajpai	Department of Applied Chemistry, Amity School of Applied Sciences, Amity University Uttar Pradesh, Lucknow Campus, Lucknow, 226028, U.P., India
Satish Kumar Singh	Department of Chemistry, Government College Daman (Affiliated to Veer Narmad South Gujarat University, Surat), Daman, U. T. of Dadra & Nagar Haveli and Daman & Diu, India
Simer Kaur	Department of Chemistry, University of Delhi, Delhi, India

T. Vinodkumar	Department of Chemistry, Central Council for Research in Ayurvedic Sciences (CCRAS), CSMCARI, Arumbakkam, Chennai, 600106, India
Trapti Aggarwal	Department of Chemistry, University of Delhi, Delhi, India
Tarnija Midha	Department of Chemistry, University of Delhi, Delhi, India
Urvashi Garg	Department of Chemistry & Biochemistry, Sharda School of Engineering & Science, Sharda University, Greater Noida, UP, 201310, India
Vishwa K. Patel	Department of Chemistry, Government College Daman (Affiliated to Veer Narmad South Gujarat University, Surat), Daman, U. T. of Dadra & Nagar Haveli and Daman & Diu, India
Vijay Bahadur	Department of Science, Alliance University, Central Campus, Chikkahadage Cross, Chandapura-Anekal, Main Road, Bengaluru, Karnataka, 562106, India

CHAPTER 1

Recent Methods for the Synthesis and Characterization of Solid-base Catalysts

Ruchi S. Pathak[1], **Abha R. Vyas**[1] and **Radhika N. Kachhadiya**[1,*]

[1] *Department of Pharmaceutical Chemistry, L. J. Institute of Pharmacy, L. J. University, Ahmedabad – 382210, Gujarat, India*

Abstract: Solid-base catalysts have acquired significant attention due to their essential role in various chemical processes, including catalytic transformations, environmental remediation, and energy conversion. This chapter highlights a comprehensive overview of current approaches to the synthesis, characterization, innovative methods, and techniques employed in the development of solid-base catalysts. A wide array of methodologies has emerged, ranging from traditional techniques to cutting-edge approaches, facilitating the design and optimization of solid-base catalysts. The synthesis section discusses novel approaches such as sol-gel hydrothermal, nanoparticle immobilization, impregnation, metal-organic framework, fly ash technique, template-assisted techniques, and conventional methods like generation of the basic site by pretreatment at high temperature. Each technique offers unique advantages in controlling catalyst morphology, composition, and surface properties. Furthermore, recent developments in characterization techniques like X-ray diffraction, Fourier-transform infrared spectroscopic, NMR, Temperature programmed desorption, microscopic and surface analysis methods such as indicator method, *etc.*, have enabled detailed insights into the physicochemical properties and active sites of catalysts and structure-activity relationships governing the catalytic performance of solid-base materials. Moreover, computational methods play an essential role in predicting and optimizing the catalytic performance of these materials. By summarizing these recent methodologies, this chapter aims to provide valuable insights into the advancements in solid-base catalyst development, paving the way for enhanced catalytic efficiency and sustainability in various chemical processes.

Keywords: Morphological analysis, Solid-base catalysts, Spectroscopic techniques, Structural analysis, Synthetic methods.

INTRODUCTION

Solid base catalysts are materials that can donate electrons or are capable of abstracting protons from reactant molecules based on their types, like Bronsted-

* **Corresponding Author Radhika N. Kachhadiya:** Department of Pharmaceutical Chemistry, L. J. Institute of Pharmacy, L. J. University, Ahmedabad – 382210, Gujarat, India; E-mail: radhika24patel6@gmail.com

Ravi Tomar, Vivek Mishra, Kapil Arya & Garima Pandey (Eds.)

Lowry Bases, Arrhenius bases, and Lewis bases. Metal oxides, Hydroxides, Carbonates, and supported alkali/alkaline earth metals are examples of Solid base catalysts used for many base-catalyzed reactions [1-4]. Solid base catalysts can be synthesized through conventional methods like basic site generation by pretreatment at high temperatures, precipitation methods, nanoparticle immobilization, and impregnation. Recently, several advanced methods like metal-organic frameworks [5,6], sol-gel, hydrothermal method, one-pot synthesis, [7, 8], and Fly ash have been widely used for the synthesis of solid base catalysts [8-14].

It is crucial to determine the main characteristics of surfaces to predict their potential as base catalysts. This can be easily achieved by having control over crucial parameters which optimize the catalytic process. The number of basic sites, its origin, and its strength are the main factors that define the basicity of any solid base catalyst. Over the past decade, significant advancement has been made in the methods pertaining to the characterization of solid base catalysts. The structural analysis of the base catalyst uses the Brunauer Emmett-Teller technique and X-ray diffraction analysis. Thermal analysis is conducted using Thermogravimetric Analysis (TGA), Differential Scanning Calorimetry (DSC), Differential Thermal Analysis (DTA), Temperature-Programmed Reduction (TPR), and Temperature-Programmed Desorption (TPD) techniques. Spectroscopy techniques include FT-IR and NMR spectroscopy.

This chapter illustrates the impact of new supports, on the one hand, recent catalyst synthesis methods and characterization techniques of solid base catalysts are also discussed.

Synthetic Methods of Solid Base Catalysts

Generation of the Basic Site by Pretreatment at High Temperature

Usually, the surfaces of basic materials are layered with water and CO_2 in the air. Because of that reason, they do not exhibit their catalytic activities. The basic site on the surface of various materials can be generated by high-temperature pretreatment. The high-temperature pretreatment removes surface species like water, CO_2, and O_2 and allows the materials to exhibit basic properties and promote base-catalyzed reactions. The temperature required for the removal of adsorbed material is the decomposition temperature of hydroxide, carbonate, and peroxide.

The nature of the generated basic site depends on the severity of pretreatment. The molecules that are covering the surface are desorbed when the pre-treatment temperature increases, which is depicted in Fig. (**1**). For example, MgO pretreated at a temperature below 723 K has relatively very little activity for migration of double bond of 1- butene, but when the temperature rises above 723 K, it shows great activity that is up to a maximum of 873 K [15].

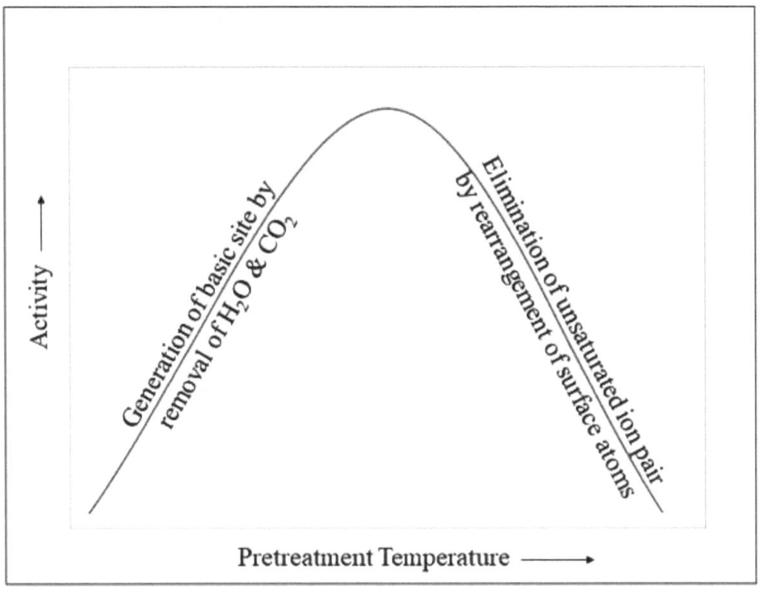

Fig. (1). Effect of pretreatment temperature on basic site.

Coluccia and Tench presented a model to study the impact of pretreatment temperature on metal oxide surface, specifically MgO, which is shown in Fig. (**2**). On the surface, there are different coordination numbers of MgO ion couples after high-temperature application to remove adsorbed molecules. Low coordination number sites are located close to edges, corners, and surfaces with a high Miller index. An incease in the pretreatment temperature generates different basic sites corresponding to ion pairs with varying coordination numbers. The most reactive site is the three-fold Mg^{2+}-three-fold O^{2-} ion pair, which is very unstable and has a tendency to reorganize or vanish at high temperatures. Despite its great adsorption of carbon dioxide and water, this pair is also very reactive.

Precipitation

The precipitation method for solid base catalysts involves the formation of solid catalysts by inducing the precipitation of metal ions or compounds from a solution. This process starts with creating a uniform mixture, followed by changes in temperature and the addition of acidic or alkaline solutions. Through chemical reactions, soluble ions in the solution interact with another reagent to produce insoluble solid compounds. Factors like temperature, pH, composition of raw materials, and the solvent used impact the formation of the solid precipitate. The process unfolds in four stages: Precursor mixing and supersaturation, nucleation, crystal formation, and aggregation [16, 17].

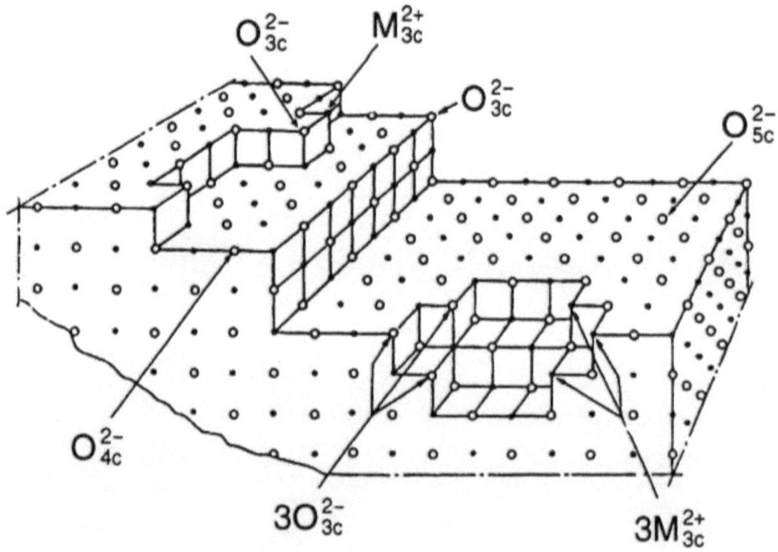

Fig. (2). MgO surface model presented by Coluccia and Tench [1].

$$\frac{dN}{dt} = \beta\ exp\left[\frac{-16\pi\sigma^3\ \vartheta^2}{3(kT)^3 In^2\ s}\right]$$

βis a pre-exponential term, σ the solid /fluid interfacial energy, ϑ is solid molecular volume, T is temperature, and *s* the supersaturation.

Reaction can be simplified as:

$$\frac{dN}{dT} = \beta\ \exp\left(\frac{-A}{In^2\ s}\right)$$

A represents the interfacial energy parameter.

Although nucleation and crystal development often occur together, they can be analyzed independently. The separation of these processes is crucial for producing mono-dispersed crystals. Nucleation is particularly challenging in multicomponent systems because it begins with the formation of clusters that can develop spontaneously by adding monomers until they reach a critical size. Smaller clusters will likely break apart, but bigger clusters will keep growing. Nucleation is similar to a chemical reaction, requiring the overcoming of an energy barrier. Therefore, it is essential to achieve a certain level of supersaturation to trigger spontaneous crystallization [17].

The primary focus of this approach is to ensure a consistent pH value throughout the precipitation process. Precipitates such as hydroxides and carbonates are commonly utilized due to their low toxicity, minimal solubility, and simple breakdown.

Co-precipitation

Co-precipitation is a standard technique for synthesizing catalyst precursors, enabling the uniform distribution of multiple metal cations in a solution. This method facilitates the formation of multiple catalyst precursors through rapid solidification, producing highly dispersed, well-mixed, and homogeneous catalysts after calcination or reduction [18].

As part of this method, salts of an active metal and a base are mixed to make a solid precursor containing both components in a single step. This technique is effective for producing catalysts with high metal mass-to-volume ratios by inducing substantial supersaturation, allowing simultaneous nucleation of all components. The merging of highly concentrated solutions results in a material that cannot dissolve completely, leading to the precipitation of most precursors. Precise control is essential to avoid issues from temperature gradients, poor mixing, or concentration gradients. Spray-drying and constant sequential precipitation are two new techniques that can help solve these real problems [19, 20]. It is also used to produce catalysts like nickel alumina for steam reforming or iron copper potassium for Fischer-Tropsch synthesis [21]. In contrast, many studies have targeted $Cu/ZnO/Al_2O_3$ catalysts for making methanol. The active sites of the catalyst are made up of metallic copper step sites that are close to zinc. A usual ratio of 70:30 Cu/Zn is used. Copper, zinc, and aluminum nitrates are mixed to make the mixture, and then alkali (bi)-carbonates are used to precipitate the mixture [22]. Once the particles have been precipitated and aged, they need to be washed to get rid of any remaining nitrates, sodium, or potassium that could cause them to sinter and stick together during the next thermal process, which would reduce their surface area. Titration tests show that the best way to participate is at a steady pH level, with steps of precipitation going from Cu to Zn [23]. The ideal building block for producing functional methanol catalysts is zincian malachite [24]. The zincian malachite precursor undergoes a calcination process after aging, resulting in CuO and ZnO particles. CuO particle size increased with increasing calcination severity, according to research on Cu/Zn/Al coprecipitates. The use of microwave irradiation as a substitute for conventional heat sources in the precipitation and aging process of Cu/Zn/Al at a molar ratio of 64:32:4 and pH 7.5–8 is also reported [25].

Coprecipitation creates a solid precursor from active metal salts and support, suited for high metal mass-to-volume ratios. Substantial supersaturation causes component nucleation and precipitation. Controlling temperature, mixing, and concentration gradients is essential. Innovative methods like spray-drying and continual consecutive precipitation solve these problems.

Deposition Precipitation

Deposition precipitation is a method where metal compounds, such as metal hydroxides, are formed from a precursor solution due to changes in pH, temperature, or evaporation. Unlike co-precipitation, deposition precipitation occurs in the presence of a support material. By gradually increasing the concentration of the new chemical, excessive substance synthesis in the solution can be prevented. The support aids in preferred precipitation by reducing the surface free energy of small nuclei or stabilizing the precipitate, lowering the energy barrier for nucleation. Nucleation occurs solely on the support surface under specific conditions, acting as a catalyst for nucleation rather than in the surrounding fluid. The core principle of the deposition precipitation method is based on nucleation and growth theory, where the solubility of a solid in a solution is controlled by the thermodynamic equilibrium between the stability of the solid phase and the dissolved components.

Energy change, ΔG_{tot}:

$$\Delta G_{tot} = 3/4\pi r^3 .\Delta\mu_s 1 + 4\pi r^2 .\gamma$$

r = radius of the particle

$\Delta\mu_s 1$= difference in thermodynamic potential of solid and solute

γ = surface free energy of the solid in contact with solution or support

The process of nucleation of metal species usually takes place by adjusting the pH to create compounds that have limited solubility. One should be cautious when starting nucleation by introducing a precipitant, such as an alkaline solution, to avoid localized concentrations beyond the threshold of supersaturating, which may result in extensive precipitation. As a result, approaches of homogeneous deposition precipitation are frequently recommended, as they cause precipitation to occur evenly throughout the whole reaction vessel. One technique entails the introduction of urea at a temperature of 363 K, which undergoes a gradual decomposition, resulting in the release of OH− ions and a progressive elevation of the pH. Alternatively, commonly used methods involve immediately reducing metal ions that are dissolved in a solvent to obtain the metal or gradually decreasing the pH by progressively eliminating ligands such as ammonia through the process of evaporation [26, 27]. The DP technique provides three primary methodologies for catalyst preparation. First, a precipitating agent is combined with a solution that contains the precursor of the active component, while the support is suspended in this combination. Alternatively, the precursor solution may be introduced into an alkali solution containing a support suspension. Ultimately, the initial solution can be mixed with an alkaline substance to produce colloidal components, which are subsequently administered onto the surface of the support [28]. Precipitation methods were created to overcome the constraints of impregnation procedures. A study was conducted where hexachloroplatinic acid was utilised to apply a maximum of 60wt % Pt onto carbon black by a deposition method that involved the degradation of urea. The Pt particle size before reduction was influenced by adjusting the urea/Pt molar ratio, where larger ratios resulted in smaller particles due to the influence of electrostatic charge effects. The size of the particles was affected by the reduction circumstances. The use of $NaBH_4$ led to the formation of bigger particles, whereas ethylene glycol produced smaller and uniformly distributed Pt particles. A further investigation revealed that manipulating the initial solution pH had an impact on the dimensions of 40 wt % PtRu particles following reduction, wherein a lower pH led to an increase in particle size [29, 30]. Cobalt nitrate and urea solution deposition precipitation has been explored for cobalt deposition on Carbon Nanofiber (CNF) supports and silica gel [31, 32]. When the cobalt loading was at 15 wt %, cobalt hydrosilicates were the main component of silica. However, due to weak precursor contact, nucleation on CNF was limited, resulting in the formation of larger cobalt particles (40 nm on silica, 25 nm on CNF) after reduction. Improved uniform deposition was accomplished by utilising cobalt carbonate and a high starting pH, which caused precipitation through ammonia evaporation at a temperature of 363 K. This approach yielded smaller particles, measuring 8 nm on CNF and 13 nm on silica [33].

Researchers opt for the DP method to address challenges encountered with the adsorption method in catalyst preparation. Two of these difficulties are (i) the limited ability of some metal compounds to adsorb on support surfaces, as in the case of inorganic Fe(III) salts depositing on SiO_2 or Al_2O_3, and (ii) the contamination of the surfaces of dispersed metallic particles with chloride ions as a result of the adhesion of a few chloro complexes of noble metals. This contamination, as seen in the adsorption of gold chloride on oxide supports, can act as catalytic poisons in various reactions. Nevertheless, the deposition of a gold precursor in the form of hydroxide particles leads to a substantial decrease in the detrimental impact of chlorine atoms [34, 35].

Template-assisted Techniques

In template-assisted techniques, organic or inorganic molecules with specific shapes and functionalities are used as molds in which the catalyst precursor materials assemble during synthesis. These templates are designed carefully so that they can tailor specific applicant solid base catalysts. There are two templating methods involved in this technique: Hard templating method and soft templating method are described in Fig. (**3**). The hard templating method contains a rigid scaffold to hold the material and the template material is removed after synthesis. Soft templating involves the self-assembly of precursor molecules around the template, which is subsequently removed or retained depending on the desired catalyst morphology. The advantages of this technique are that it can control surface area, pore size distribution, and active sites of catalyst which increase selectivity and catalytic activity of solid base catalysts.

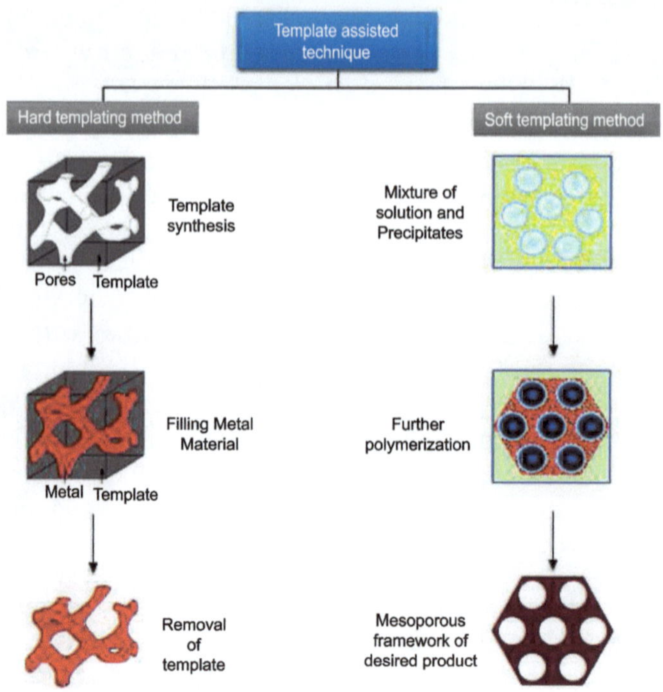

Fig. (3). Types of template-assisted technique [36, 37].

Impregnation Method

The impregnation technique involves the deposition of dissolved aqueous precursors onto oxide supports. In this, a precursor solution containing active catalyst elements came into contact with the solid support. After a specific duration, this process is followed by the drying of access-absorbed liquid and, at last, calcination to activate the catalyst [38, 39]. There are two types of Impregnation methods: 1) Dry Impregnation and 2) Wet Impregnation. In dry Impregnation, the volume of solution used is equal to or slightly less than the pore volume of the support. Conversely, Wet impregnation requires excess solution and higher temperatures for catalyst activation. In both methods, the temperature affects the viscosity of the solution and solubility of the precursor [40]. Another disadvantage of this method is the non-uniform distribution of active species, which can lead to reduced catalytic activity and selectivity.

The metal salt solution interacts with the solid oxide support through adsorption, ion exchange, or precipitation, followed by thermal treatment to form the targeted solid base catalyst (Fig. **4**). The bromide, chloride, fluoride, iodide, hydroxide, and nitrate of potassium were used for impregnation of alumina support. Dried NaX zeolites are prepared by using the impregnation method, where NaX zeolites are soaked in an aqueous solution of potassium hydroxide for a specific time, followed by drying and calcination for activation [42]. The impregnation method is also used in the preparation of alumina-loaded solid base catalysts using potassium nitrate [43].

Sol-gel Method

Sol-gel processing is a technique used to create porous materials by solidifying them from a solution phase [44]. The process entails creating stable colloidal solutions, referred to as “sol”, first. Subsequently, the colloidal particles undergo anisotropic condensation, resulting in the formation of polymeric chains intertwined with the by-products of condensation. When an external solvent is not used, this process leads to the creation of either a “lio-” or “hydrogel” or a “monolith”. After the by-products are removed by washing, the solvent is removed, resulting in the creation of “xero-” or “aerogels,” depending on the drying process used. Each of them has a unique primary particle structure and packing arrangement (texture) [45, 46]. The sol-gel approach involves the creation of a stable colloidal mixture, known as a sol, which consists of finely distributed highly pure solid particles. The initial substance experiences partial condensation and hydrolysis to produce the sol, which then undergoes polycondensation to transform into a gel. The liquid contained in the gel is removed through the process of evaporative drying, resulting in the production of a dried product known as a xerogel. Subsequently, the xerogel characteristics were enhanced by heating [47].

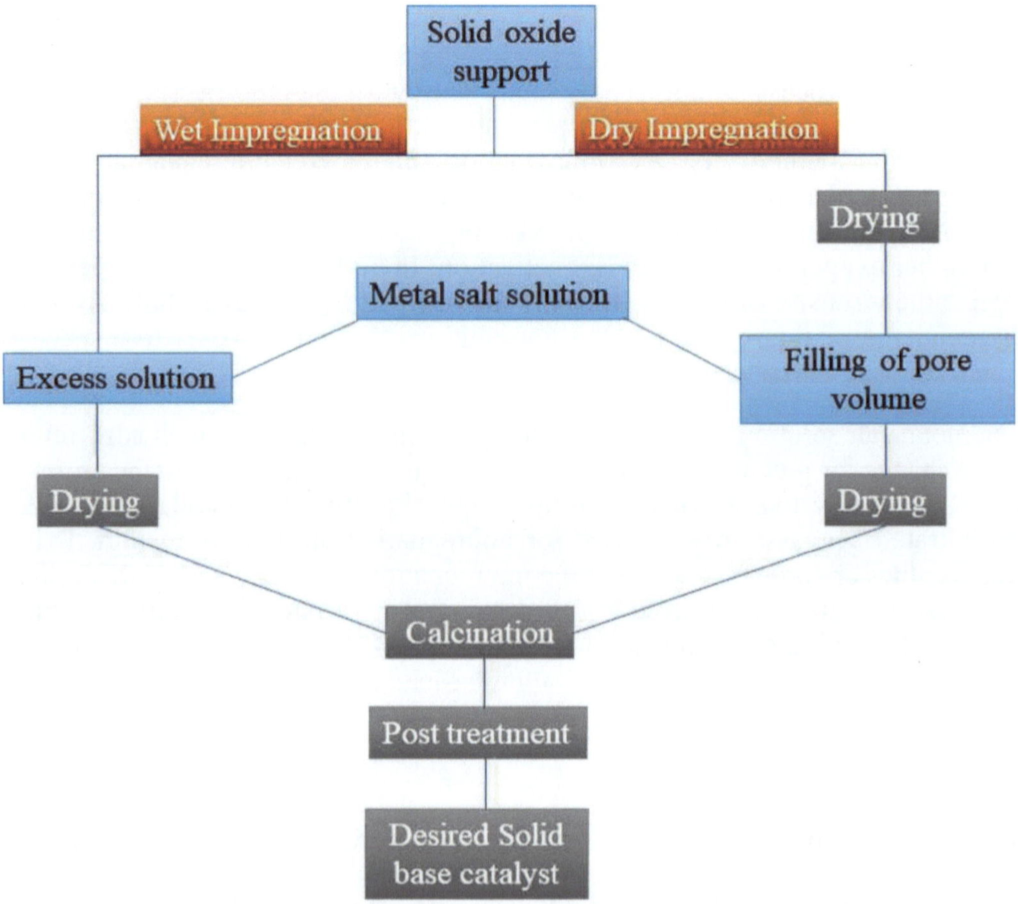

Fig. (4). Flow & types of Impregnation technique for the synthesis of solid base catalyst [41].

Understanding the unique characteristics of the sol-gel method is essential to separate it from other methods of hardening materials from liquids. The method starts when dispersed molecular precursors condense, resulting in the production of a transparent colloidal solution. Furthermore, in the next gelation phase, these colloidal particles combine into chains of polymers through chemical interaction between reactive groups on their surfaces. This hinders the clumping together of micelles due to their uniform distribution. Subsequently, the desolvation stage is employed to form porous solids, such as xerogels or aerogels, and the specific drying mode chosen determines the outcome. Sol-gel processing provides a flexible approach to creating solid catalytic materials using different methods [8-48]. The procedure entails the solidification of a colloidal solution, which is then followed by the removal of solvent to produce three distinct categories of substances. Bulk uniphase materials refer to xerogels or aerogels that are composed of either a single metal or many metals. In bulk multiphasic materials, molecular components or condensed phases are confined within the gel matrix by being trapped between the polymeric chains or by being co-gelated from a mixed colloidal solution. Porous coating and nanometric films can be created by inducing gelation in a thin layer of colloidal solution on a substrate. This process involves a series of steps, as shown in Fig. (**5**).

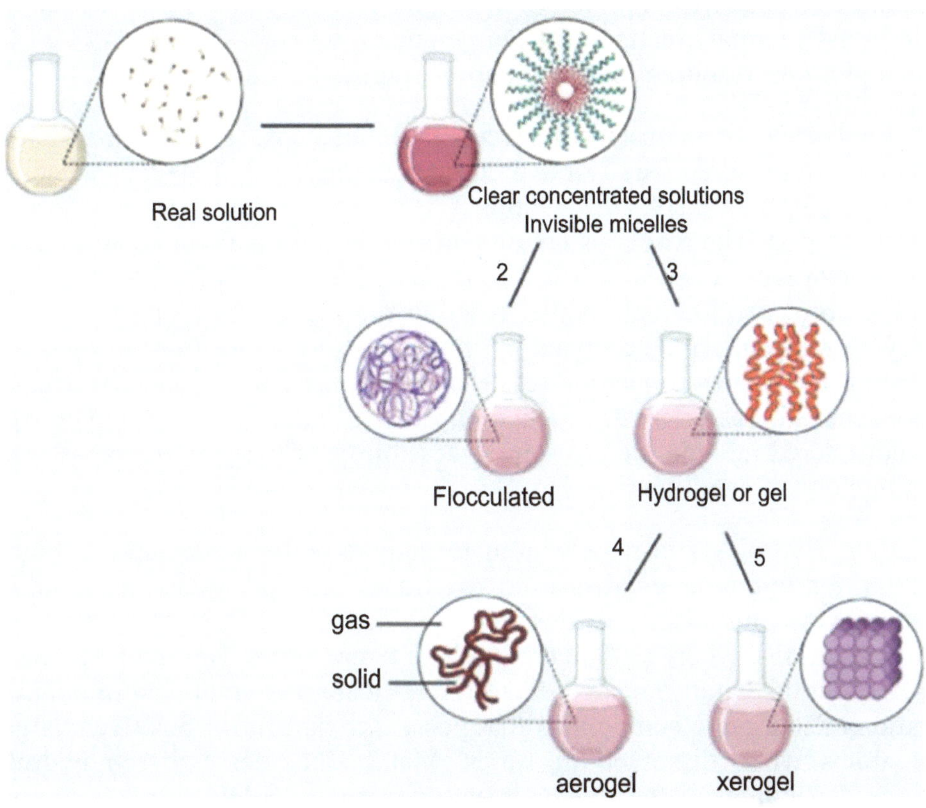

Fig. (5). Atypical scheme of sol-gel processing in preparation of solid materials 1-colloidization, 2-flocculation, 3- gelation, 4- supercritical fluid processing, and 5- drying by evaporation.

The sol-gel approach is recommended due to its positive impact on the uniformity, structural surface, and characteristics of catalysts. The end product is influenced by several parameters, including the concentration of precursors, the properties of the solvent, the pH level, the concentration and type of additives, and the duration of ageing.

The sol-gel method was employed to synthesize a ZrO_2-SiO_2 solid catalyst utilising, acetylacetone, tetraethoxysilane, zirconium n-propoxide, and isopropyl alcohol. Acetylacetone and isopropyl alcohol were initially combined in precise amounts to create a blended solvent. The solvent was subsequently employed to dilute zirconium n-propoxide, with the molar percentage of the combined zirconium n-propoxide and solvent varying from 0 to 1. Tetraethoxysilane was introduced into the solution and agitated for an interval of 15 minutes. Afterward, distilled water was used to initiate hydrolysis and create a gel. The gel was then kept undisturbed at room temperature for a duration of one week, earlier dried at a temperature of 333 K for a period of 48 hours. The dehydrated gel was subjected to additional thermal treatment at temperatures between 773 to 1173 K for a period of 4 hours, resulting in the formation of the solid catalyst [49].

The sol-gel processing consists of several essential steps: (i) activating dissolved molecular precursors, (ii) Activated precursors are polycondensed into nanoclusters to make a sol, (iii) gelation, (iv) aging, (v) washing, (vi) drying, and (vii) stabilization. The reactions taking place in the sol-gel process, as well as the required additives and circumstances to regulate the characteristics of the emerging porous solid, such as pH, reaction time, concentrations, temperature, and drying method, are determined by the chemical composition of the selected precursors. The choice of method relies on the capacity to regulate the activation and condensation reaction rates, which impact the material texture. Other aspects to be considered include the cost and accessibility of the precursors, as well as their simplicity of handling.

Activation: Hydrolysis is widely used to activate sol-gel precursors, which is necessary for initiating condensation reactions. This procedure is utilised for metal alkoxides dissolved in alcoholic solutions or metal ions dissolved in aqueous solutions, where parameters such as temperature, cation charge density and pH affect the balance of hydrolysis. Acidification is employed to induce the activation of inorganic compounds that possess a significant positive charge on metal atoms when dissolved in water. Metal alkoxides undergo hydrolysis reactions depending on the metal's electronegativity. Silica provides the most favourable circumstances for regulated hydrolysis. Alkoxides can be activated to yield partially hydroxylated species through nonhydrolytic mechanisms, such as heat intermolecular hydroxylation or particular chemical reactions.

Polycondensation process entails the reaction of active precursors to produce colloidal solutions composed of oligomeric species. Compounds with hydroxyl groups as intermediates undergo condensation reactions through olation, oxolation, or alcoxolation pathways, which are catalysed by acidic or basic conditions. Si-O-Si bridges are formed as a result of these reactions. The acidification of aqueous solutions containing high-valence metal anionic precursors reduces the electrostatic repulsion, which in turn enhances the condensation process. Aprotic condensation happens in nonhydrolytic processing, which involves interactions between alkoxides of different metals or with metal esters and chlorides. The pace of condensation is influenced by the ratios of the precursors, temperature, the presence of a catalyst, and the structure of the R group. Metal cation complexes react with chelating agents and polyalcohols by acid-catalyzed polyesterification. As the hydrolysis/polycondensation degree increases, water molecules and hydroxyl ligands are lost. This process terminates polycondensation once it has produced dimers, tetramers, or oligomers that include more than 10 metal atoms.

Gelation/Aging/Washing refers to the transformation of a liquid sol into an elastic gel network. This is usually accomplished by gradually adding a base or hydrolyzing urea. Gelation can be induced in metal salt solutions by adding proton scavengers or chelating agents. The gel structure is influenced by the extent of ionic character in metal-oxygen bonds and the equilibrium between hydrolysis and condensation rates. Following gelation, the process advances to aging, which encompasses polycondensation, syneresis, coarsening, and phase change, ultimately resulting in the optimum gel density.

Gel desolvation is accomplished by either evaporating the solvent or using supercritical drying. Xerogels are produced through the process of solvent evaporation, which includes desolvation, pore opening, and the final evaporation of the liquid film. Supercritical drying is a method that creates aerogels by maintaining the texture of a wet gel without the use of capillary forces. To cleanse the sample, 30 ml of high-purity water is introduced to the glass frit contents, agitated for several minutes with a magnetic stirrer, and subsequently extracted. The quantity of washing cycles and water volume affects catalytic performance. It is crucial to thoroughly eliminate nitrate and sodium whilst preventing the precipitate from reforming into a sol through re-peptization.

In the case study presented, neither $NaNO_3$ nor gerhardtite was identified by XRD analysis, and elementary analysis did not detect sodium. The washing process requires careful removal of nitrate and sodium to prevent the formation of compounds prone to sudden decomposition [50]. Following the elimination of by-products through washing, the removal of the solvent yields either 'xerogels' or 'aerogels', contingent upon the drying method employed. These resultant gels exhibit distinct characteristics in terms of primary particle structure and arrangement [51, 52]. The stabilization of xero- and aerogels is achieved through heat treatment, which improves their stability, modifies their surface, and induces crystallographic transformations. These processes have an impact on the pore geometry and surface area of the materials.

A novel catalyst with nanoscale dimensions has been created, comprising a ferrite core coated with a double hydroxide layer. The procedure entails the dissolution of magnesium nitrate hexahydrate and aluminum nitrate nonahydrate in water, after which the addition of magnesium ferrite to the solution using a rotating mill. Gel formation occurred after the addition of sodium carbonate and sodium hydroxide to maintain a pH of 11 during mixing. The gel was left undisturbed, thereafter washed, filtered, and dried before undergoing calcination to produce the catalyst [53].

Sol-gel technique is widely employed for fabricating solid catalytic materials in many applications. It allows for the production of different materials, such as large-scale single or multiple metallic oxide catalysts, mixed multiphasic catalysts containing trapped catalytic components, catalytic coatings, and membranes. Furthermore, it is utilized in the fabrication of prefabricated porous solids such as silica and alumina, which function as catalyst supports. Sol-gel processes offer a lot of adaptability in manipulating the pore and phase structures through the adjustment of precursor characteristics and preparation conditions. Nevertheless, this adaptability requires meticulous choice of chemical pathways and processing parameters according to the intended catalytic substance.

Hydrothermal Technique

A hydrothermal process (Fig. **6**) involves subjecting a consistent aqueous solution containing precursors and alkali to raised temperature and pressure within an autoclave for a defined period of time. The chemicals undergo dissolution and subsequent recrystallization under these circumstances. The resultant precipitate is gathered *via* filtration, rinsed, and subsequently dehydrated. When reactants are autoclaved, ions or molecules split due to temperature differences between the top and bottom sections. Crystallization occurred in dissolved compounds, as ions or molecules travelled to low temperatures [54].

A multi autoclave system with Teflon blocks and 100 reaction chambers allows up to 1000 hydrothermal syntheses per experiment. This method and automated X-ray powder diffraction efficiently identified zeolite crystalline structures in Na_2O-SiO_2-Al_2O_3 and Na_2O-Li_2O-Cs_2O-tetramethylammonium phase diagrams. Similar methods were utilized to examine alum phosphate molecular sieves and optimize zeolite beta production. High-throughput experimentation discovered new zeolite structures like ITQ-30 and ITQ-33, proving their efficacy in synthesis and characterization [55-57]. Pulverized magnesium sources are placed in a

Teflon-coated autoclave with solvents like ethylenediamine solution or aqueous ammonia and heated to 453 K for 20 hours in the hydrothermal process. After cooling, magnesium hydroxide is filtered, washed, and dried before being calcined at 553–723 K. Mixing ammonium carbonate solution with aqueous magnesium nitrate solution, hydrothermal treatment, centrifuging, washing, and drying produce magnesium carbonate hydrates. Finally, magnesium carbonate hydrates are calcined at 773 K for 6 hours to produce magnesium oxide.

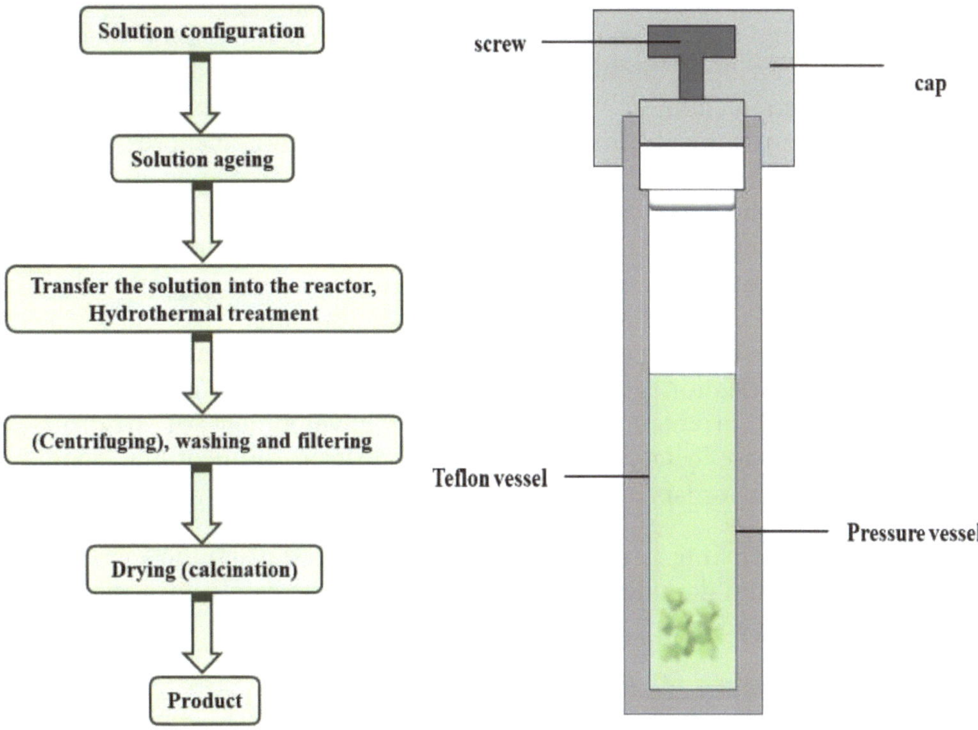

Fig. (6). Series of steps involved in hydrothermal process and typical equipment used in the process [54].

Hydrothermal technologies have many advantages over conventional ones, including low reaction temperature. Hydrothermal processing generates pure, well-crystallized minerals without thermal treatment. High temperatures are not needed for hydrothermal synthesis. Hydrothermal methods may synthesize nano-structured materials at 413 – 473 K, while solid-state processes require 1073 K. This feature can produce nano- or sub micro-metric materials [58]. Three mixtures, including equal amounts of ZrO_2/CeO_2, ZrO_2/TiO_2, and ZrO_2/La_2O_3, were synthesized using the hydrothermal method. These mixtures were thoroughly characterized. The inclusion of lanthanum led to the formation of a two-phase system comprising oxychlorides, accompanied by a little enhancement in surface area and the presence of robust basic sites. The incorporation of cerium oxide resulted in the formation of a heterogeneous solid solution that exhibited a greater surface area and notable presence of both acidic and basic sites, particularly Bronsted acidic sites. The inclusion of titanium oxide resulted in the formation of a mesoporous amorphous phase, which exhibited a considerably greater surface area in comparison to pure ZrO_2 [59].

Sonochemical Method

In the sonochemical method, nanoparticles are prepared using sonotrode, which acts as both an ultrasound emitter and electrode. This method operates based on the cavitation phenomenon, in which the rapid formation and collapse of bubbles within a liquid create localized zones of high temperature and pressure. These intense conditions facilitate the formation of highly reactive species, promoting the synthesis of solid base catalysts with unique properties. One of the main advantages of the sonochemical method is that it has precise control over particle size. Current density and sonication time may affect the particle size. Sonochemical method enhances catalytic performance by utilizing the morphology and surface area of the crystals. Crystal size reduces with a decrease in temperature, sonication gap, and high current density [60]. Sonochemical method provides a promising avenue for the development of efficient and sustainable solid base catalysts for various industrial applications, including biodiesel production, organic synthesis, and environmental remediation.

Previously, hydroxyapatite zirconia catalyst was prepared using the sonochemical method [61]. Titanium dioxide-palladium nanocomposite catalyst is also prepared by ultrasonication irradiation/ sonochemical method [62].

Vapor Phase Deposition Method

The vapor phase deposition method involves the coating of a thin layer of the catalyst material onto a support. In this method, the starting material is vaporized and carried by a carrier gas to the reaction zone, where it reacts and forms the desired catalyst material. Vapor phase deposition methods are generally categorized into two categories such as Chemical Vapor Deposition (CVD) and Physical Vapor Deposition (PVD). It is dependent on the coating process, which chemical reactions or physical impacts may drive. Target elements from the sample are supplied as volatile molecules for chemical vapor deposition. These molecules react chemically with the substrate or other precursor residues to generate the desired coating film. Conversely, a sample containing the target elements is subjected to a number of operations during physical vapor deposition, like electric discharge, resulting in the emission of atoms that subsequently condense onto the substrate surface [63].

Fig. (7) Illustrates both PVD and CVD processes. Vapor phase deposition is used to synthesize a variety of solid base catalysts like metal oxides, zeolites and supported alkali metals. For example, Titanium/silicon dioxide [64], silicon dioxide/calcium oxide [65], and barium zirconate [66] solid catalysts are synthesized using the vapor phase deposition method.

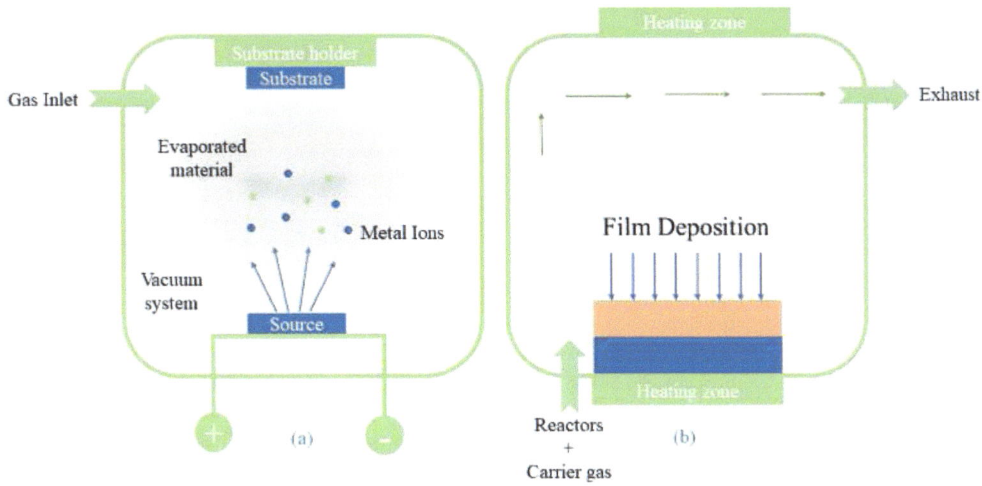

Fig. (7). Schematic diagram of (a) physical vapor deposition & (b) chemical vapor deposition.

Metal-Organic Frameworks (MOFs)

This is a novel synthetic technique in which Metal-Organic Frameworks (MOFs) are used as a precursor for the synthesis of solid base catalysts, which involves the conversion of MOF structures into metal oxide-based catalysts through thermal decomposition. MOFs are porous and crystalline solids of metal ions or clusters connected by organic ligands with high versatility in terms of composition and structure (Fig. **8**) [67]. MOFs generally produce effective catalytic active sites, which provide ways that reduce the effort required by researchers to develop novel organic processes.

The flexible and durable skeleton, large surface area, regular porosity, high pore volume, varied chemical capabilities, and affordable synthesis are important benefits of using MOFs [69]. Lanthanide Metal-Organic Frameworks (Ln-MOFs) [70] and Zinc Metal-Organic Frameworks (Zn-MOF) [71] are examples of solid base catalysts that are synthesized by MOFs.

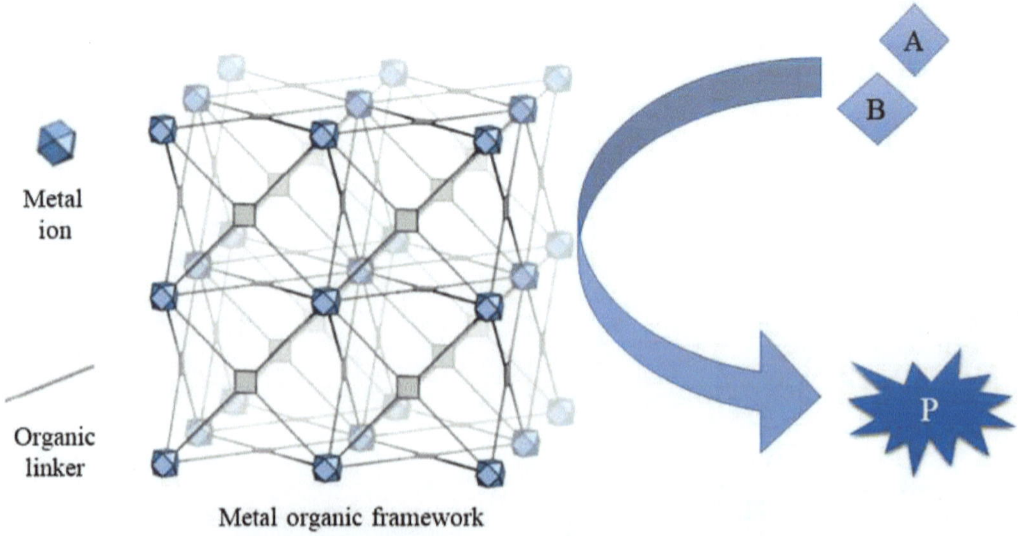

Fig. (8). Schematic diagram of metal-organic frameworks (MOFs) [68].

One-Pot Synthesis

Multiple synthetic stages can be integrated into a single reaction vessel using the streamlined one-pot synthesis method. Because it avoids time-consuming separation procedures and the purification of intermediate chemical compounds, this synthetic technique is favored by chemists. This technique also maximizes chemical yield. The one-pot synthesis can be carried out in three different ways: Domino or cascade reactions, Multicomponent Reactions (MCRs), and One-Pot Stepwise Synthesis (OPSS), which is shown in Fig. (**9**) [72].

A single-operation reaction with multiple chemical transformations is called a cascade, domino, or tandem reaction. No more reactants, reagents, or catalysts are added to the reaction system after the reaction starts. Although MCR uses multiple reactants, it is still a single-operation reaction. Because of the several operating steps, an OPSS enables the stepwise introduction of reactants, reagents, and catalysts, as well as the modification of reaction conditions at each stage.

The one-pot technique is used to synthesize various types of solid base catalysts, for example, Potassium-functionalized mesoporous γ-alumina [73], mesoporous Ca/SBA-15, which is used for the transesterification of sunflower oil [74], calcium-incorporated MCM-41 base catalyst, which is used for transesterification of palm olein [75] and potassium-loaded Mg-Al oxide as the solid super base catalyst [76].

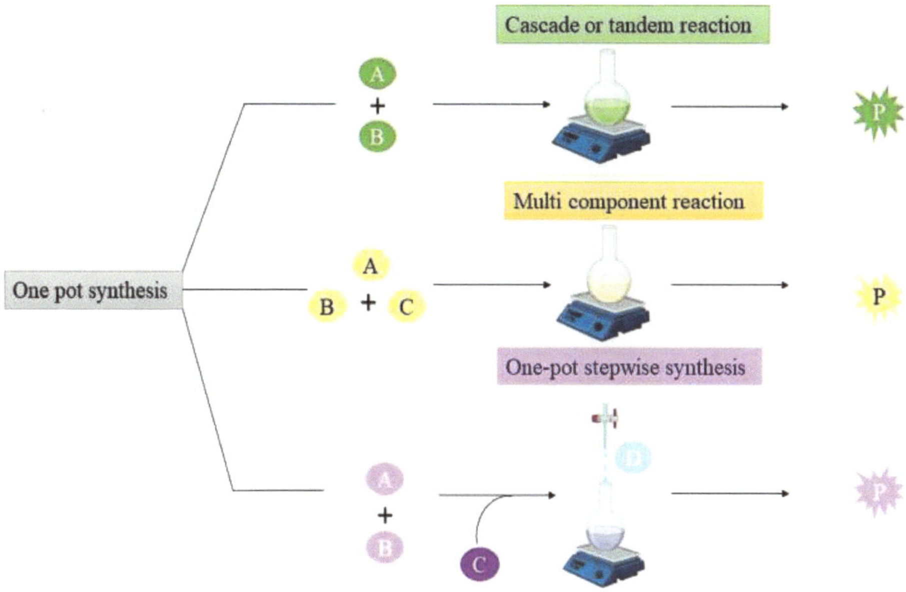

Fig. (9). Types of one pot synthetic method Where; A, B, C, D = substrates, P = desired product.

Fly Ash-Derived Solid Base Catalysts

Fly ash is a waste product of industries that is a by-product of coal combustion. It has drawn interest lately as a possible starting point for the development of solid base catalysts. Fly ash is a porous substance with a high specific surface area and good adsorption qualities that make it a useful catalyst carrier and adsorbent. The complex compounds in fly ash are mainly dominated by metal oxides. So, raw fly ash can be directly used as a solid base catalyst or modify the fly ash by incorporating metal oxides [77].

Another way is the conversation of fly ash into zeolites, for example, Ni-Cu and Ni-Co-Modified Fly Ash Zeolite Catalysts. Fly ash can be used as a solid base catalyst by activating it chemically or thermally at high temperatures [79]. Modified fly ash as an SBC is depicted in Fig. (**10**). Solid base catalysts synthesized from fly ash play a crucial role in many catalytic reactions, including transesterification for biodiesel production, aldol condensation, and Knoevenagel condensation.

Melt Infiltration

In Melt infiltration (Fig. **11**), a molten precursor is infused into a porous support material, where it solidifies within the pores of the support material to produce the active phase of the catalyst. Mesoporous silica, alumina, and zeolites are usually used as support materials because they possess thermal stability and high surface area. This technique is used as an alternate technique of impregnation and drying [80]. Melt infiltration is dependent on the capillary action and wettability of porous support material. Solvent-free capillary forces can be used to introduce metal salts that melt before they decompose into a porous support. After solidification, the catalyst may undergo additional post-treatment steps, such as calcination and activation at high temperature [81].

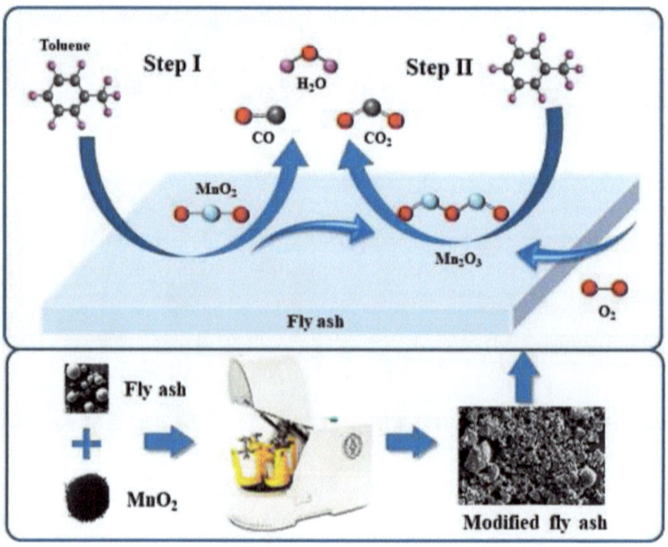

Fig. (10). Modification of fly ash by metal oxide for use as a solid base catalyst [78].

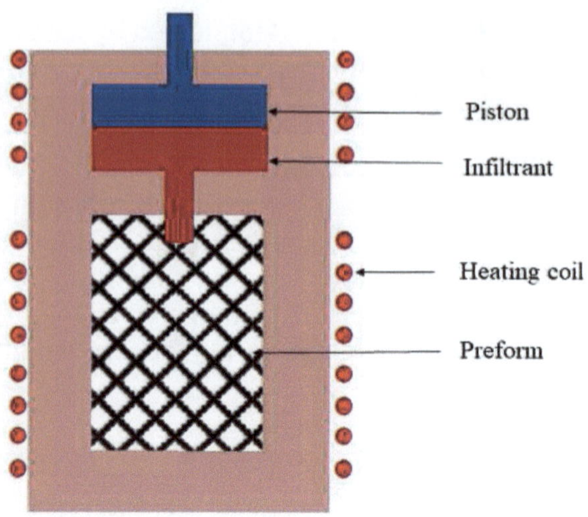

Fig. (11). Melt Infiltration Assembly [82].

Structure Analysis of Solid Base Catalysis

Brunauer Emmett-Teller technique

The BET technique is widely utilized to measure the precise surface area created by pores at an atomic level. This is achieved through the adsorption of an unreactive gas. As most gaseous and solid interact weakly, the solid surface is cooled using cryogenic liquid to induce the adsorption [83]. The solid surface is maintained at isothermal conditions while the pressure and concentration of absorbing gas are systemically raised. As the pressure increases, a greater number of molecules adhere to the surface. The volume of absorbed gas can justify the number of molecules adhered to the surface in the monolayer (Fig. **12**). Brunauer, Emmet and Teller developed the BET equation which was published in 1938 by Brunauer. It is still the most commonly employed approach to quantify the amount of gas atoms utilized to create a monolayer on the solid pore surface [84]. The BET equation, as mentioned below, demonstrates the correlation between the quantity of gas molecules absorbed (X) at the specific relative pressure (P/P_0) forming a monolayer (Xm). Here, C represents the secondary component that is associated with the heat of absorption [8].

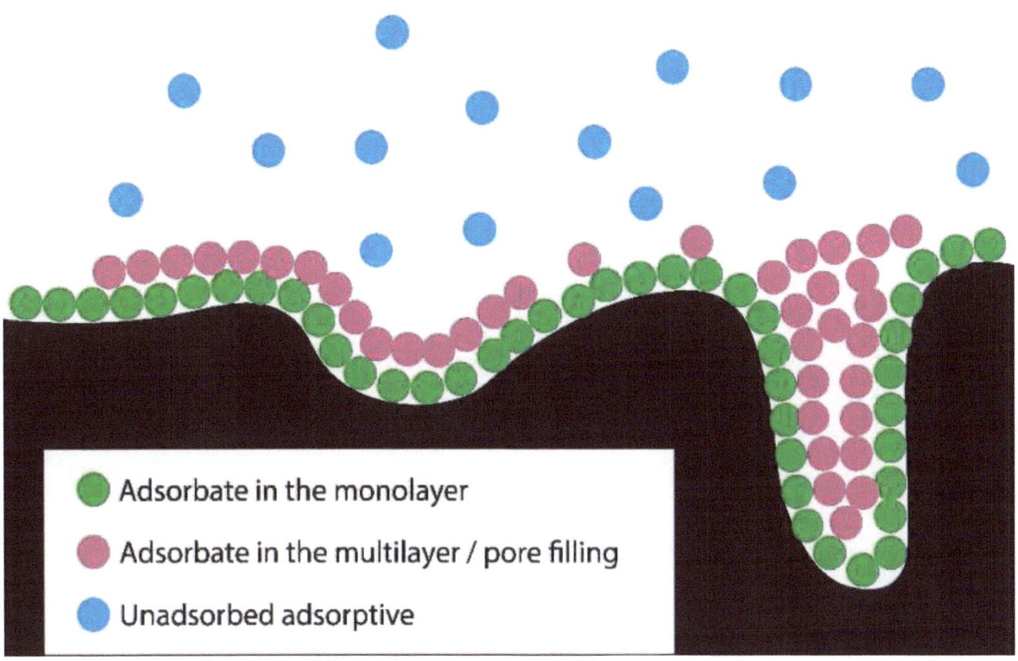

Fig. (12). Monolayer formation on the solid base catalyst [82].

$$\frac{1}{X\left[\left(\frac{P_0}{P}\right)-1\right]}=\frac{1}{X_mC}+C-\frac{1}{X_mC}\left(\frac{P}{P_0}\right)$$

X-ray Diffraction

X-ray diffraction is the process of scattering the electromagnetic waves with the particular wavelengths within the atomic structure. The scattering process can be understood as the incoming wave stimulating an electron to emit radiation of the same wavelength but in a different direction than the incoming wave. Multiple secondary waves generated due to many electrons in the crystal give rise to interference phenomenon. Interference at the atomic level is responsible for the absence of 'zero intensity' diffraction pattern, also known as diffuse background. Instead, a much stronger diffraction occurs from the periodic arrangement of atoms in lattice, resulting in distinct and sharp maxima known as a diffraction pattern. Geometric analysis enables the collection of the crystal's structural information [85].

XRD is an essential technique used for the design, development and production of catalyst, which provides valuable insights into the overall structure and composition of solid catalyst materials [86]. The average size of crystallite can be calculated using Debye Scherrer formula,

$$D_v = K\lambda/\beta\cos\theta$$

Where,

Dv = Average particle size,

λ = wavelength of radiation

β = full width at half maximum of radiation peak

θ = Bragg's angle (angle of diffraction)

The catalytic activity of a catalyst is directly related to its defects (nanostructures). Active site of the catalyst is not considered part of the bulk system and is characterized by its non-equilibrium nature. XRD, being the most sensitive technique to identify the defects, can easily be employed to detect and analyze the active site of the catalyst. The fundamental advantage of XRD for crystal characterization lies in its ability to be used at high pressure and the potential to minimize radiation damage and interference from beam chemistry by selecting wavelengths with minimal interaction cross-section when synchrotron is used as the light source. To attain these benefits, a radiation source must be accessible. Although many synchrotron sources offer significant potential for catalysis-related powder diffraction experiments, limitations such as current beam time allocation practices and the lack of dedicated stations capable of conducting and monitoring realistic chemical reactions often hinder these opportunities. In future, the latest radiation sources, such as free electron lasers and improved synchrotron sources, will allow the elucidation of the nanostructure of a working catalyst without beam damage as a consequence of a short time exposure. The increasing demand for comprehension of catalyst production and operation will stimulate the advancement of several strategies for the characterization of solid catalysts during their functioning.

Thermal Analysis and Colorimetric Techniques

The term Thermal Analysis (TA) stands for the analysis of fluctuation in sample's characteristics caused by changes in temperature, whereas colorimetry refers to the measurement of heat changes. Table **1** summarizes the techniques used to determine the related property of the sample by measuring its time and temperature.

Table 1. Compilation of various thermal techniques and their application.

Physical property	Name of Technique	Application
Temperature	DTA (Differential Thermal Analysis)	Phase change, decomposition, Redox reactions, dehydration, desorption, *etc.*
Heat flow	DSC (Differential scanning Calorimetry)	Phase change, decomposition, hydrogenation, kinetics, catalysis, Redox reactions, *etc.*
Heat	Colorimetry	Redox reactions, absorption, adsorption, heat capacity, kinetics, thermodynamics, *etc.*
Mass	TGA (Thermogravimetric Analysis)	Redox reactions, absorption, adsorption, kinetics, decomposition, *etc.*
Length, Volume	TMA (Thermomechanical Analysis)	Expansion, phase change, sintering, *etc.*

Differential Thermal Analysis (DTA)

As defined by International Confederation for Thermal Analysis and Colorimetry (ICTAC), the Differential Thermal Analysis (DTA) technique involves measuring the temperature difference between a sample and a reference material while they undergo the same controlled temperature variation in a controlled atmosphere. It involves mounting the thermocouples differently in sample (Te) and reference material (Tr). The assembly is kept in either hot and cold controlled temperature and the temperature difference between the sample and reference material is measured using a thermocouple (Fig. **13**).

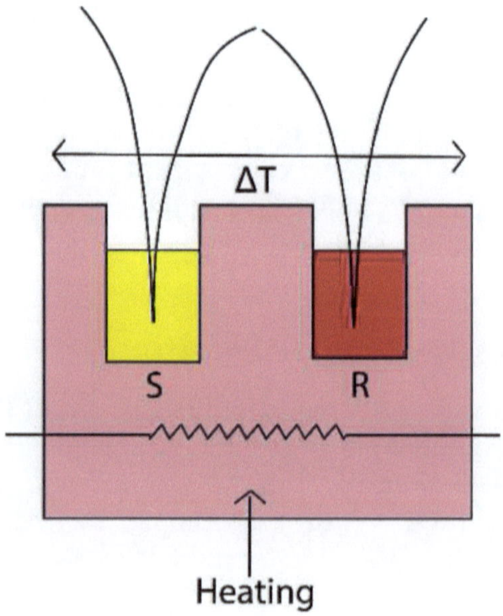

Fig. (13). Assembly depicting the principle of DTA [85].

DTA detectors employ various thermocouples that are selected as per the specific temperature range that is being studied (Fig. **14**). Crucible prevents its interaction with the sample and facilitates the passage of heat. Alumina, graphite, tungsten and platina are the most commonly used materials for crucibles. In a nutshell, DTA is a versatile technique with a wide range of applications in material science, particularly in the field of catalyst research and development [87].

Differential Scanning Colorimetry (DSC)

DSC is a crucial method in thermal analysis that provides accurate information regarding the thermal properties of the material. It is based on the fundamental idea of measuring the amount of heat transferred between a sample and its surrounding while the sample is subjected to controlled changes in temperature. The difference between the heat flow between sample and reference in a controlled temperature environment is recorded through DSC equipment, which allows the precise analysis of phase transition, reactions, and thermal characteristics with high sensitivity and accuracy [10-84].

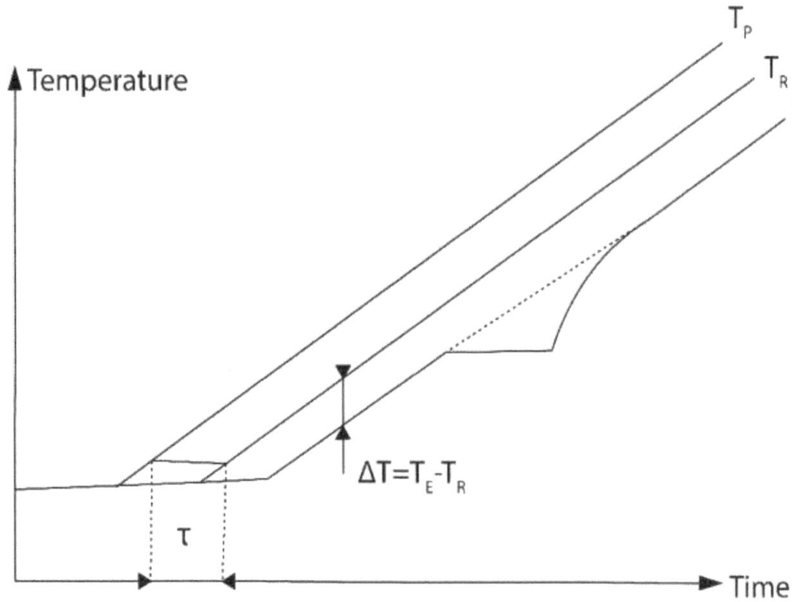

Fig. (14). Recording of the different temperatures as per DTA principle [85].

DSC can be performed using multiple operating modes, such as isothermal and scanning temperature modes. Different modes offer specific benefits in terms of experiment design and data collection. The mathematical equation describes the heat flow, thermal resistance and temperature differences. The characteristic equation for colorimetric measurement can be written as follows:

$$dh/\mathrm{dt} = -\mathrm{dq}/\mathrm{dt} + (C_s\text{-}C_r)\mathrm{dTp}/\mathrm{dt}\text{-}\mathrm{R}C_s d^2q/dt^2$$

Where,

dh/dt = Heat flux produced by the transformation of the sample or the reaction

Cs = heat capacity of the sample including container

Cr = heat capacity of the reference including container

R = thermal resistance

The diversity in the DSC instruments is due to various types of detectors, which create a huge range of DSC instruments, including heat flux DSC, power compensated DSC and Calvet DSC. All of these instruments utilize distinct detection methods to detect the tiny change in the heat flux. DSC detectors, ranging from plate-based sensors to advanced fluxmeter components, in detail, capture the complex interaction of thermal energies. This leads to the precise analysis of various materials with exceptional accuracy.

The most crucial aspect of DSC testing is the crucible selection which depends upon the material compatibility, gas interactions, and experimental requirements. DSC is a technique that facilitates the investigation of surface interactions and catalytic mechanisms allowing for a detailed understanding of the complexities of catalytic processes with the most accuracy.

DSC, by accurately measuring the heat capacity of a material, enables the calculation of specific heat capacity in various experimental settings and its mathematical analysis. It also allows the analysis of changes in heat capacity that occur with varying temperatures, thus providing significant information about the thermodynamic properties of material in several scientific fields. In a nutshell, DSC is an essential instrument in thermal analysis technique, providing unmatched capabilities in the assessment of material characteristics and thermal processes. The multiple operation modes, precision instrumentation, and rigorous mathematical foundation of the instrument have made the DSC a pioneering research method in various scientific disciplines.

The Calorimetric Technique

Calorimetry is the scientific method used to calculate the amount of heat transferred between a sample and its surroundings during specific types of processes such as physical, chemical, or biological. It offers valuable insights about thermodynamic characteristics and reaction kinetics, making it an essential tool in a wide range of research and industrial applications.

The calorimetric principle is explained in Fig. (**15**). The calorimetry process involves placing a sample in a vessel (A) and surrounding it with a thermostat jacket (B) to regulate the temperature of the surroundings and measure the flow of heat. The temperature of the sample (T_i) and the same jacket (T_j) differ depending on the calorimetric mode, which ultimately affects the heat transfer and creates the difference in temperature between the sample and its surroundings. Numerous calorimetric principles such as isothermal, isoperiodic, adiabatic and scanning calorimetry are employed to investigate the desired properties. In Table **2**, types of calorimetry have been listed, which cater to different types of measurements.

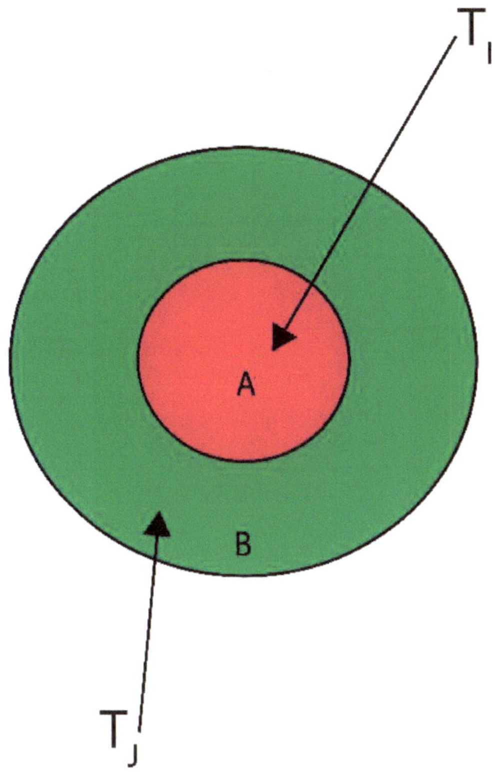

Fig. (15). Explanation of calorimetric principle [10].

Table 2. Types of calorimetric techniques.

Types of Calorimetry	Types of Measurement
Isothermal calorimetry	Instrument ensures a constant temperature for the sample and calculates heat flow by comparing temperature difference between the sample and its surroundings.
Isoperibolic calorimetry	Jacket temperature is constantly maintained and sample temperature is allowed to fluctuate and observation is made based on temperature variation related to sample.
Adiabatic calorimetry	Sample and jacket temperature are kept constant. The difference in temperature is observed based on the changes in the sample due to endothermic or exothermic reaction.
Calvet calorimetry	Uses thermocouple cylindrical detectors to measure the heat flow in both isothermal and scanning modes.
Titration calorimetry	Uses various solutions to measure the heat of interaction.
Flow calorimetry	Measures the heat flow in the moving liquid.

Apart from this, the C80 microcalorimeter by the brand name SEATRAM is a very versatile tool in the catalytic investigation as it offers various modes such as isothermal, scanning, gas adsorption, liquid adsorption, reaction and pressure calorimetry. For different types of applications, it is equipped with suitable vessels to carry out the experiment as shown in Table **3**.

Table 3. Types of vessels for C80 microcalorimetric measurements.

Vessel type for C80 microcalorimeter	Types of measurement
Standard batch vessel	For isothermal and scanning colorimetry. Suitable for the heat capacity determination of solid and liquid samples.
High pressure vessel	For high pressure studies including gas hydrate formation and supercritical gas adsorption.
Mixing vessels	To study liquid-liquid and solid-liquid interactions.
Percolation vessels	To study interaction during wetting, adsorption or percolation through solid powders.
Flow mixing calorimetry	To study enthalpies of absorption in gas absorption processes such as CO_2 capture.

Thermogravimetric Technique

Thermogravimetric Analysis (TGA) is the technique that monitors the changes in the mass of a sample as a function of time or temperature upon its heating at a controlled temperature. It can be performed under several controlled atmospheres, including inert, reactive, corrosive or humid conditions, in order to investigate the interaction of the sample with different environments. TGA can showcase two main types of transformations: (1) Reduction in mass (such as dehydration, desorption, and breakdown) and (2) Increase in mass (such as adsorption and reaction). Based upon this, a derivative curve (DTG) can be calculated, which offers valuable information on the kinetics of the transformation by analysing the rate at which the mass changes.

A thermobalance is an essential component of TGA, including a controlled environment cabinet and a very sensitive balance module-a furnace. The balances utilised in TGA measures change in mass by using a compensating current. Null position balance is the most commonly used one. TGA can be performed using many approaches, such as mono TGA (uses a single furnace for sample) or symmetrical TGA (uses 2 furnaces, sample and reference each). Moreover, TGA can also conducted at high pressure and as a corrosive TGA. Crucibles made from various materials like silica, alumina, platinum and graphite can be selected based on the temperature range and the characteristics of the sample to prevent any chemical interaction between the crucible and the sample. The buoyancy has a huge impact on the mass measurement when temperatures are low. Hence, to resolve this issue, blank tests and other related symmetrical thermo balances are carried out. Combining TGA with DTA and DSC amplifies the result significantly. This helps in understanding the properties of the sample more effectively.

Spectroscopic Analysis

UV-Vis-NIR Spectroscopy

UV-vis-NIR spectroscopy is an effective method for characterizing solid catalysts, as it provides information on their electronic transitions. The oxidation state and coordination environments of metal through metal-centered and charge-transfer transitions can be easily understood through this approach. It helps in evaluating the dispersion of supported oxide moieties through charge transfer transitions and determining the particle size and shape of dispersed coinage metals *via* surface plasmon resonance. It also helps in the research of semiconductor band gaps *via* absorption edge investigations and identification of surface functional groups through examination of overtone and combination vibrational modes. Absorbates like probes, reactants, products, and reaction intermediates can be identified if they possess recognizable chromophoric groups or significant vibrational overtone and combination modes [88].

However, the usage of UV-vis-NIR spectroscopy in catalysis research is limited due to various issues including inadequate absorption or overlapping absorption bands. Quantification is significantly more complicated for solid catalyst samples examined *via* diffuse reflectance spectroscopy than for the direct transmission investigation for a non-scattering medium. Keeping aside its limitation, an important advantage of UV-vis-INR spectroscopy is its capacity to acquire spectra of solid samples amongst gas or liquid phases over a board temperature range, achieving a time resolution of under one second. This makes it highly suitable for investigating catalysts under working conditions, enabling real-time studies of their functional behavior [89, 90].

IR Spectroscopy

Infrared (IR) absorption spectroscopy is the most basic method used to analyse catalyst, especially in studying the chemosorption of catalytically significant substances on supported metals. Initially, transmission mode was utilised to examine species like carbon monoxide, which are absorbed onto different metal particles that usually spread over supports with high surface areas. Advanced findings noted by Eischens identified the distinct peaks in the C-O stretching vibrational region, which offered valuable information about the bonding mechanism between CO bonds and metal surfaces. Later, Blyholder's model clarified the bonding mechanism between absorbates and metals. IR spectroscopy uses the chemical specificity of vibrational frequencies to investigate the catalytic surroundings. Molecular vibrations in the mid-IR range exhibit distinct spectra known as fingerprint regions for each species, facilitating their identification and characterization of bonding. FTIR has recent advances that broaden its range of the catalysis field application. Different setups can create different conditions that can support various catalysts and their transmission, such as in-situ or operando modes. The non-intrusive nature of FTIR enables its application in gas and liquid phase interactions. This helps in understanding the details about the molecules involved and the surface locations they interact with. It also helps in the determination of kinetic parameters [91].

However, some challenges, such as the intricate nature of organic reactant spectra and low signal intensities, are the major remain as the major concern which needs to be addressed. Quantum mechanics calculations have facilitated peak assignments; however, further enhancement is still needed for the accuracy of vibrational mode assignments. Besides, factors like source intensity, detector sensitivity and setup designs are the current areas being addressed, which can help in kinetic and microscopy studies.

In a nutshell, FTIR is the most commonly used characterization technique in catalysis because of its versatility, accessibility and cost-effectiveness. It remains a dynamic and effective method for understanding the intricacies of catalysis system, offering valuable insights into progressively more complicated systems.

Raman Spectroscopy

Similar to IR spectroscopy, Raman spectroscopy provides information about the rotational and vibrational energy states of molecules. The main difference between both techniques lies in how light interacts with molecules. In crystalline solids, atomic vibrations occur as phase-related movements, causing different vibrational modes termed phonons. Based on the presence or absence of a net dipole moment, phonons can be categorized as acoustic or optical [92]. This vibration also causes inelastic light scattering phenomena. Upon interaction of electromagnetic radiation with an atom, absorption, reflection, or scattering can take place. The majority of scattered radiation retains its frequency, a phenomenon referred to as Rayleigh scattering. In Raman scattering, an incoming photon raises a molecule from its ground state to its virtual state. The virtual state subsequently relaxes, releasing photons with different energy. If the photon possesses lower energy, it's known as the Stokes scattering phenomenon, whereas, if the photon possesses higher energy, it is called anti-stokes scattering phenomenon [93, 94]. These energy differences occur *via* the following fundamental steps:

- Photon absorption results in the creation of a virtual electron-hole pair.
- The virtual electron-hole pair interacts with vibrational mode of molecule, resulting in the creation or destruction of phonon and the formation of second virtual electron-hole pair.
- The subsequent electron-hole pair recombines, leading to the emission of Raman-scattered photon.

The Raman scattering phenomenon adheres to the fundamental principles of energy and momentum conversation. The shape and intensity of the Raman bands are governed by Raman scattering transition moment, which is defined by Raman scattering tensor. The Raman scattering, developed as Taylor series of phonons, can be written as follows:

ε (Q,ω)= ε (ω)+(dε (ω))/dQ Q+1/22 (dε (ω))/(dQ^2)+•••

Where, ε is the frequency dependent dielectric constant, ω is the frequency of incident light and Q is phonon normal coordinate. The series of derivatives describes the Raman tensor of first and second order and so on.

Raman spectroscopy faces several challenges that can influence its precision and use. A major concern is laser heating, which may modify temperature sensitive material, resulting in dehydration, phase transition, reduction or disintegration which can compromise the validity of results. To overcome this issue, low laser strength (<10 mW), conventional spectrometers containing holographic filters and cooled CCD detectors, inert gas cooling and methods such as sample rotation, cylindrical lens focusing, and distributed laser energy are utilized [95].

NMR Technique

In situ, solid state NMR (Nuclear Magnetic Resonance) spectroscopy is the most commonly used method for investigating the structure and movements of absorbed reactants, intermediate, and products on solid catalyst surfaces. The use of MAS (Magic Angle Spinning) allows the detection of active sites and observation of chemical processes inside NMR rotors can be made. These studies are carried out using either batch-like or continuous flow conditions. Although economical and easily accessible, batch-type studies showcase minimal similarity to the actual catalytic reaction as compared to the continuous flow study. Advancement in solid state NMR techniques in real time environment has greatly enhanced the comprehension of heterogeneous catalysis. MAS NMR simultaneously can be used with Density Functional Theoretical (DFT) computations to detect intermediates on catalysts and understand the reaction processes. The addition of laser-enhanced MAS NMR in the in-situ flow mode enables the examination of reaction kinetics within the nano space providing valuable information about the rate of reaction under actual settings. *In situ*, NMR MAS can also be integrated with GC or UV/Vis spectroscopy to analyze the catalytic process [96].

Some potential areas to improve the sensitivity of MAS NMR include techniques such as Dynamic Nuclear Polarisation (DNP), direct NMR measurements of low nuclei during reactions using ultra-high field magnets, and advanced NMR sequences. MAS NMR can also be combined with quantum chemical calculations and simulations to gain a better understanding of the reaction mechanisms and kinetics of solid catalysts.

Microscopic Techniques

The scanning electron microscope is widely used to examine the morphological and chemical composition of nanomaterials. The signal generated from the interaction between the electron beam and the sample is responsible for producing SEM images. The 3D representation of sample pictures can be generated due to the extensive depth of field used in SEM. Transmission Electron Microscopy (TEM) is extensively used to ascertain the crystal structure and shape of nanomaterial. It has higher resolution than light-based imaging techniques, facilitating the creation of 2D projection images of inside structures. SEM generates images by detecting secondary and back scattered electrons, whereas TEM generates images through transmitted electrons. Apart from this, the High-Resolution Transmission Electron Microscopy (HRTEM) technique is employed to examine atomic structures [97]. One primary limitation of SEM is that it captures relatively low-resolution images as compared to TEM which makes it less effective for studying nanoscale features of catalysts. SEM can provide valuable information on morphology and surface structure; however, it cannot reliably capture the distribution of smaller catalytic sites or distinguish between surface elements with similar atomic numbers. TEM, on the other side, provides atomic resolution but needs extensive sample preparation which may alter the native state of the catalyst. The vacuum conditions required for both SEM and TEM might also cause structural changes or desorption of surface species, limiting the applicability of these techniques *in situ* studies. Also, solid base catalysts involve lightweight elements such as oxygen, nitrogen, and alkali metals, which are challenging to detect with high precision using electron microscopy due to contrast to weak electron scattering. All these limitations of the techniques indicate that SEM and TEM with other characterization tools can provide a more holistic understanding of solid base catalysts [92-98].

CONCLUSION

In conclusion, the development and refinement of synthetic methods for the synthesis of solid base catalysts is a significant advancement in catalysis. Various recent approaches, including the impregnation method, sol-gel process, template-assisted techniques, nanoparticle immobilization, melt infiltration, fly ash technique, sono chemical method, MOFs, and hydrothermal synthesis, have shown their effectiveness in the development of catalysts with desirable properties such as high surface area, strong basic sites, desired pore size distribution, and robust stability. The particular application requirements and the necessary catalyst qualities determine which approach is best.

Further research ought to focus on enhancing these methods in order to boost solid base catalyst's activity, selectivity, scalability, reproducibility, and durability as well as investigating environmentally friendly and sustainable synthesis strategies. The development of these catalysts may further contribute to more sustainable and effective chemical processes in an assortment of industrial applications, such as the synthesis of fine chemicals, the generation of biodiesel, and environmental remediation.

The characterization of solid base catalysts is crucial for a complete understanding of their catalytic properties and the enhancement of their efficiency in diverse chemical portfolios. The prerequisite for determining the basicity of solid base catalysis includes the quantity, source, and potency of the basic sites. In the past decade, there has been a significant improvement in the techniques used to describe and examine these catalysts. The background includes an examination of their structure, studying thermal properties, and analysis of the same using various types of spectroscopy. In the current chapter, we have described various characterization methods such as BET, XRD, DTA, DSC, NMR, and FTIR. BET and XRD techniques are used for structural analysis as both can provide valuable information about the surface area, pore size, and crystalline structure of catalysts. The BET method uses the adsorption isotherm phenomenon to accurately quantify surface area at the atomic scale which is crucial for calculating the active surface area available for the catalytic processes. XRD, on the other hand, gives a clear idea about the understanding of the crystalline structure and the average particle size of the catalysts. The method also helps understand the defect in sites that usually serve as the active sites for the catalytic events. Thermal analysis techniques, DTA and DSC help in understating the thermal stability and its phase change due to the temperature. DSC measures the heat flow linked to the thermal transition and gives precise insights into the phase alterations and reaction rates.

Whereas, DTA uses temperature difference between sample and reference material to detect phase transitions and chemical reactions occurring within the catalyst. FTIR and NMR indicate the structural changes in the catalyst upon examination of chemical surroundings. NMR combined with MAS makes the direct study of structural and dynamic processes possible in its respective environment. This technique yields a more comprehensive understanding of the active sites and other intermediate reactions occurring on solid catalysts.

Future advancements in the characterization of solid base catalysis hold more refined and precise methods, such as DNP in NMR and the utilization of modern synchrotron radiation sources for XRD. These techniques will improve the ability to investigate the complex interaction of catalytic processes and will provide a comprehensive reaction mechanism with exceptional features. All in all, the incorporation of sophisticated characterization methods is necessary to keep up with the advancements in catalysis processes. These techniques are reliable and provide in-depth information about the structural, thermal, and chemical characteristics of solid base catalysts, which helps in the improvement of catalytic processes leading to the development of its wide array of applications.

REFERENCES

[1] Mishra, V.; Cho, J.K.; Shin, S.H.; Suh, Y.W.; Kim, H.S.; Kim, Y.J. Ruthenium-Na_2CO_3-catalyzed one-pot synthesis of ring-hydrogenated carbamates from aromatic amines and organic carbonates under H2. *Appl. Catal. A Gen.*, **2014**, *487*, 82-90. [http://dx.doi.org/10.1016/j.apcata.2014.09.013]

[2] Yadav, N.; Mudgal, D. Catalytic applications of cerium-based material. *In Bentham Books- Cerium-based material: synthesis, properties, and applications,* **2023**. Available from: https://benthambooks.com/book/9789815080087/preface/

[3] Oh, S.G.; Mishra, V.; Cho, J.K.; Kim, B-J.; Kim, H.S.; Suh, Y-W.; Lee, H.; Park, H.S.; Kim, Y.J. One pot catalytic NO_2 reduction, ring hydrogenation, and N-alkylation from nitroarenes to generate alicyclic amines using Ru/C-$NaNO_2$. *Catal. Commun.*, **2014**, *43*(1), 79-83. [http://dx.doi.org/10.1016/j.catcom.2013.09.012]

[4] Kaushal, A.; Bhukya, J.; Yadav, A.K.; Mishra, V.; Kumar, D. In(III)-catalyzed C3-alkylation of indoles with 2H-chromen-2-ol derivatives. *Tetrahedron Lett.*, **2023**, *129*, 154742. [http://dx.doi.org/10.1016/j.tetlet.2023.154742]

[5] Dwivedi, B.; Mishra, P.K. Chiral metal-organic frameworks for asymmetrical catalysis In: *Metal-Organic Frameworks (MOFs) as Catalysts*; , **2022**; pp. 10-1007. 19.

[6] Chatterjee, S.; Jeevanandham, S.; Mukherjee, M.; Vo, D.V.N.; Mishra, V. Significance of re-engineered zeolites in climate mitigation – A review for carbon capture and separation. *J. Environ. Chem. Eng.*, **2021**, *9*(5), 105957. [http://dx.doi.org/10.1016/j.jece.2021.105957]

[7] Pratap Singh, R.; Shee, S.; Soni, M.; Kale, A.; Shukla, L.; Ghoshal, T.; Mishra, V. One pot synthesis of carboxylic acid, ester, and amide derivatives from iodoarenes in aqueous solubilized systems using sodium oleate as surfactant at room temperature. *Tetrahedron Lett.*, **2024**, *137*, 154930. [http://dx.doi.org/10.1016/j.tetlet.2024.154930]

[8] Mishra, V.; Mudgal, D.; Singh, R.P.; Yadav, N.; Bharti, T. Exploring the catalytic efficiency of

copper-doped magnetic carbon aerogel for the coupling reaction of isatin oxime with phenylboronic acid derivatives. *SynOpen,* **2023**, *7*(4), 570-579.
[http://dx.doi.org/10.1055/a-2182-7757]

[9] Pratap Singh, R.; Ghoshal, T.; Mishra, V. Trifluoromethylation strategies of alcohols and phenols. *Asian J. Org. Chem.,* **2024**, *13*(11), e202400179.
[http://dx.doi.org/10.1002/ajoc.202400179]

[10] Ono, Y.; Hattori, H. *Characterization of solid base catalysts.,* **2011**, 11-68.
[http://dx.doi.org/10.1007/978-3-642-18339-3_2]

[11] Mudgal, D.; Yadav, N.; Mishra, V. Nickel-doped magnetic carbon aerogel derived from xanthan gum: A competent catalyst for the degradation of single and binary dye-based water pollutants. *Environ. Sci. Pollut. Res. Int.,* **2024**, *31*(25), 36748-36760.
[http://dx.doi.org/10.1007/s11356-024-33605-2] [PMID: 38750276]

[12] Kumar, M.; Verma, S.; Mishra, V.; Reiser, O.; Verma, A.K. Visible-light-accelerated copper-catalyzed [3 + 2] cycloaddition of *N* -tosylcyclopropylamines with alkynes/alkenes. *J. Org. Chem.,* **2022**, *87*(9), 6263-6272.
[http://dx.doi.org/10.1021/acs.joc.2c00491] [PMID: 35476544]

[13] Mishra, V.; Aggarwal, S.; Pandey, S. Porous polymer for heterogeneous catalysis. *Porous Polymer Science and Applications*; CRC Press, **2022**, pp. 91-120.
[http://dx.doi.org/10.1201/9781003169604-6]

[14] Panda, S.K.; Aggarwal, I.; Kumar, H.; Prasad, L.; Kumar, A.; Sharma, A.; Vo, D.V.N.; Van Thuan, D.; Mishra, V. Magnetite nanoparticles as sorbents for dye removal: a review. *Environ. Chem. Lett.,* **2021**, *19*(3), 2487-2525.
[http://dx.doi.org/10.1007/s10311-020-01173-9]

[15] Hattori, H. Solid base catalysts: Generation, characterization, and catalytic behavior of basic sites. *J. Jpn. Petrol. Inst.,* **2004**, *47*(2), 67-81.
[http://dx.doi.org/10.1627/jpi.47.67]

[16] Jambhulkar, D.K.; Ugwekar, R.P.; Bhanvase, B.A.; Barai, D.P. A review on solid base heterogeneous catalysts: preparation, characterization and applications. *Chem. Eng. Commun.,* **2022**, *209*(4), 433-484.
[http://dx.doi.org/10.1080/00986445.2020.1864623]

[17] de Jong, K.P. *Synthesis of Solid Catalysts*; John Wiley & Sons, **2009**.
[http://dx.doi.org/10.1002/9783527626854]

[18] Behrens, M. Coprecipitation: An excellent tool for the synthesis of supported metal catalysts – From the understanding of the well known recipes to new materials. *Catal. Today,* **2015**, *246*, 46-54.
[http://dx.doi.org/10.1016/j.cattod.2014.07.050]

[19] Kaluza, S.; Schröter, M.K.; Naumann d'Alnoncourt, R.; Reinecke, T.; Muhler, M. High surface area zno nanoparticles *via* a novel continuous precipitation route. *Adv. Funct. Mater.,* **2008**, *18*(22), 3670-3677.
[http://dx.doi.org/10.1002/adfm.200800457]

[20] Kaluza, S.; Behrens, M.; Schiefenhövel, N.; Kniep, B.; Fischer, R.; Schlögl, R.; Muhler, M. A novel synthesis route for Cu/ZnO/Al_2O_3 catalysts used in methanol synthesis: combining continuous consecutive precipitation with continuous aging of the precipitate. *ChemCatChem,* **2011**, *3*(1), 189-199.
[http://dx.doi.org/10.1002/cctc.201000329]

[21] Shroff, M.D.; Kalakkad, D.S.; Coulter, K.E.; Kohler, S.D.; Harrington, M.S.; Jackson, N.B.; Sault, A.G.; Datye, A.K. Activation of precipitated iron fischer-tropsch synthesis catalysts. *J. Catal.,* **1995**, *156*(2), 185-207.
[http://dx.doi.org/10.1006/jcat.1995.1247]

[22] Behrens, M.; Schlögl, R. How to prepare a good Cu/ZnO catalyst or the role of solid state chemistry

for the synthesis of nanostructured catalysts. *Z. Anorg. Allg. Chem.,* **2013**, *639*(15), 2683-2695. [http://dx.doi.org/10.1002/zaac.201300356]

[23] Behrens, M.; Brennecke, D.; Girgsdies, F.; Kißner, S.; Trunschke, A.; Nasrudin, N.; Zakaria, S.; Idris, N.F.; Hamid, S.B.A.; Kniep, B.; Fischer, R.; Busser, W.; Muhler, M.; Schlögl, R. Understanding the complexity of a catalyst synthesis: Co-precipitation of mixed Cu,Zn,Al hydroxycarbonate precursors for Cu/ZnO/Al2O3 catalysts investigated by titration experiments. *Appl. Catal. A Gen.,* **2011**, *392*(1-2), 93-102. [http://dx.doi.org/10.1016/j.apcata.2010.10.031]

[24] Zander, S.; Seidlhofer, B.; Behrens, M. *In situ* EDXRD study of the chemistry of aging of co-precipitated mixed Cu,Zn hydroxycarbonates – consequences for the preparation of Cu/ZnO catalysts. *Dalton Trans.,* **2012**, *41*(43), 13413-13422. [http://dx.doi.org/10.1039/c2dt31236k] [PMID: 23007170]

[25] Li, Z.; Yan, S.; Fan, H. Enhancement of stability and activity of Cu/ZnO/Al2O3 catalysts by microwave irradiation for liquid phase methanol synthesis. *Fuel,* **2013**, *106*, 178-186. [http://dx.doi.org/10.1016/j.fuel.2012.11.003]

[26] de Jong, K.P.; Geus, J.W. Production of supported silver catalysts by liquid-phase reduction. *Appl. Catal.,* **1982**, *4*(1), 41-51. [http://dx.doi.org/10.1016/0166-9834(82)80288-1]

[27] Lok, C.M. Novel highly dispersed cobalt catalysts for improved fischer-tropsch productivity.*Studies in surface science and catalysis*; Elsevier, **2004**, 147, pp. 283-288. [http://dx.doi.org/10.1016/S0167-2991(04)80065-X]

[28] Vayenas, C.G. *Catalysis and Electrocatalysis at Nanoparticle Surfaces*; Macel Dekker, Inc: New York, **2003**.

[29] Fang, B.; Chaudhari, N.K.; Kim, M.S.; Kim, J.H.; Yu, J.S. Homogeneous deposition of platinum nanoparticles on carbon black for proton exchange membrane fuel cell. *J. Am. Chem. Soc.,* **2009**, *131*(42), 15330-15338. [http://dx.doi.org/10.1021/ja905749e] [PMID: 19795863]

[30] Kim, M.S.; Lim, S.; Chaudhari, N.K.; Fang, B.; Bae, T.S.; Yu, J.S. Effect of pH on electrocatalytic property of supported PtRu catalysts in proton exchange membrane fuel cell. *Catal. Today,* **2010**, *158*(3-4), 354-360. [http://dx.doi.org/10.1016/j.cattod.2010.04.037]

[31] Deraz, N.M. The comparative jurisprudence of catalysts preparation methods: I. precipitation and impregnation methods. *J. Ind. Environ. Chem,* **2018**, *2*(1), 19-21.

[32] Bezemer, G.; Radstake, P.; Koot, V.; Vandillen, A.; Geus, J.; Dejong, K. Preparation of Fischer–Tropsch cobalt catalysts supported on carbon nanofibers and silica using homogeneous deposition-precipitation. *J. Catal.,* **2006**, *237*(2), 291-302. [http://dx.doi.org/10.1016/j.jcat.2005.11.015]

[33] Eschemann, T.O.; Bitter, J.H.; de Jong, K.P. Effects of loading and synthesis method of titania-supported cobalt catalysts for Fischer–Tropsch synthesis. *Catal. Today,* **2014**, *228*, 89-95. [http://dx.doi.org/10.1016/j.cattod.2013.10.041]

[34] Zanella, R.; Giorgio, S.; Henry, C.R.; Louis, C. Alternative methods for the preparation of gold nanoparticles supported on TiO_2. *J. Phys. Chem. B,* **2002**, *106*(31), 7634-7642. [http://dx.doi.org/10.1021/jp0144810]

[35] Zanella, R.; Louis, C.; Giorgio, S.; Touroude, R. Crotonaldehyde hydrogenation by gold supported on TiO_2: structure sensitivity and mechanism. *J. Catal.,* **2004**, *223*(2), 328-339. [http://dx.doi.org/10.1016/j.jcat.2004.01.033]

[36] Li, H.; Wang, L.; Wei, Y.; Yan, W.; Feng, J. Preparation of templated materials and their application to typical pollutants in wastewater: A review. *Front Chem.,* **2022**, *10*(April), 882876.

[http://dx.doi.org/10.3389/fchem.2022.882876] [PMID: 35480393]

[37] Eslek Koyuncu, D.D.; Okur, M. Removal of AV 90 dye using ordered mesoporous carbon materials prepared *via* nanocasting of KIT-6: Adsorption isotherms, kinetics and thermodynamic analysis. *Separ. Purif. Tech.,* **2021**, *257*, 117657.https://doi.org/https://doi.org/10.1016/j.seppur.2020.117657 [http://dx.doi.org/10.1016/j.seppur.2020.117657]

[38] Schwarz, J.A.; Contescu, C.; Contescu, A. Methods for preparation of catalytic materials. *Chem. Rev.,* **1995**, *95*(3), 477-510. [http://dx.doi.org/10.1021/cr00035a002]

[39] Perego, C.; Villa, P. Catalyst preparation methods. *Catal. Today,* **1997**, *34*(3-4), 281-305. [http://dx.doi.org/10.1016/S0920-5861(96)00055-7]

[40] Shahinuzzaman, M.; Yaakob, Z.; Ahmed, Y. Non-sulphide zeolite catalyst for bio-jet-fuel conversion. *Renew. Sustain. Energy Rev.,* **2016**, *2017*(December), 1-10. [http://dx.doi.org/10.1016/j.rser.2017.01.162]

[41] De Vos, Y.; Jacobs, M.; Van Der Voort, P.; Van Driessche, I.; Snijkers, F.; Verberckmoes, A. Development of stable oxygen carrier materials for chemical looping processes—A review. *Catalysts,* **2020**, *10*(8), 926. [http://dx.doi.org/10.3390/catal10080926]

[42] Xie, W.; Huang, X.; Li, H. Soybean oil methyl esters preparation using NaX zeolites loaded with KOH as a heterogeneous catalyst. *Bioresour. Technol.,* **2007**, *98*(4), 936-939. [http://dx.doi.org/10.1016/j.biortech.2006.04.003] [PMID: 16740389]

[43] Vyas, A. P.; Subrahmanyam, N.; Patel, P. A. Production of biodiesel through transesterification of jatropha oil using KNO_3 / Al_2O_3 solid catalyst. *Fuel,* **2009**, *88*, 625-628. [http://dx.doi.org/10.1016/j.fuel.2008.10.033]

[44] Schüth, F. *Handbook of Porous Solids*; No Title, **2002**. [http://dx.doi.org/10.1002/9783527618286]

[45] Baiker, A. Towards molecular design of solid catalysts.*Studies in Surface Science and Catalysis*; Elsevier, **1996**, 101, pp. 51-62. [http://dx.doi.org/10.1016/S0167-2991(96)80214-X]

[46] Pierre, A.C.; Pajonk, G.M. Chemistry of aerogels and their applications. *Chem. Rev.,* **2002**, *102*(11), 4243-4266. [http://dx.doi.org/10.1021/cr0101306] [PMID: 12428989]

[47] Lee, D.W.; Yoo, B.R. Advanced metal oxide (supported) catalysts: Synthesis and applications. *J. Ind. Eng. Chem.,* **2014**, *20*(6), 3947-3959. [http://dx.doi.org/10.1016/j.jiec.2014.08.004]

[48] Mudgal, D.; Yadav, N.; Singh, J.; Srivastava, G.K.; Mishra, V. Xanthan gum-based copper nano-magnetite doped carbon aerogel: A promising candidate for environmentally friendly catalytic dye degradation. *Int. J. Biol. Macromol.,* **2023**, *253*(Pt 7), 127491. [http://dx.doi.org/10.1016/j.ijbiomac.2023.127491] [PMID: 37852396]

[49] Zhan, Z.; Zeng, H.C. A catalyst-free approach for sol–gel synthesis of highly mixed ZrO_2–SiO_2 oxides. *J. Non-Cryst. Solids,* **1999**, *243*(1), 26-38. [http://dx.doi.org/10.1016/S0022-3093(98)00810-2]

[50] Bems, B.; Schur, M.; Dassenoy, A.; Junkes, H.; Herein, D.; Schlögl, R. Relations between synthesis and microstructural properties of copper/zinc hydroxycarbonates. *Chemistry,* **2003**, *9*(9), 2039-2052. [http://dx.doi.org/10.1002/chem.200204122] [PMID: 12740852]

[51] Solsona, B.; Garcia, T.; Jones, C.; Taylor, S.; Carley, A.; Hutchings, G. Supported gold catalysts for the total oxidation of alkanes and carbon monoxide. *Appl. Catal. A Gen.,* **2006**, *312*, 67-76. [http://dx.doi.org/10.1016/j.apcata.2006.06.016]

[52] Khoudiakov, M.; Gupta, M.C.; Deevi, S. Au/Fe2O3 nanocatalysts for CO oxidation: A comparative study of deposition–precipitation and coprecipitation techniques. *Appl. Catal. A Gen.*, **2005**, *291*(1-2), 151-161.
[http://dx.doi.org/10.1016/j.apcata.2005.01.042]

[53] Zhang, H.; Qi, R.; Evans, D.G.; Duan, X. Synthesis and characterization of a novel nano-scale magnetic solid base catalyst involving a layered double hydroxide supported on a ferrite core. *J. Solid State Chem.*, **2004**, *177*(3), 772-780.
[http://dx.doi.org/10.1016/j.jssc.2003.09.009]

[54] Yang, G.; Park, S.J. Conventional and microwave hydrothermal synthesis and application of functional materials: A review. *Materials (Basel)*, **2019**, *12*(7), 1177.
[http://dx.doi.org/10.3390/ma12071177] [PMID: 30978917]

[55] Pescarmona, P.P.; Rops, J.J.T.; van der Waal, J.C.; Jansen, J.C.; Maschmeyer, T. High-speed experimentation techniques applied to the study of the synthesis of zeolites and silsesquioxanes. *J. Mol. Catal. Chem.*, **2002**, *182-183*, 319-325.
[http://dx.doi.org/10.1016/S1381-1169(01)00494-0]

[56] Pescarmona, P.P.; van der Waal, J.C.; Maxwell, i.e.; Maschmeyer, T. A new, efficient route to titanium-silsesquioxane epoxidation catalysts developed by using high-speed experimentation techniques. In: *Angew. Chemie-German Ed*; , **2001**; 113, pp. (4)762-764.

[57] Corma, A.; Díaz-Cabañas, M.J.; Jordá, J.L.; Martínez, C.; Moliner, M. High-throughput synthesis and catalytic properties of a molecular sieve with 18- and 10-member rings. *Nature*, **2006**, *443*(7113), 842-845.
[http://dx.doi.org/10.1038/nature05238] [PMID: 17051215]

[58] Raza, A.; Azam, A.; Saeed, M.S.; Ahsan, M.; Qayyum, F.; Yaseen, M. Hydrothermal synthesis and characterization of co0. 5Zn0. $5Fe_2O_4$ nano-material and evaluation of its photo-catalytic activity under visible light irradiation. *Dig. J. Nanomater. Biostruct.*, **2016**, *11*(4), 1289-1298.

[59] Caillot, T.; Salama, Z.; Chanut, N.; Cadete Santos Aires, F.J.; Bennici, S.; Auroux, A. Hydrothermal synthesis and characterization of zirconia based catalysts. *J. Solid State Chem.*, **2013**, *203*, 79-85.
[http://dx.doi.org/10.1016/j.jssc.2013.04.005]

[60] Sáez, V.; Mason, T. J. *Sonoelectrochemical Synthesis of Nanoparticles.*, **2009**, 4284-4299.
[http://dx.doi.org/10.3390/molecules14104284]

[61] Ekka, B.; Nayak, S.R.; Achary, L.S.K.; Kumar, A.; Mawatwal, S.; Dhiman, R.; Dash, P.; Patel, R.K. Department of Chemistry SC. *Biochem. Pharmacol.*, **2018**.
[http://dx.doi.org/10.1016/j.jece.2018.09.026]

[62] Karuppasamy, L.; Chen, C.; Anandan, S.; Wu, J.J. ScienceDirect sonochemical reduction method for synthesis of TiO_2 e Pd nanocomposites and investigation of anode and cathode catalyst for ethanol oxidation and oxygen reduction reaction in alkaline medium. *Int. J. Hydrogen Energy*, **2018**, **•••**, 1-14.
[http://dx.doi.org/10.1016/j.ijhydene.2018.08.080]

[63] Pedersen, H.; Elliott, S. D. *Studying chemical vapor deposition processes with theoretical chemistry.*, **2014**.
[http://dx.doi.org/10.1007/s00214-014-1476-7]

[64] Li, K.; Lin, P.; Lin, S. *Preparation of Ti/SiO_2 catalysts by chemical vapor deposition method for olefin epoxidation with cumene hydroperoxide,* **2006**, *301*, 59-65.
[http://dx.doi.org/10.1016/j.apcata.2005.11.012]

[65] Matsuhashi, H.; Kawamura, A. Preparation of a novel solid base catalyst of CaO covered with SiO_2. *Catal. Today*, **2012**, *185*(1), 236-240.
[http://dx.doi.org/10.1016/j.cattod.2011.08.015]

[66] Seo, M.; Kim, S.Y.; Kim, Y.D.; Park, E.D.; Uhm, S. Highly stable barium zirconate supported nickel oxide catalyst for dry reforming of methane: From powders toward shaped catalysts. *Int. J. Hydrogen*

Energy, **2018**, *43*(24), 11355-11362.
[http://dx.doi.org/10.1016/j.ijhydene.2018.03.181]

[67] Cirujano, F.G.; Martin, N.; Wee, L.H. Design of hierarchical architectures in metal–oganic frameworks for catalysis and adsorption. *Chem. Mater.,* **2020**, *32*(24), 10268-10295.
[http://dx.doi.org/10.1021/acs.chemmater.0c02973]

[68] Naser, S.A.E.; Badmus, K.O.; Khotseng, L. Synthesis, properties, and applications of metal organic frameworks supported on graphene oxide. *Coatings,* **2023**, *13*(8), 1456.
[http://dx.doi.org/10.3390/coatings13081456]

[69] Gangu, K. K.; Jonnalagadda, S. B. *A Review on metal-organic frameworks as congenial heterogeneous catalysts for potential organic transformations.,* **2021**, *9*, 1-12.
[http://dx.doi.org/10.3389/fchem.2021.747615]

[70] Huang, J.; Li, C.; Tao, L.; Zhu, H.; Hu, G. Synthesis, characterization and heterogeneous base catalysis of amino functionalized lanthanide metal-organic frameworks. *J. Mol. Struct.,* **2017**, *1146*, 853-860.
[http://dx.doi.org/10.1016/j.molstruc.2017.06.045]

[71] Gu, J.M.; Kim, W.S.; Huh, S. Size-dependent catalysis by DABCO-functionalized Zn-MOF with one-dimensional channels. *Dalton Trans.,* **2011**, *40*(41), 10826-10829.
[http://dx.doi.org/10.1039/c1dt11274k] [PMID: 21931931]

[72] Ma, X.; Zhang, W. Recent developments in One-Pot Stepwise Synthesis (OPSS) of small molecules. *iScience,* **2022**, *25*(9), 105005.
[http://dx.doi.org/10.1016/j.isci.2022.105005] [PMID: 36093043]

[73] Sun, L. B.; Yang, J.; Kou, J. H.; Gu, F. N.; Chun, Y.; Wang, Y.; Zhu, J. H. *One-Pot synthesis of potassium-functionalized mesoporous g -alumina : A solid superbase **,* **2008**, 3418-3421.
[http://dx.doi.org/10.1002/anie.200800034]

[74] Sun, H.; Han, J.; Ding, Y.; Li, W.; Duan, J.; Chen, P.; Lou, H.; Zheng, X. One-pot synthesized mesoporous Ca/SBA-15 solid base for transesterification of sunflower oil with methanol. *Appl. Catal. A Gen.,* **2010**, *390*(1-2), 26-34.
[http://dx.doi.org/10.1016/j.apcata.2010.09.030]

[75] Tantirungrotechai, J.; Thananupappaisal, P.; Yoosuk, B.; Viriya-empikul, N.; Faungnawakij, K. One-pot synthesis of calcium-incorporated MCM-41 as a solid base catalyst for transesterification of palm olein. *Catal. Commun.,* **2011**, *16*(1), 25-29.
[http://dx.doi.org/10.1016/j.catcom.2011.09.002]

[76] Zhao, J.; Xie, J.; Au, C.T.; Yin, S.F. One-pot synthesis of potassium-loaded MgAl oxide as solid superbase catalyst for Knoevenagel condensation. *Appl. Catal. A Gen.,* **2013**, *467*, 33-37.
[http://dx.doi.org/10.1016/j.apcata.2013.06.054]

[77] Darmansyah, D.; You, S.J.; Wang, Y.F. Modification of recycled industrial solid waste incineration fly ash as a low-cost catalyst for nox removal. *Aerosol Air Qual. Res.,* **2024**, *24*(3), 230247.
[http://dx.doi.org/10.4209/aaqr.230247]

[78] Liu, J.; Shi, N.; Wang, T.; Yang, J.; Liu, P.F.; Zhang, Y.; Pan, W.P. Significant enhancement of VOCs conversion by facile mechanochemistry coupled MnO2 modified fly ash: Mechanism and application. *Fuel,* **2021**, *304*(June), 121443.
[http://dx.doi.org/10.1016/j.fuel.2021.121443]

[79] Jain, D.; Khatri, C.; Rani, A. Synthesis and characterization of novel solid base catalyst from fly ash. *Fuel,* **2011**, *90*(6), 2083-2088.
[http://dx.doi.org/10.1016/j.fuel.2010.09.025]

[80] Eggenhuisen, T. M.; Breejen, J. P.; Den, ; Verdoes, D.; Jongh, P. E.; De, ; Jong, K. P.; De, Fundamentals of melt infiltration for the preparation of supported metal catalysts. *The case of Co / SiO2 for Fischer-Tropsch synthesis,* **2010**, 18318-18325.

[http://dx.doi.org/10.1002/adma.201301912]

[81] Jongh, P. E.; De, ; Eggenhuisen, T. M. Melt infiltration: An emerging technique for the preparation of novel functional nanostructured materials *Adv. Mater.,* **2013**, *25*, 6672-6690. [http://dx.doi.org/10.1002/adma.201301912]

[82] S-de-la-Muela, A.M.; Cambronero, L.e.g.; Ruiz-Román, J.M. Molten metal infiltration methods to process metal matrix syntactic foams. *Metals (Basel),* **2020**, *10*(1), 149. [http://dx.doi.org/10.3390/met10010149]

[83] Agarwal, A.K.; Kadu, M.S.; Pandhurnekar, C.P.; Muthreja, I.L. Brunauer-Emmett-Teller (B.E.T.), Langmuir and freundlich isotherm studies for the adsorption of nickel ions onto coal fly ash. *Asian J. Water Environ. Pollut.,* **2016**, *13*(2), 49-53. [http://dx.doi.org/10.3233/AJW-160016]

[84] Védrine, J.C. Acid–base characterization of heterogeneous catalysts: an up-to-date overview. *Res. Chem. Intermed.,* **2015**, *41*(12), 9387-9423. [http://dx.doi.org/10.1007/s11164-015-1982-9]

[85] Sankar, G.; Thomas, J.M. *In situ* combined X-ray absorption spectroscopic and X-ray diffractometric studies of solid catalysts. *Top. Catal.,* **1999**, *8*(1/2), 1-21. [http://dx.doi.org/10.1023/A:1019188422574]

[86] Schlögl, R. X□ray diffraction. *Adv. Catal.,* **2009**, *52*(C), 273-338. [http://dx.doi.org/10.1016/S0360-0564(08)00005-9]

[87] Le Parlouër, P.. *Thermal analysis and calorimetry techniques for catalytic investigations*; , **2013**, 154, . [http://dx.doi.org/10.1007/978-3-642-11954-5_2]

[88] Rainer, D. N.; Mazur, M. *Electron Microscopy Methods for Characterisation of Zeolite Catalysts.,* **2020**. [http://dx.doi.org/10.1039/9781788019477-00151]

[89] Rizvi, A.; Mulvey, J.T.; Carpenter, B.P.; Talosig, R.; Patterson, J.P. A close look at molecular self-assembly with the transmission electron microscope. *Chem. Rev.,* **2021**, *121*(22), 14232-14280. [http://dx.doi.org/10.1021/acs.chemrev.1c00189] [PMID: 34329552]

[90] Gao, S.; Wang, L.; Zhao, C.; Wang, A.; Lang, X.; Liu, Z.; Wang, W. A review of the synergistic effect of multi-coordination crystal fields on electrocatalysts. *Mater. Chem. Front.,* **2021**, *5*(18), 6718-6734. [http://dx.doi.org/10.1039/D1QM00658D]

[91] Zaera, F. New advances in the use of infrared absorption spectroscopy for the characterization of heterogeneous catalytic reactions. *Chem. Soc. Rev.,* **2014**, *43*(22), 7624-7663. [http://dx.doi.org/10.1039/C3CS60374A] [PMID: 24424375]

[92] Imelik, B.; Vedrine, J.C. *Catalyst Characterization: Physical Techniques for Solid Materials*; Springer Science & Business Media, **2013**.

[93] Lercher, J.A.; Jentys, A. Infrared and raman spectroscopy for characterizing zeolites.*Studies in surface science and catalysis*; Elsevier, **2007**, 168, pp. 435-476.

[94] Bañares, M.A.; Mestl, G. Structural characterization of operating catalysts by raman spectroscopy. *Adv. Catal.,* **2009**, *52*, 43-128. [http://dx.doi.org/10.1016/S0360-0564(08)00002-3]

[95] Loridant, S. Raman spectroscopy as a powerful tool to characterize ceria-based catalysts. *Catal. Today,* **2021**, *373*, 98-111. [http://dx.doi.org/10.1016/j.cattod.2020.03.044]

[96] Zhang, W.; Xu, S.; Han, X.; Bao, X. *In situ* solid-state NMR for heterogeneous catalysis: A joint experimental and theoretical approach. *Chem. Soc. Rev.,* **2012**, *41*(1), 192-210. [http://dx.doi.org/10.1039/C1CS15009J] [PMID: 21743940]

[97] Su, D.S.; Zhang, B.; Schlögl, R. Electron microscopy of solid catalysts--transforming from a challenge to a toolbox. *Chem. Rev.,* **2015**, *115*(8), 2818-2882. [http://dx.doi.org/10.1021/cr500084c] [PMID: 25826447]

[98] Guo, F.; Peng, Z.G.; Dai, J.Y.; Xiu, Z.L. Calcined sodium silicate as solid base catalyst for biodiesel production. *Fuel Process. Technol.,* **2010**, *91*(3), 322-328. [http://dx.doi.org/10.1016/j.fuproc.2009.11.003]

CHAPTER 2

Recent Advances in Supported Solid Phase Catalysis for Organic Synthesis

Monika Kamboj[1,*], **Sangeeta Bajpai**[1] and **Garima Pandey**[2]

[1] *Department of Applied Chemistry, Amity School of Applied Sciences, Amity University Uttar Pradesh, Lucknow Campus, Lucknow-226028, U.P., India*

[2] *Department of Chemistry, SRM Institute of Science and Technology, Delhi NCR Campus, Modinagar Ghaziabad, U.P., India*

Abstract: Solid-phase-supported catalysis has become a key method in organic synthesis over the last two decades, offering significant advantages in expediting both synthesis and purification. This chapter will explore the types of solid phases used in metal catalysis and the various approaches employed for organic transformations.

Keywords: Catalysis, Organic synthesis, Polymer catalyst, Solid-phase.

INTRODUCTION

R.B. Merrifield pioneered solid-phase organic synthesis and developed this technique for peptide synthesis in the 1960s [1]. In 1984, he received a Nobel Prize in chemistry for his pioneering work [2]. Since then, this field has had unbelievable interest [3 - 11]. Although the concept of solid phase-supported catalysis for organic synthesis was conceived in the 1940s, it gained importance after the Merrifield solid-phase peptide synthesis. Solid phase-supported catalysis for organic synthesis involves one reactant in an immobilized phase. It has become an important method for organic transformation due to its ability to expedite synthesis and purification and the removal of immobilized products easily by simple filtration [12]. Organic transformations, typically completed in multiple steps, can be carried out in a single step using solid-phase-supported catalysis. Another advantage of this technique is reduced side reactions, ease of handling, low susceptibility to moisture, cost-effectiveness, stability, high surface area, and reusability. These supported catalysts are stable because they are chemically bound to the solid support, allowing for easy removal after the

* **Corresponding author Monika Kamboj:** Department of Applied Chemistry, Amity School of Applied Sciences, Amity University Uttar Pradesh, Lucknow Campus, Lucknow-226028, U.P., India; E-mail: mkamboj@lko.amity.edu

Ravi Tomar, Vivek Mishra, Kapil Arya & Garima Pandey (Eds.)

reaction [13 - 15]. Moreover, they are utilized in industries for carrying redox reactions and carbon-carbon bond formation. These distinctive features make them ideal catalysts for facilitating green synthetic transformations, as they address issues such as pollution, chemical disposal, and sustainability [16 - 18]. Fig. (**1**) depicts the advantage of a solid-supported catalyst.

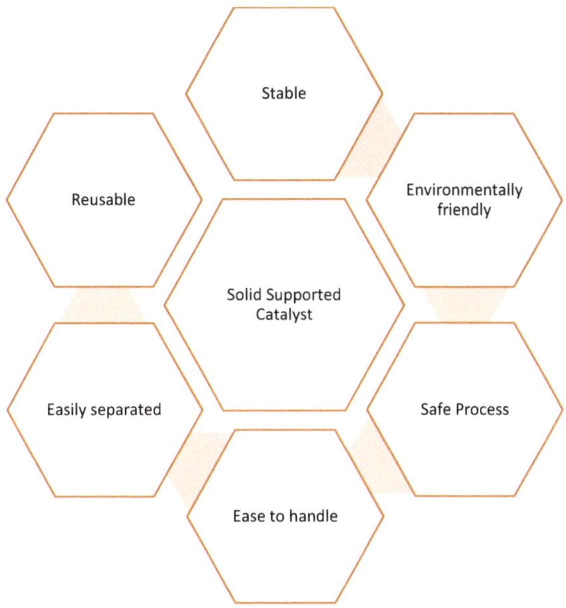

Fig. (1). Advantage of solid-supported catalyst.

These distinctive features make them ideal catalysts for green synthesis, as they help mitigate pollution, waste disposal, and sustainability challenges. Table **1** summarizes the various solid support catalysts used for organic transformation.

Various Types of Solid Supported Transition Metal Catalyst

Transition metal catalysts like palladium, ruthenium, rhodium, cobalt, and copper are supported on various heterogeneous solid surfaces and are used for various organic transformations. The noteworthy advantages of immobilized catalysts on solid support that make them a versatile and innovative tool in organic chemistry include reusability, ease of handling, and isolation from products. The different types of transition metal catalysts supported on solid phase and the various approaches used for organic transformations are discussed below:

Reaction Catalyzed by Solid Supported Palladium

The transition metal, palladium, is most extensively used as a catalyst in organic reactions. The Pd complexes are successful in homogeneous catalysis. However, removing the catalyst from the reaction mixture is difficult, limiting its use. This problem is overcome by using solid-supported palladium. The notable advantage of solid-supported palladium catalysts is that they are easy to prepare, recyclable, and have no leaching of Pd from support. These features increase the efficiency of solid-supported palladium catalysts. Palladium has the ability to change its oxidation state easily, thereby making it a versatile catalyst. Also, it shows high tolerance towards various functional groups and can withstand the reaction conditions. It also shows high selectivity, thereby facilitating the desired bond formation. Some of the organic reaction catalysts by solid-supported palladium are discussed herein:

Table 1. Different types of solid support for catalysis.

Types of Solid Support in Catalysis	Features	Examples	References
Metal Oxides	• Provide thermal and chemical stability to the catalyst • Easy to prepare.	TiO_2, CeO_2, SiO_2.	[19 - 21]
Zeolites	• They can catalyze hydrocarbon reactions due to the presence of Bronsted acid site on them. • Effective catalyst for gas phase reaction above 300 °C • Highly porous and stable.	Aluminosilicate material.	[22 - 24]
Carboneous Material	• They are carbon nanostructures • Porous and high surface area • The surface of carbonaceous material can be modified.	Graphene, graphite, Black carbon, Graphite oxide, Activated Carbon and carbon fibre.	[25 - 27]
Polymeric compounds	• They are employed for a metal catalyst • Improves stability of catalyst by preventing sintering.	Polyorganophosphazenes, styrene/divinylbenzene copolymers, dendrimers, fibers, and biopolymers.	[28, 29]

In Suzuki–Miyaura C-C cross-coupling reaction, aryl halides and arylboronic acids react to form biphenyls (40-99%). The efficient catalytic system involves a Pd complex with ligands having phosphorous and nitrogen. These ligands stabilize the complex and play a crucial role in controlling its activity and selectivity. Polymer-supported Pd catalyst (Fig. **2**) with N-heterocyclic carbenes ligand finds application in Suzuki–Miyaura cross-coupling reactions—the product obtained in moderate to excellent yields. Nie *et al.* demonstrated that the catalyst is recovered and reused readily for further transformations up to four cycles with the same catalytic power [30] (Scheme **1**).

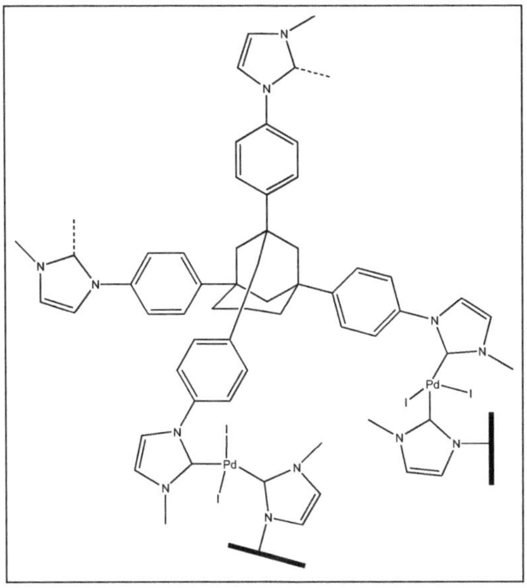

Fig. (2). Polymer-supported Pd catalyst (Pd cat).

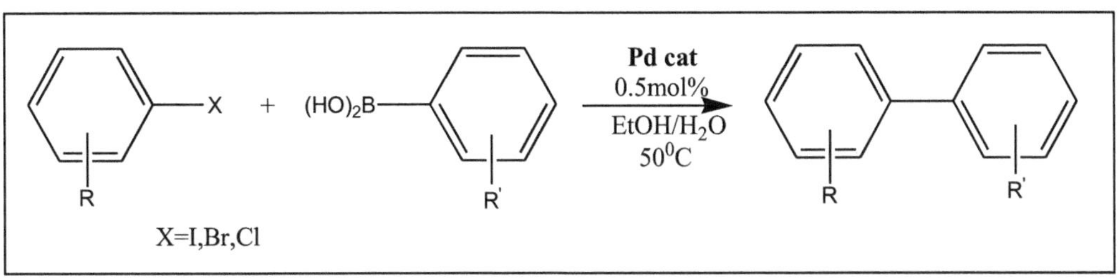

Scheme (1). Suzuki–Miyaura C-C cross-coupling reaction using polymer-supported Pd catalyst.

Formation of diaryls from aryl bromides by Suzuki–Miyaura reactions in a toluene/ethanol/water solvent mixture can be successfully carried out using a Rasta Resin-supported Pd catalyst (RRPd catalyst) (Fig. (**3**). Polymeric linear chains of Rasta resin act as a ligand for solid support that immobilizes the Pd catalyst. The catalyst can be recovered and reused for up to 5 cycles with the same efficiency [31] (Scheme **2**).

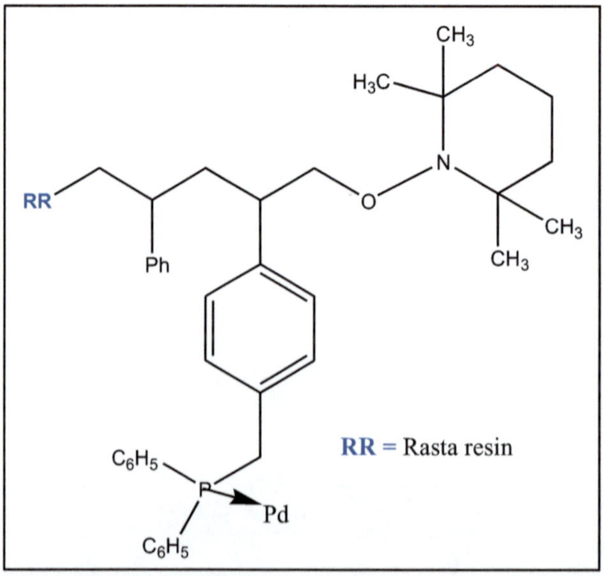

Fig. (3). Rasta Resin-supported Pd catalyst (RRPd catalyst).

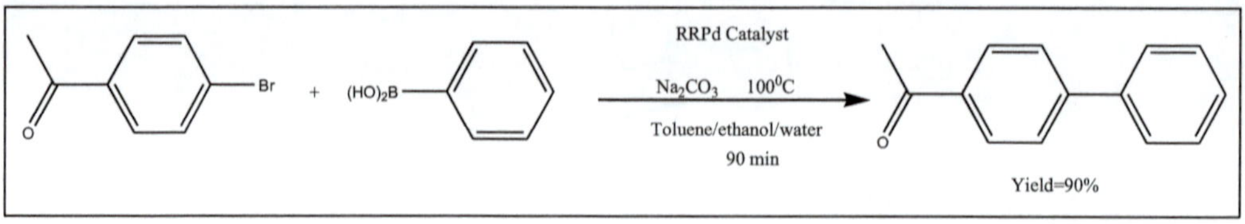

Scheme (2). Formation of Diaryls by Suzuki–Miyaura reactions using a Rasta Resin-supported Pd catalyst (RRPd catalyst).

Suzuki cross-coupling reactions can be successfully carried out using bimetallic copolymers containing palladium catalyst *i.e.*, palladium-lanthanide copolymers in DMF/water solvent system [32]. The lanthanides enhance the reductive elimination step of the cross-coupling reaction due to its electropositive tendency (Scheme **3**).

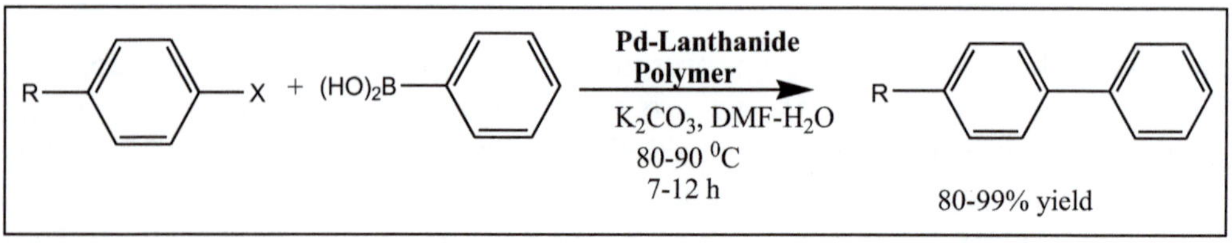

Scheme (3). Suzuki cross-coupling reactions catalyzed by Pd-Lanthanide Polymer.

K. Ebitani and researchers have reported heterogeneous organozinc copolymer catalyst functionalized with amine to support graft palladium. This system shows exceptional catalytic efficiency in Suzuki–Miyaura coupling and Mizoroki–Heck coupling reactions [33]. Shū Kobayashi *et al.* reported Polysilane/Bone Charcoal-supported Pd catalyst (PSi/BC-Pd) for effective catalytic hydrogenation of aliphatic nitro compounds. The hydrogenation of nitro compounds to amine was successful under mild conditions, with good yield. The catalyst retains its activity even after 5 days without the leaching of Pd. The catalyst was taken in the SUS column and packed with celite-3. The SUS column was preheated to 30 °C after passing ethanol solvent through it. 15 mL of H2 gas per minute was introduced into the column, followed by 0.2 M ethanolic solution of aliphatic nitroalkanes at the rate of 0.2 mL per minute under the continuous flow of hydrogen gas [34]. (Scheme **4**).

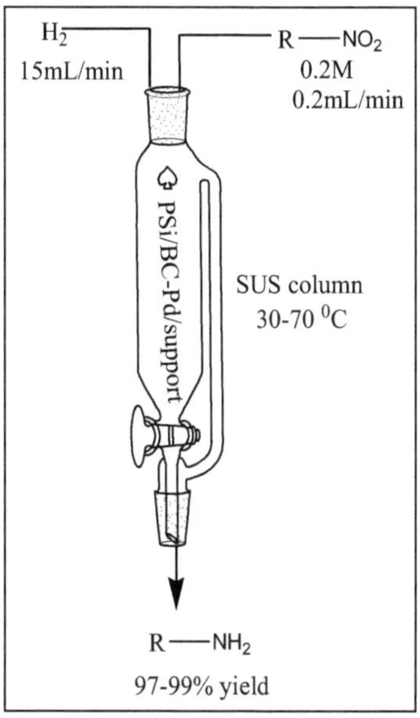

Scheme (4). Catalytic hydrogenation of aliphatic nitro compounds by Polysilane/Bone Charcoal-supported Pd catalyst (PSi/BC-Pd).

Reaction catalyzed by Solid supported Ruthenium

Various organic transformations are successful with ruthenium catalysts. Ruthenium also has the potential to change its oxidation state (-2 to +8) by accepting or donating electrons, thereby acting as a good catalyst. It is stable in air and solvents. It shows a high tendency to complexes with various ligands. It is highly effective and selective in various oxidative organic transformations. Solid-supported Ru- catalysts are versatile tools for such reactions. Various organic polymers extensively support various transition metals used in heterogeneous catalysis. Herein, some of the organic transformations catalyzed by Ru- supported catalysts are discussed:

X Cai *et al.* reported the synthesis of a phosphorous-rich polymer with a micro-mesoporous structure through Friedel-Crafts alkylation of tris(2-thienyl) phosphine and thiophene [35]. The polymer with a micro-mesoporous structure provides solid support to Ru nanoparticles (NPs), thereby immobilizing NPs. This Ru-polymer catalyst shows catalytic activity in reducing nitroarenes to amine-arenes by using $NaBH_4$ as a reductant (Scheme **5**).

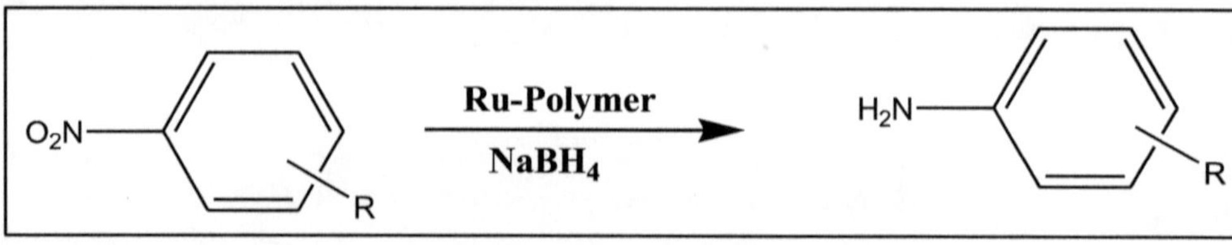

Scheme (5). Catalytic reduction of Nitroarenes by Ru-phosphorous-rich polymer.

Molla *et al.* reported a heterogeneous catalyst (Fig. **4**), a polystyrene-supported ruthenium (III) complex for hydrogenating ketones to alcohols using isopropanol (as the hydrogen donor) and KOH. Both aliphatic and aromatic ketones can be reduced to alcohols in open air by this catalyst [36]. The catalyst can be recycled and reused with the same catalytic activity for up to six cycles. Also, it shows stability to air and moisture (Scheme **6**).

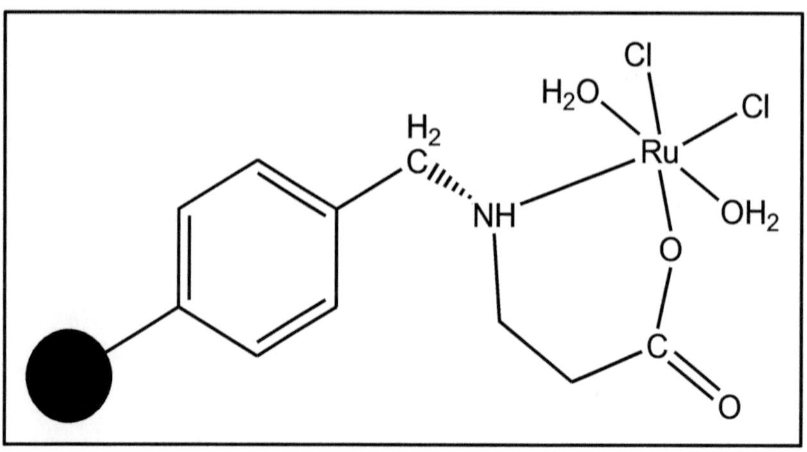

Fig. (4). Polystyrene-supported ruthenium(III) catalyst.

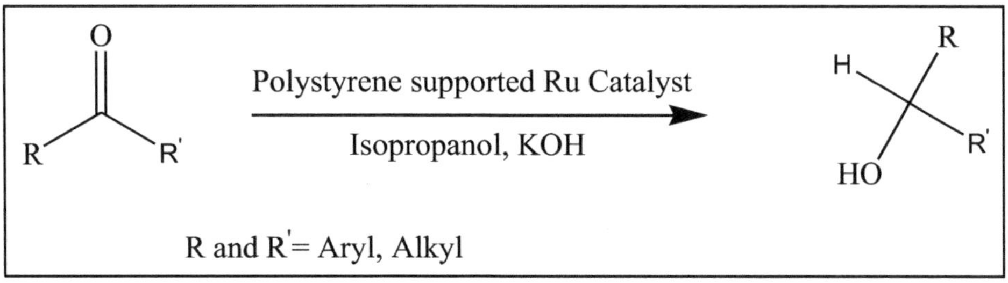

Scheme (6). Reduction of ketones to alcohol by using Polystyrene supported Ru Catalyst.

Reaction Catalyzed by Solid Supported Rhodium

Rhodium shows a variable oxidation state from 0 to +6. It can easily donate and accept electrons in its partially filled d orbital. Rhodium complexes in +1 and +3 oxidation states are widely used as catalysts. As a catalyst, it easily brings out various organic reactions by forming unstable intermediate complexes with the reactant molecules. It shows high-temperature range stability. All these properties make it a good catalyst.

Xia *et al.* reported a heterogeneous catalyst, a highly porous organic copolymer immobilizing rhodium (Rh-POC) (Fig. **5**). This catalyst is efficiently used for the regioselective hydroformylation of olefin with its remarkable reusability of up to 10 runs. The hydroformylation of olefin (1 hexane) was carried out by passing H_2 and CO (10 bar) in the presence of 5mg of catalyst for 5 hr at 90 ^{0}C [37] (Scheme 7).

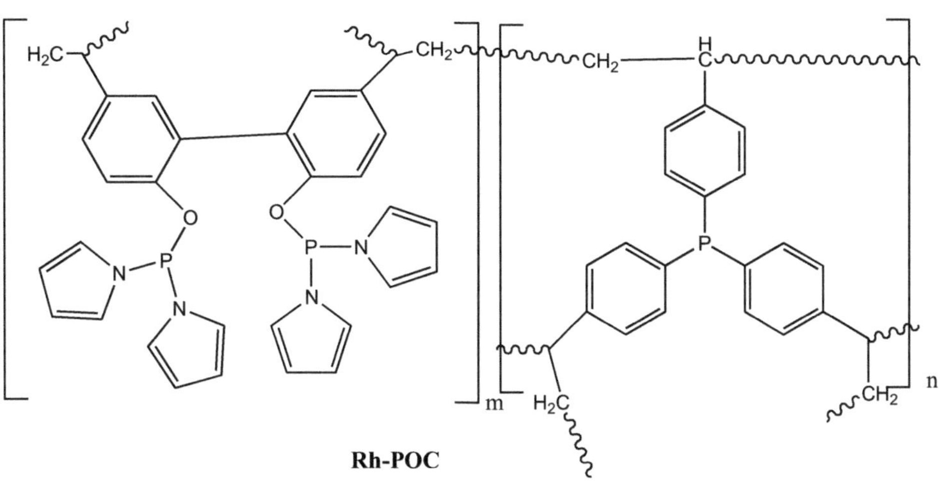

Fig. (5). Porous organic copolymer immobilizing rhodium (Rh-POC).

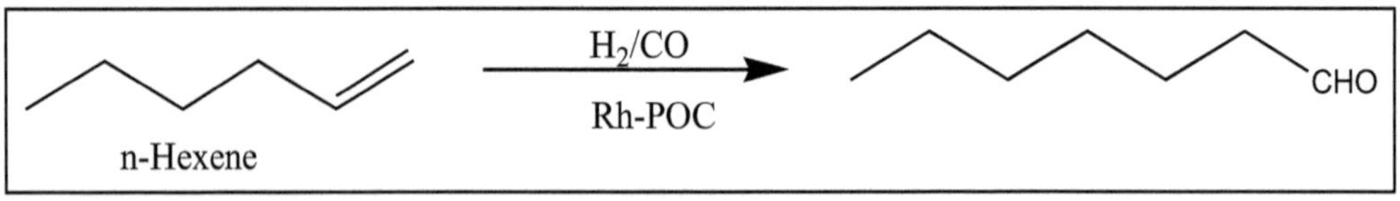

Scheme (7). Hydroformylation of olefin catalyzed by Rh-POC catalyst.

Liu *et al.* also developed the heterogenous catalyst, which is a highly porous organic polymer containing heteroatoms nitrogen and phosphorus that immobile Rh complexes on its surface and efficiently catalyst hydroformylation of olefin with CO and H_2 [38].

Ball *et al.* have reported the synthesis of an efficient Rh (II) catalyst supported on a solid tripodal peptide for the cyclopropanation reaction [39]. The catalyst shows remarkable advantages and can be reused after separation from the reaction mixture. The cyclopropanation reaction between phenyldiazoacatate and alkenes was catalyzed successfully using Rh (II) catalyst supported on a solid tripodal peptide (Scheme **8**).

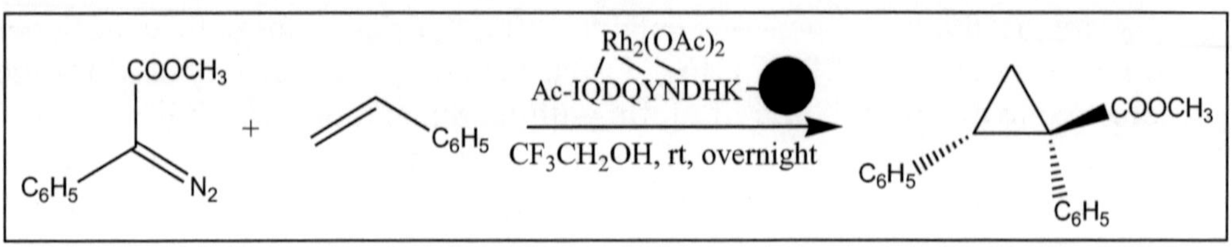

Scheme (8). Cyclopropanation reaction catalyst by Rh (II) catalyst supported on a solid tripodal peptide.

Rosenberg *et al.* synthesized a novel and effective Rh-containing heterogeneous catalyst with an organic-inorganic hybrid surface containing poly(allylamine) covalently attached to a silica gel surface. This Rh-containing heterogeneous catalyst effectively catalyzes hydroformylation of unsaturated hydrocarbons [40].

Campos *et al.* synthesize a novel and stable heterogeneous catalyst of Rh(I) diphenylphosphine complexes immobilized on organic polymer resin for alkene hydrogenation [41]. The catalyst (Fig. **6**) was successful in chemoselective hydrogenation of styrene (Scheme **9**). The electron-attracting groups in styrene enhance the catalytic power of the catalyst, whereas electron-donating groups decrease its activity. The activity of the catalyst remained intact till 7 repeated runs and Rh metal was not leached during this condition.

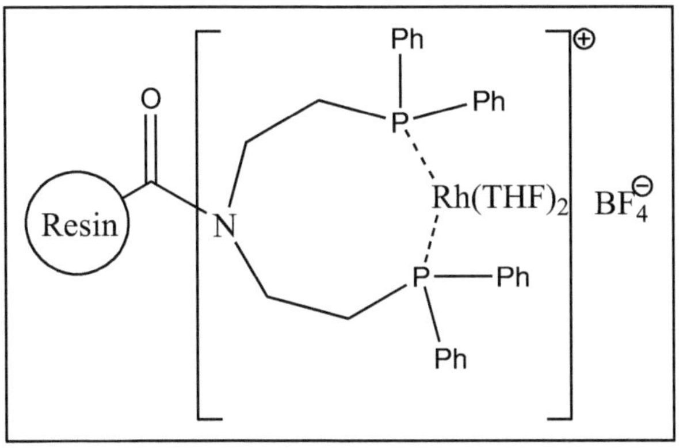

Fig. (6). Resin supported Rh(I) catalyst.

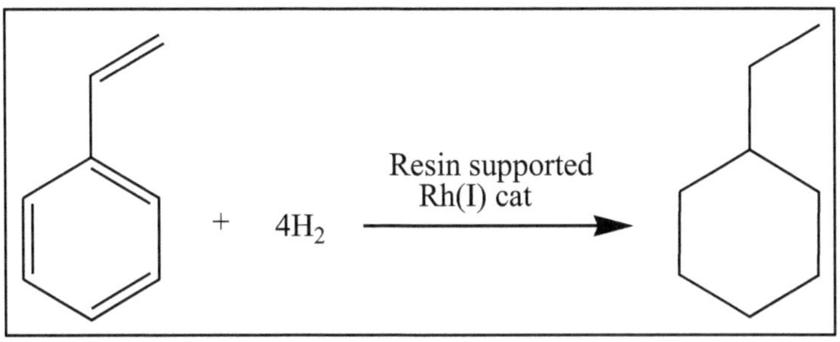

Scheme (9). Hydrogenation of styrene using Resin supported Rh(I) catalyst.

Reaction Catalyzed by Solid Supported Copper

Copper shows an oxidation state of 0,+1 and +2. The catalytic power of copper is dependent on its oxidation state. This flexibility in the oxidation state allows it to participate in reactions involving one or two electron transfers making it a versatile catalyst in organic transformation. On the solid support, copper is dispersed on its surface, thereby providing a large surface area with more active sites for the reaction to occur, enhancing its catalytic ability. Primarily, copper-based solid-supported catalysts have attracted wide interest and found vast applications, possibly because of their high natural abundance, low cost, toxicity, and relatively easy procedures for preparation. Solid supports like hydroxyapatite, cellulose acetate, and titanium dioxide support copper for catalysis.

Salah-Eddine Stiriba and researchers prepared an eco-friendly biopolymer-supported copper heterogeneous catalyst with cellulose acetate backbone (Cu-CA) [42]. The Cu-CA catalyst was used at room temperature for the selective cycloaddition reactions of substituted alkynes and azides to form 1,4-disubstituted-1,2,3-triazoles in an aqueous medium (Scheme **10**).

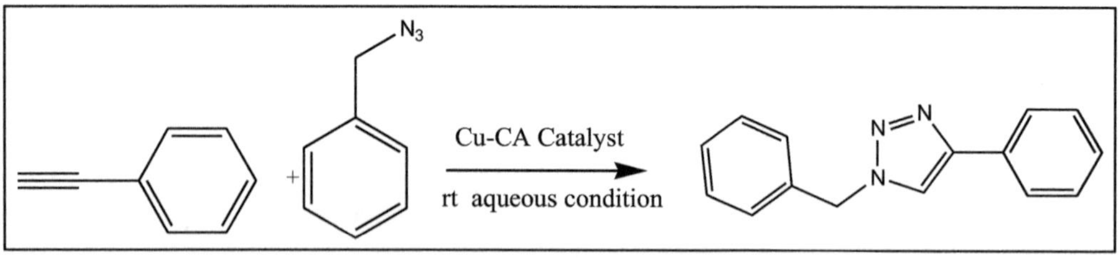

Scheme (10). Cycloaddition reaction catalyzed by biopolymer-supported copper (Cu-CA).

Hydroxyapatite has a high ion exchange capacity and serves as a good support for ions, both cations and anions. Ranu *et al.* reported a hydroxyapatite supported CuI (Cu-HAP) catalyst for the efficient cyanation of styrenyl bromides by using $K_4Fe(CN)_6$ [43]. With this simple and efficient protocol, a high yield of cinnamonitriles was obtained for the first time (Scheme **11**).

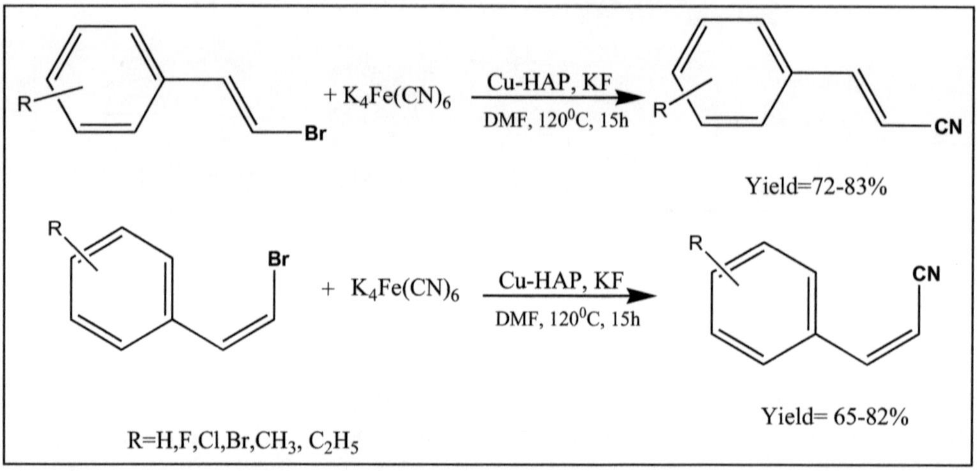

Scheme (11). Cyanation of styrenyl bromides using hydroxyapatite-supported CuI (Cu-HAP) catalyst.

Yu *et al.* synthesized $(NH_4)_4[CuMo_6O_{18}(OH)_6]$ catalyst having copper supported by an inorganic ligand, polyoxometalates for catalyzing alkyne coupling reactions. The catalytic power of the catalyst was effective for six consecutive runs. A high yield of conjugated 1,3- diynes was synthesized by coupling symmetric and asymmetric alkynes with this non-toxic and green Cu catalyst [44] (Scheme **12**).

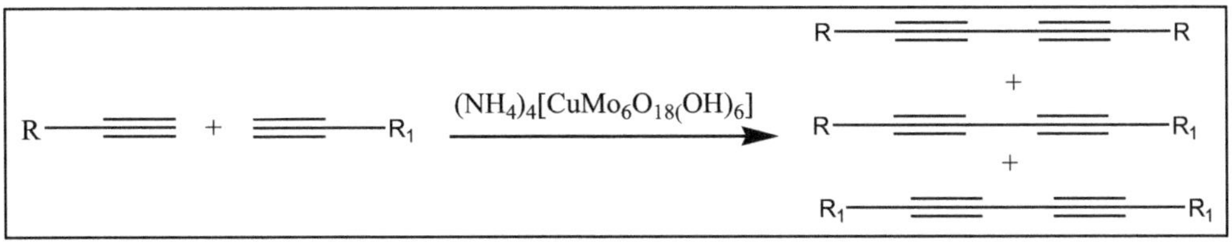

Scheme (12). Coupling of terminal alkyne catalyzed by $(NH_4)_4[CuMo_6O_{18}(OH)_6]$ catalyst.

Reaction Catalyzed by Solid Supported Cobalt

Cobalt also shows variable oxidation states, but the most common is +2 and +3. It has the capacity to exchange electrons with a variety of substances, thereby participating in various organic transformations as a catalyst. When cobalt is dispersed on a porous solid support, it provides a large surface area with more active sites. The reactant molecules then interact with these active sites of the cobalt catalyst; thereby enhancing the catalytic ability of cobalt.

Yamada *et al.* developed a heterogenous catalyst, poly(4-vinyl pyridine) cobalt (II) chloride (PVP-$CoCl_2$) (Fig. **7**) [45]. Cyclotrimerization of terminal aryl alkynes was reported with this polymer-supported-cobalt-catalyst in a regioselective manner. The catalyst was recycled for up to 3 cycles (Scheme **13**).

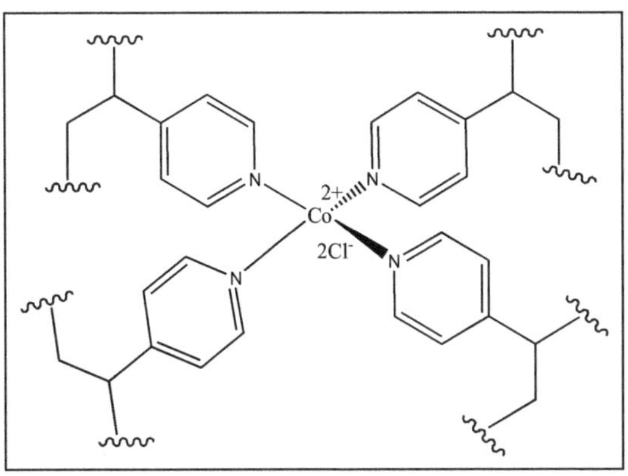

Fig. (7). Poly(4-vinyl pyridine) cobalt (II) chloride (PVP-$CoCl_2$) catalyst.

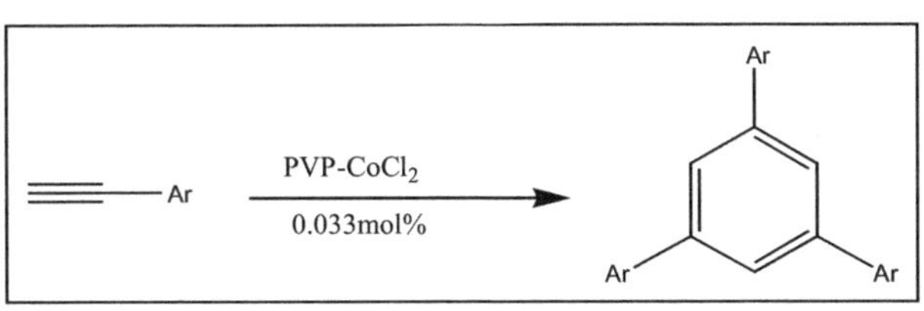

Scheme (13). PVP-$CoCl_2$ catalyzed cyclotrimerization of terminal aryl alkynes.

Błaż and Pielichowski synthesized an 8-hydroxyquinoline complex heterogeneous catalyst with cobalt (II) supported by the polymer (Fig. **8**). This polymer-supported co-catalyst oxidizes alkene under mild conditions in the presence of molecular oxygen as an oxidant [46] (Scheme **14**).

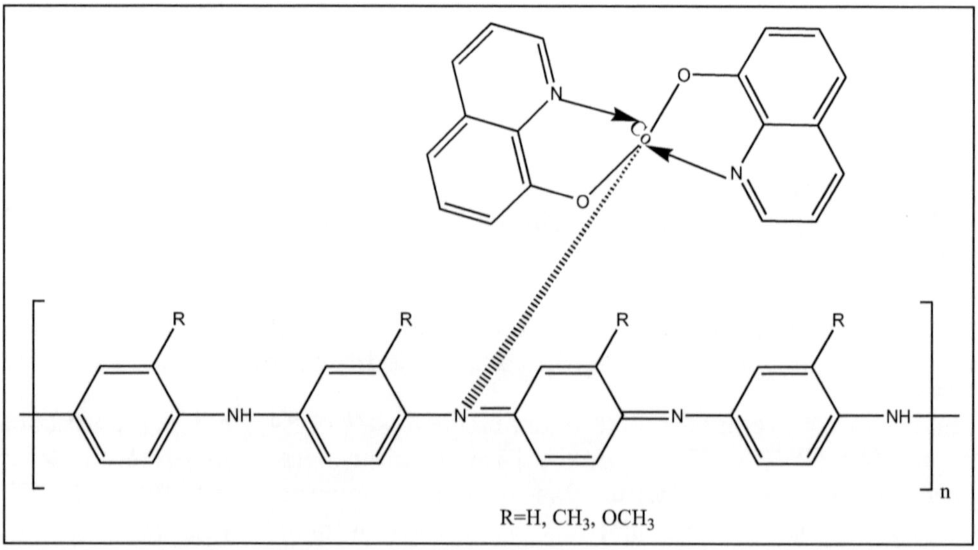

Fig. (8). Cobalt (II) 8-hydroxyquinoline complex supported by polymer.

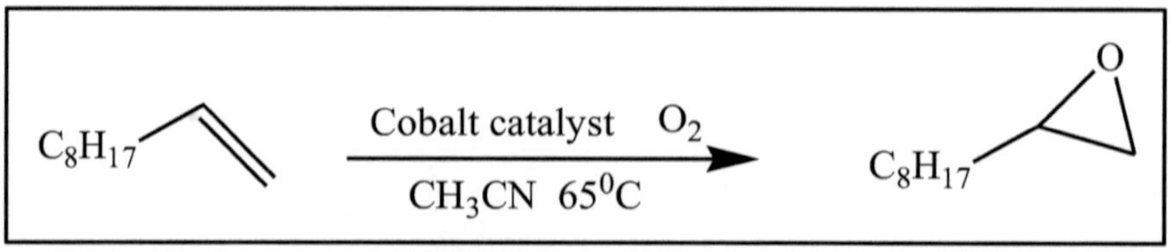

Scheme (14). Oxidation of alkene catalyst by Cobalt (II) catalyst.

CONCLUSION

Solid-supported transition metals are useful, sustainable heterogeneous catalysts that are alternatives to homogeneous ones. The heterogenous solid-supported catalysts play useful catalytic roles in synthetic organic reactions due to their ability to accelerate synthesis and purification. These catalysts can be recycled, reused, and easily removed from the reaction mixture. The present work was focused on the catalytic activities of an important class of transition metals like palladium, ruthenium, rhodium, cobalt, and copper catalysts supported on heterogeneous solid surfaces to be useful to the scientific community working on heterogeneous catalysis. Further advancement in this field will expand the horizon of heterogeneous catalysis.

REFERENCES

[1] Merrifield, R.B. Solid phase peptide synthesis. I. the synthesis of a tetrapeptide. *J. Am. Chem. Soc.,* **1963**, *85*(14), 2149-2154.
[http://dx.doi.org/10.1021/ja00897a025]

[2] Larsen, B.D.; Christensen, D.H.; Holm, A.; Zillmer, R.; Nielsen, O.F. The Merrifield peptide synthesis studied by near-infrared Fourier-transform Raman spectroscopy. *J. Am. Chem. Soc.,* **1993**, *115*(14), 6247-6253.
[http://dx.doi.org/10.1021/ja00067a044]

[3] Posner, G.H. Organic reactions at alumina surfaces. *Angew. Chem. Int. Ed. Engl.,* **1978**, *17*(7), 487-496.
[http://dx.doi.org/10.1002/anie.197804871]

[4] Mckillop, A.; Young, D.W. Organic synthesis using supported reagents - Part I. *Synthesis,* **1979**, *1979*(6), 401-422, 481-500.
[http://dx.doi.org/10.1055/s-1979-28699]

[5] Cornelis, A.; Laszlo, P. Clay-supported copper(II) and iron(III) nitrates: Novel multi-purpose reagents for organic synthesis. *Synthesis,* **1985**, *1985*(10), 909-918.
[http://dx.doi.org/10.1055/s-1985-31382]

[6] Parish, R.V. *Preparative chemistry using supported reagents*; Laszlo, P., Ed.; Academic Press Inc., **1987**.

[7] Smith, E.K. *Solid supports and catalysts in organic synthesis*; Ellis Horwood, **1992**.

[8] Miceli, M.; Frontera, P.; Macario, A.; Malara, A. Recovery/reuse of heterogeneous supported spent catalysts. *Catalysts,* **2021**, *11*(5), 591.
[http://dx.doi.org/10.3390/catal11050591]

[9] Alazemi, A.M.; Dawood, K.M.; Al-Matar, H.M.; Tohamy, W.M. Efficient and recyclable solid-supported Pd(II) Catalyst for microwave-assisted suzuki cross-coupling in aqueous medium. *ACS Omega,* **2022**, *7*(33), 28831-28848.
[http://dx.doi.org/10.1021/acsomega.2c01809] [PMID: 36033663]

[10] Adeyeye Nafiu, S.; Ajeebi, A.M.; Alghamdi, H.S.; Aziz, M.A.; Nasiruzzaman Shaikh, M. Solid-supported catalysts for organic functional group transformations. *Asian J. Org. Chem.,* **2023**, *12*(4), e202300051.
[http://dx.doi.org/10.1002/ajoc.202300051]

[11] Munnik, P.; de Jongh, P.E.; de Jong, K.P. Recent developments in the synthesis of supported catalysts. *Chem. Rev.,* **2015**, *115*(14), 6687-6718.
[http://dx.doi.org/10.1021/cr500486u] [PMID: 26088402]

[12] Crich, D.; Smith, M. Solid-phase synthesis of β-mannosides. *J. Am. Chem. Soc.,* **2002**, *124*(30), 8867-8869.
[http://dx.doi.org/10.1021/ja011406u] [PMID: 12137540]

[13] Hirao, T.; Higuchi, M.; Ikeda, I.; Ohshiro, Y. A novel synthetic metal catalytic system for dehydrogenative oxidation based on redox of polyaniline. *J. Chem. Soc. Chem. Commun.,* **1993**, *194*(2), 194.
[http://dx.doi.org/10.1039/c39930000194]

[14] Pratap Singh, R.; Ghoshal, T.; Mishra, V. Trifluoromethylation strategies of alcohols and phenols. *Asian J. Org. Chem.,* **2024**, *13*(11), e202400179.
[http://dx.doi.org/10.1002/ajoc.202400179]

[15] Mishra, V.; Mudgal, D.; Singh, R.P.; Yadav, N.; Bharti, T. Exploring the catalytic efficiency of copper-doped magnetic carbon aerogel for the coupling reaction of isatin oxime with phenylboronic acid derivatives. *SynOpen,* **2023**, *7*(4), 570-579. [http://dx.doi.org/10.1055/a-2182-7757]

[16] Bhamra, R.; Dani, S.; Burnard, K. Resilience: The concept, a literature review and future directions. *Int. J. Prod. Res.,* **2011**, *49*(18), 5375-5393. [http://dx.doi.org/10.1080/00207543.2011.563826]

[17] Mishra, V.; Cho, J.K.; Shin, S.H.; Suh, Y.W.; Kim, H.S.; Kim, Y.J. Ruthenium-Na2CO3-catalyzed one-pot synthesis of ring-hydrogenated carbamates from aromatic amines and organic carbonates under H2. *Appl. Catal. A Gen.,* **2014**, *487*, 82-90. [http://dx.doi.org/10.1016/j.apcata.2014.09.013]

[18] Keeble, B.R. The Brundtland report: 'Our common future'. *Med. War,* **1988**, *4*(1), 17-25. [http://dx.doi.org/10.1080/07488008808408783]

[19] Promhuad, P.; Sawatmongkhon, B.; Theinnoi, K.; Wongchang, T.; Chollacoop, N.; Sukjit, E.; Tunmee, S.; Tsolakis, A. Effect of metal oxides (CeO_2, ZnO, TiO_2, and Al_2O_3) as the support for silver-supported catalysts on the catalytic oxidation of diesel particulate matter. *ACS Omega,* **2024**, *9*(17), 19282-19294. [http://dx.doi.org/10.1021/acsomega.4c00218] [PMID: 38708233]

[20] Dagle, V. L.; Flake, M. D.; Lemmon, T. L.; Lopez, J. S.; Kovarik, L.; Dagle, R. A. Effect of the SiO2 support on the catalytic performance of Ag/ZrO2/SiO2 catalysts for the single-bed production of butadiene from ethanol *Appl. Cat. B,* **2018**, *236*, 576-587. [http://dx.doi.org/10.1016/j.apcatb.2018.05.055]

[21] Mi, R.; Li, D.; Hu, Z.; Yang, R.T. Morphology effects of CeO_2 nanomaterials on the catalytic combustion of toluene: A combined kinetics and diffuse reflectance infrared fourier transform spectroscopy study. *ACS Catal.,* **2021**, *11*(13), 7876-7889. [http://dx.doi.org/10.1021/acscatal.1c01981]

[22] Stocker, K.; Ellersdorfer, M.; Lehner, M.; Raith, J.G. Characterization and utilization of natural zeolites in technical applications. *BHM Berg- Huttenmann. Monh.,* **2017**, *162*(4), 142-147. [http://dx.doi.org/10.1007/s00501-017-0596-5]

[23] Martins, A.; Nunes, N.; Carvalho, A.P.; Martins, L.M.D.R.S. Zeolites and related materials as catalyst supports for hydrocarbon oxidation reactions. *Catalysts,* **2022**, *12*(2), 154. [http://dx.doi.org/10.3390/catal12020154]

[24] Xu, Z.X.; Liu, L.; Zhang, J. Synthesis of metal–organic zeolites with homochirality and high porosity for enantioselective separation. *Inorg. Chem.,* **2016**, *55*(13), 6355-6357. [http://dx.doi.org/10.1021/acs.inorgchem.6b00865] [PMID: 27315193]

[25] Chutia, A. Adsorption of transition metal catalysts on carbon supports: A theoretical perspective. *Johnson Matthey Technol. Rev.,* **2022**, *66*(1), 4-20. [http://dx.doi.org/10.1595/205651322X16212512135401]

[26] Yazici, M.S. Electrochemical responses of flexible graphite in H2SO4. *Turk. J. Chem.,* **2019**, *43*(5), 1398-1406. [http://dx.doi.org/10.3906/kim-1906-19]

[27] Mironenko, R.M.; Belskaya, O.B.; Likholobov, V.A. Carbon black as a support in palladium catalysts for hydrogenation of organic compounds. *Solid Fuel Chem.,* **2020**, *54*(6), 362-367. [http://dx.doi.org/10.3103/S0361521920060087]

[28] Panziera, N.; Pertici, P.; Barazzone, L.; Caporusso, A. M.; Vitulli, G.; Salvadori, P.; Borsacchi, S.; Geppi, M.; Veracini, C. A.; Gianmario Martra, L. B. MVS-derived palladium nanoparticles deposited on polydimethylphosphazene as recyclable catalysts for Heck-type reactions: Preparation, structural study, and catalytic activity *J. Cat.,* **2007**, *246*(2), 351-361.

[29] Caporusso, A.; Innocenti, P.; Aronica, L.; Vitulli, G.; Gallina, R.; Biffis, A.; Zecca, M.; Corain, B. Functional resins in palladium catalysis: Promising materials for Heck reaction in aprotic polar solvents. *J. Catal.*, **2005**, *234*(1), 1-13.
[http://dx.doi.org/10.1016/j.jcat.2005.05.008]

[30] Xia, W.; Huang, L.; Huang, X.; Wang, Y.; Lu, C.; Yang, G.; Chen, Z.; Nie, J. Main-chain NHC-palladium polymers based on adamantane: Synthesis and application in Suzuki–Miyaura reactions. *J. Mol. Catal. Chem.*, **2016**, *412*, 93-100.
[http://dx.doi.org/10.1016/j.molcata.2015.11.020]

[31] Diebold, C.; Becht, J.-M.; Lu, J.; Toy, P. H.; Le Drian, C. An efficient and reusable palladium catalyst supported on a Rasta resin for Suzuki—Miyaura cross-couplings. *ChemInform,* **2012**, *43*(27).
[http://dx.doi.org/10.1002/chin.201227078]

[32] You, L.; Zhu, W.; Wang, S.; Xiong, G.; Ding, F.; Ren, B.; Dragutan, I.; Dragutan, V.; Sun, Y. High catalytic activity in aqueous heck and Suzuki–Miyaura reactions catalyzed by novel Pd/Ln coordination polymers based on 2,2′-bipyridine-4,4′-dicarboxylic acid as a heteroleptic ligand. *Polyhedron,* **2016**, *115*, 47-53.
[http://dx.doi.org/10.1016/j.poly.2016.04.032]

[33] Choudhary, H.; Nishimura, S.; Ebitani, K. Tailored design of palladium species grafted on an amino functionalized organozinc coordination polymer as a highly pertinent heterogeneous catalyst. *J. Mater. Chem. A Mater. Energy Sustain.*, **2014**, *2*(43), 18687-18696.
[http://dx.doi.org/10.1039/C4TA03889D]

[34] Saito, Y.; Ishitani, H.; Kobayashi, S. Catalytic hydrogenation of aliphatic nitro compounds with polysilane/bone charcoal□supported palladium catalysts under continuous-flow conditions. *Asian J. Org. Chem.*, **2016**, *5*(9), 1124-1127.
[http://dx.doi.org/10.1002/ajoc.201600279]

[35] Cai, X.; Nie, J.; Yang, G.; Wang, F.; Ma, C.; Lu, C.; Chen, Z. Phosphorus-rich network polymer supported ruthenium nanoparticles for nitroarene reduction *Mater. Lett.,* **2019**, *240*, 80-83.
[http://dx.doi.org/10.1016/j.matlet.2018.12.140]

[36] Molla, R.A.; Roy, A.S.; Ghosh, K.; Salam, N.; Iqubal, M.A.; Tuhina, K.; Islam, S.M. Polymer anchored ruthenium complex: A highly active and recyclable catalyst for one-pot azide–alkyne cycloaddition and transfer-hydrogenation of ketones under mild conditions. *J. Organomet. Chem.*, **2015**, *776*, 170-179.
[http://dx.doi.org/10.1016/j.jorganchem.2014.11.007]

[37] Jia, X.; Liang, Z.; Chen, J.; Lv, J.; Zhang, K.; Gao, M.; Zong, L.; Xie, C. Porous organic polymer supported rhodium as a reusable heterogeneous catalyst for hydroformylation of olefins. *Org. Lett.*, **2019**, *21*(7), 2147-2150.
[http://dx.doi.org/10.1021/acs.orglett.9b00459] [PMID: 30854859]

[38] Lan, Y.; Yun, D.; Xia, C.; Qian, B.; Liu, J. Containing nitrogen and phosphorus porous organic polymer-supported rhodium catalyst for hydroformylation of olefins. *Catal. Lett.*, **2024**, *154*(7), 3911-3923.
[http://dx.doi.org/10.1007/s10562-024-04602-4]

[39] Sambasivan, R.; Zheng, W.; Burya, S.J.; Popp, B.V.; Turro, C.; Clementi, C.; Ball, Z.T. A tripodal peptide ligand for asymmetric Rh(II) catalysis highlights unique features of on-bead catalyst development. *Chem. Sci. (Camb.),* **2014**, *5*(4), 1401-1407.
[http://dx.doi.org/10.1039/C3SC53354A]

[40] Gorbunov, D.; Safronova, D.; Kardasheva, Y.; Maximov, A.; Rosenberg, E.; Karakhanov, E. New heterogeneous rh-containing catalysts immobilized on a hybrid organic–inorganic surface for hydroformylation of unsaturated compounds. *ACS Appl. Mater. Interfaces,* **2018**, *10*(31), 26566-26575.
[http://dx.doi.org/10.1021/acsami.8b02797] [PMID: 29979868]

[41] Campos, C.H.; Belmar, J.B.; Jeria, S.E.; Urbano, B.F.; Torres, C.C.; Alderete, J.B. Rhodium(I) diphenylphosphine complexes supported on porous organic polymers as efficient and recyclable catalysts for alkene hydrogenation. *RSC Advances,* **2017**, *7*(6), 3398-3407.
[http://dx.doi.org/10.1039/C6RA26104C]

[42] Stiriba, S.E.; Bahsis, L.; Benhadria, E.; Oudghiri, K.; Taourirte, M.; Julve, M. Cellulose acetate-supported copper as an efficient sustainable heterogenous catalyst for azide-alkyne cycloaddition click reactions in water. *Int. J. Mol. Sci.,* **2023**, *24*(11), 9301.
[http://dx.doi.org/10.3390/ijms24119301] [PMID: 37298251]

[43] Saha, D.; Adak, L.; Mukherjee, M.; Ranu, B.C. Hydroxyapatite-supported Cu(I)-catalysed cyanation of styrenyl bromides with $K_4[Fe(CN)_6]$: An easy access to cinnamonitriles. *Org. Biomol. Chem.,* **2012**, *10*(5), 952-957.
[http://dx.doi.org/10.1039/C1OB06467C] [PMID: 22130731]

[44] Deng, M.; Xie, Y.; Zhou, J.; Shi, D.; Han, S.; Wu, L.; Yu, H. The inorganic ligand-supported copper catalyst catalyzing alkyne coupling reactions. *Eur. J. Org. Chem.,* **2024**, *27*(21), e202301238.
[http://dx.doi.org/10.1002/ejoc.202301238]

[45] Sen, A.; Sato, T.; Ohno, A.; Baek, H.; Muranaka, A.; Yamada, Y.M.A. Polymer-supported-coba-t-catalyzed regioselective cyclotrimerization of aryl alkynes. *JACS Au,* **2021**, *1*(11), 2080-2087.
[http://dx.doi.org/10.1021/jacsau.1c00360] [PMID: 34841419]

[46] Błaż, E.; Pielichowski, J. Polymer - supported cobalt (II) catalysts for the oxidation of alkenes. *Molecules,* **2006**, *11*(1), 115-120.
[http://dx.doi.org/10.3390/11010115] [PMID: 17962751]

CHAPTER 3

Advancing Biomass Utilization: Chemical Conversions, Catalysis, and Energy Applications

Jaya Tuteja[1,*]

[1] *School of Basic Science, Galgotias University, India*

Abstract: The use of biomass for chemical changes and catalysis is one of the causes that may mitigate the depletion of fossil fuels and environmental concerns. Biomass, which is a product of photosynthesis and the only carbon-based renewable resource, can be engineered into useful liquid fuels and other fine chemicals. In this regard, this innovation is of high importance in the promotion of carbon neutralization and sustainable green industrial development. Biomass catalytic conversion encompasses several steps in the chemical processes oriented towards the production of useful chemicals and fuels with less energy effort. The search for new materials and technologies for catalysts is an important factor in increasing the efficiency of biomass processing. Innovative catalysts can be a game changer for new processes, including lignin degradation, cellulose hydrolysis, biofuel production, *etc.*, in the efforts to provide cleaner energy. Research is currently being carried out to develop catalysts that are both highly selective and environmentally friendly and improve the conditions under which biomass is converted. It aims to offer a complete understanding of the chemical processes that underlie bioenergy use, while providing innovative ideas for surface and interface designs. Ultimately, converting biomass into forms amenable for use besides supporting energy security can assist in addressing some of the urgent social ills that come with reliance on fossil fuel energy sources.

Keywords: Biomass, Heterogeneous catalysts, Solid catalysis, Sustainability.

INTRODUCTION

Fossil fuels are the major energy source to meet the energy demands worldwide. It is impossible to think about life without fossil fuel resources. Urbanization has led to an increase in the consumption of fossil fuel resources at a much larger scale than their production, which has made it difficult to sustain or rely on natural fossil fuel resources. This urges a strong demand to look for other sustainable and renewable alternatives for energy sources to meet global energy demands. Various growing renewable energy sources are being employed and are

* **Corresponding author Jaya Tuteja:** School of Basic Science, Galgotias University, India; Tel: +91-9599495856; E-mail: jayatuteja89@gmail.com

Ravi Tomar, Vivek Mishra, Kapil Arya & Garima Pandey (Eds.)

under research, such as electric, hydroelectric, wind, and solar energy. However, the uses of these resources are limited and they cannot be employed to generate value-added chemicals as they lack carbon content in them. Thus, to meet this criterion, alternatives that are carbon-rich and can be converted to fine chemicals need to be explored. One such great source of carbon from renewable energy sources is biomass. The governments of various countries have motivated industrial companies to work in this area in order to shift from fossil fuel resources to ones that can be generated by biomass resources [1, 2]. Biomass is the only carbon source available in nature that can be converted to produce various chemical compounds; thus, it has gained tremendous attention from researchers in past decades [3]. The only barrier to utilizing it as a precursor for fuel compounds is its effective transformation as it possesses high oxygen content and is quite complex in nature. Thus, effective catalytic routes are required to convert the complex oxygenated species to petrochemical products [4 - 7]. The same approach that is applied to convert fossil fuel resources to petrochemical products cannot be applied to process biomass resources to produce value-added commodities as they differ in their starting materials. In the processing of fossil fuel precursors, isomerization and alkylation are major steps, while for carrying out biomass deoxygenation, hydrolysis and hydrogenolysis are important reactions to start with [6] (Fig. **1**).

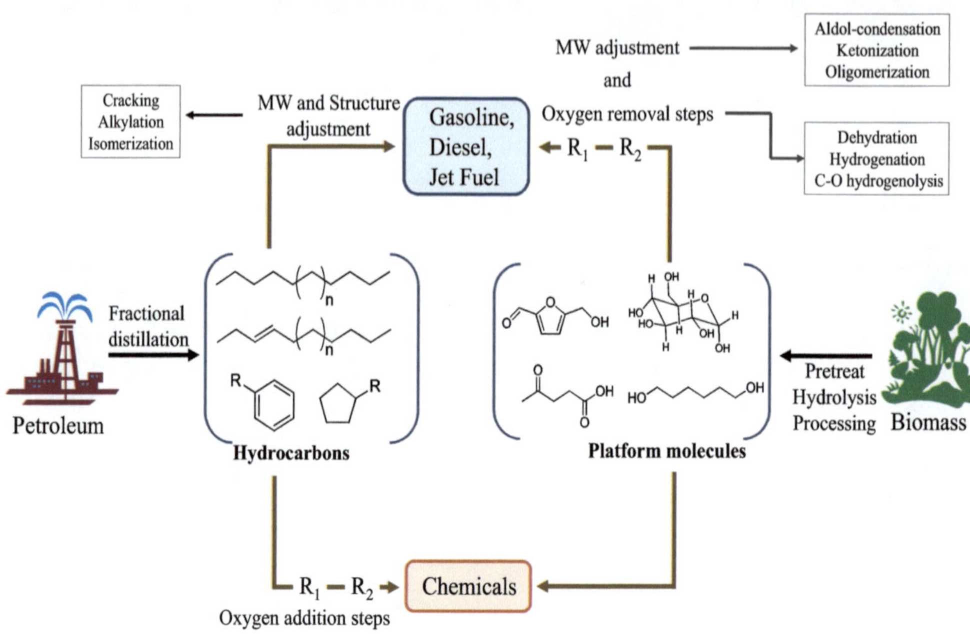

Fig. (1). Important reactions to process value-added products from fossil fuel resources and biomass resources.

To use biomass precursors as potential fuel precursors, the primary step involves breaking down their complex structures into simpler units. Biomass, specifically lignocellulosic biomass, is basically composed of 3 units known as cellulose, hemicellulose, and lignin [8 - 11]. Upon simplifying the complex biomass units *via* pre-treatment (ball milling and hydrolysis), these are sent to biorefinery units where they are converted to materials, energy, and fuel, as required. Upon the destruction of biomass after its preliminary treatment, it is converted into platform molecules based on the utility. These platform molecules or chemical building blocks that carry plenty of functional groups are further processed to a range of value-added products *via* various catalytic routes [12 - 15].

Role of Solid Catalysts

Due to the importance of using renewable feedstock as mentioned above, on the same grounds, catalysis also plays a crucial role from the viewpoint of green chemistry [16]. The role of catalysts in chemical manufacturing of fine chemicals is irreplaceable. However, unfortunately, after even becoming aware of green chemistry principles and the advantages of using solid catalysts, our manufacturing companies are working on the use of homogeneous acid and base catalysts, for instance, mineral acids, like sulfuric acid, hydrochloric acid, and phosphoric acid as acid catalysts; NaOH and KOH as base catalysts; oxidizing agents, $KMnO_4$ and $K_2Cr_2O_7$; reduction using Na, Mg, and Zn metals, *etc*. The use of these chemicals in a stoichiometric ratio generates hazardous and carcinogenic effects; hence, the only alternative to prohibit or reduce the emission of hazardous substances is the use of a heterogeneous catalytic system [17 - 20].

Long back, in the year 1958, researchers began to explore the possibility of using solid materials as acids and bases for catalytic reactions [21]. The basic characteristics of solid alumina were discovered in the 1930s. The history of using solid base catalysts in chemical reactions is amazing. It has been seen that with due course of time, their significance or utilization has increased. Solid catalysts are materials that possess catalytic active sites on their surfaces and are used to catalyze various chemical reactions. Unlike traditional liquid or homogeneous catalysts, solid catalysts are heterogeneous and exist in a solid phase. These catalysts have gained attention in various industrial and laboratory applications due to their advantages, such as ease of separation, reusability, and stability. In the 1990s, numerous metal oxides, such as MgO, CaO, SrO, BaO, LaO, and ZrO were identified as effective solid base catalysts [22, 23]. After the implications of metal oxides as base catalysts, further innovations have led to the emergence of layered double hydroxides and hydrotalcites as solid base catalysts, possessing unique properties [24 - 26].

It is interesting to note that solid acid catalysts were recognized earlier for industrial applications in the petroleum industry [27]. They are employed to catalyze reactions that are crucial for refining crude oil into valuable products. The hydrocarbons produced by the petroleum industry involve interactions with solid acid catalysts at every step, including catalyst cracking, hydrocracking, isomerization, alkylation, etherification, *etc*. On the contrary, solid base catalysts came into the picture almost two decades later, but were found to significantly contribute to the catalytic world owing to their versatile properties [28]. The heterogeneous base catalysts are now being used in various industrial applications as they exhibit enhanced surface area, ease of separation, and highly porous and uniform basicity. These days, heterogeneous base catalysts are used in a variety of reactions, such as dehydration, isomerization, hydrogenation, *etc* [29]. The characteristics that have made them suitable for industrial applications are their high activity, which signifies their ability to use minimum energy to transform the reactant into the desired product. High selectivity refers to the ability of solid base catalysts to minimize side reactions, ensuring the maximum conversion of reactants to desired products. The heterogeneous catalysts also offer better stability as they do not undergo deactivation, leaching, *etc* [30]. Recyclability and easy recovery are the added advantages of using a heterogeneous catalytic system for various industrial applications.

Solid base catalysis is one of the topics that have gained tremendous attention in the past 10-15 years [31 - 36]. Numerous publications can be seen in this era on the topic of heterogeneous catalysis. This chapter provides a brief summary of the recent advances in this field, along with some basic aspects related to the synthesis, development, and characterization of solid catalysts for desired transformations. The overall focus of the chapter is on sustainable development by developing solid catalysts, providing new insights into transforming the reaction from homogeneous to heterogeneous catalysis. A list of solid base catalysts known for years is provided below in Table **1**.

Table 1. Types of solid base catalysts [17].

S. No.	Solid Base Catalysts	Types	Examples
1.	Metal oxides	Alkaline metal oxides	CaO, MgO, BaO, SrO
		Other oxides	Al_2O_3, ZrO_2, ThO_2, TiO_2, Mo_2O_3
		Rare metal oxides	Sm_2O_3, La_2O_3
2	Mixed metal oxides	Double metal oxides	ZnO-Al_2O_3, MgO-TiO_2
		Layered double hydroxides	Anionic clays that contain positively charged layers of metal hydroxides.

(Table 1) cont.....

S. No.	Solid Base Catalysts	Types	Examples
3	Zeolites	Alkali ion-added zeolites	Cs_2O/Cs–X
		Alkali ion-exchanged zeolites	Na–X, Cs–X
4	Supported alkali metal compounds	Alkali metal compounds based on alumina	NaO/Al_2O_3, Na_2O/Al_2O_3
		Alkali metal compounds based on silica	NaO/SiO_2, Na_2O/SiO_2
		Alkali metal ions based on alkaline earth oxides	Na_2O/MgO
5	Clay minerals	Hydrotalcite	$Mg_6Al_2(OH)_{16}CO_3{\cdot}4H_2O$
		Chrysolite	$Mg_3(Si_2O_5)O_3(OH)_4$
		Sepiolite	$Mg_4Si_6O_{15}(OH)_2$
6	Non-oxides	KF supported on alumina	KF/Al_2O_3
		Lanthanide imide and nitride on zeolite	-
		Metal oxynitrides	ALPON, VALPON

A Few Important Chemical Reactions for the Conversion of Biomass to Commodity Chemicals

Deoxygenation Reaction:

Platform molecules derived from biomass resources include highly oxygenated species, such as carbohydrates, including glucose, sucrose, xylose, *etc*., and carboxylic acids, like succinic, lactic, levulinic, itaconic acids, *etc*. These highly oxygenated products need to be converted into deoxygenated products, like furan derivatives, such as furfural, 5-hydroxymethylfurfural, and 2,5-dimethylfuran, which serve as an alternative to biofuels. The reason behind decreasing the oxygen content is to increase the energy density along with inducing a decrease in chemical reactivity, which is necessary to consider them as potential fuel precursors. The deoxygenation of platform molecules obtained from biomass to value-added chemicals is a catalytic process that is being dominated by heterogeneous catalytic systems these days. In this process, the major reactions include decarboxylation, which removes CO and CO_2; dehydration process for removing H_2O; hydrogenolysis to break the C-O bond, *etc*. (Scheme **1**). Despite the importance of deoxygenation, it has been observed that all these reactions require hydrogen gas, which is sourced from fossil fuel resources only. Thus, to make the process more effective, the focus should not be only on developing heterogeneous catalysts, but also on the development of a heterogeneous catalytic system capable of facilitating hydrogenation reactions, thereby reducing our reliance on formic acid, which can be derived from renewable sources [38 - 40].

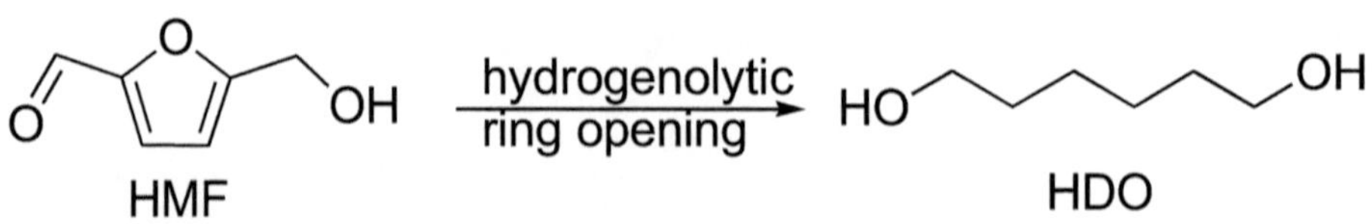

Scheme (1). Dehydration of arabinose to furfural.

The utilization of formic acid as a hydrogen source generates *in situ* hydrogen, which is absorbed on the catalyst surface and utilized in the reaction when required. Despite being an abundant resource, lignin's potential for producing value-added chemicals has not been fully explored. To date, its primary use has been in heat and power generation, as well as in the pulp and paper industry through combustion, owing to its high calorific value. Lignin is constituted by a highly complex polymeric structure of phenylpropane units. Due to its rigid nature, it resists microbial attack and cannot be depolymerized by water treatment as it repels water molecules. R. Luque *et al.* have been working on the utilization of lignin sources by developing various heterogeneous catalysts for almost a decade [41 - 43]. They have developed a microwave-assisted methodology for lignin deconstruction using MNP/Si-Al-Si (Metal Nanoparticles supported on the surface of Aluminosilicates) and employing a hydrogenated solvent as the hydrogen source [41]. The metal nanoparticles selected for this study include a combination of precious/noble metals (such as Pd, Pt, Ru, and Rh) and non-noble/transition metals (like Cu and Ni), synthesized using a dry milling approach and supported on the highly active surface of aluminosilicates [44]. The hydrogen transfer agents selected for this study include tetralin, isopropanol, and formic acid, with formic acid identified as the most effective for donating hydrogen under the reaction conditions used for lignin depolymerization [45, 46].

Hydrolysis

Hydrolysis represents an important reaction to convert lignocellulosic biomass to value-added commodities as it represents the first step in producing cellulose and hemicellulose from lignocellulosic biomass, further hydrolysis of which forms C5 and C-6 sugar units that serve as building blocks [47]. Cellulose on hydrolysis results in glucose units whereas hemicellulose on hydrolysis provides C5 sugars, like arabinose, xylose, *etc.* The acid catalysts, specifically solid acid catalysts, have gained tremendous attention for the hydrolysis of cellulose and hemicellulose. Solid acid is made up of a solid surface with highly active acidic sites (Brønsted or Lewis), which leads to the hydrolysis of cellulose. Apart from easy recovery and eco-friendliness, the important benefits of using solid catalysts include corrosion resistance to reaction vessels [48]. A list of solid acid catalysts used for hydrolysis of lignocellulosic biomass is provided in Table **2**.

Table 2. A list of solid acid catalysts for biomass transformation.

S. No.	Type	Solid Acid Catalyst	References
1.	Zeolites	HY-zeolite H-beta	[49]
2.	Heteropoly acid	$H_3PW_{12}O_{40}$	[50]
3.	Metal oxide	$HNbMoO_6$	[51]
4.	Acidic polymers	Amberlyst-15 Nafion NR-50	[52-54]
5.	Metal-supported catalysts	Pd/C Ru/CMK-3	[55]
6.	Magnetic solid acid catalysts	Fe_3O_4/SBA-SO_3H MNPs/SiO_2-SO_3H Fe_3O_4/C-SO_3H	[56 57 58]
7.	Inorganic-organic hybrid composites	Si_mC_n-SO_3H	[59]

When considering solid acid catalysts for hydrolysis, it is important to determine whether the acidity arises from Brønsted or Lewis sites. Secondly, it needs to assess whether the acidic sites are embedded on the surface or incorporated into the inner part of solid catalysts.

Catalytic Activity of Acidic Sites

Strong Brønsted sites are provided by the SO_3H group, which works by donating its H^+ to break the β-1,4-glycosidic linkage of cellulose in order to form glucose units. The commercially available catalysts, like Amberlyst-15 and Nafion NR-50, possess a sulfonic moiety on polymer beads. Inorganic solid acids, like sulfonated silica, *etc*., have exhibited remarkable activity for the hydrolysis of biomass-derived molecules [60 - 62]. The literature suggests that the density or amount of SO_3H moiety is not the only factor that depicts the catalytic activity [63], rather the location and accessibility of the sulfonic moieties play a significant role in the productivity and selectivity of the catalyst [64]. For instance, the sulfonated carbonaceous solid acid, if prepared under mild temperature conditions, has more surface acidic moieties that are easily accessible by cellulose to undergo hydrolysis. If the same carbonaceous solid acid is prepared at high-temperature conditions, the SO_3H moieties are surrounded by polymeric carbon atoms and are not easily approachable by the reactant [65]. The main concern with these strongly acidic solid acids is their selectivity; while they catalyze cellulose hydrolysis to glucose, they also promote the further conversion of glucose to HMF (5-hydroxymethylfurfural) and LA (Levulinic Acid) [66]. Thus, fine-tuning of acid sites and their location is a must. Here comes the role of weak solid acid sites (COOH) to hydrolyze the glycosidic linkages of cellulose. A 2019 report on green chemistry reported untreated carbon (without sulfonic acid treatment) to exhibit activity in converting cellulose to glucose. The study suggested that the hydroxyl and carboxyl groups on the activated carbon function as acid sites, contributing to the catalytic activity [67]. Another report by Fukoka *et al.* also provided the same result using alkali-activated carbon (K26) without acid treatment [68]. All these findings indicate the role of acidic sites, strong or weak, in the hydrolysis process of cellulose.

Dehydration

5-Hydroxymethyl Furfural (HMF) obtained from the dehydration of C-6 sugars and furfural from C-5 sugars are important building blocks for biomass (Scheme **2**). They both have been incorporated in the top 10 commodities or platform molecules derived from biomass to be converted into fuel additives or chemicals [69]. Stoichiometrically, to convert C-6 sugars into HMF, there is a need to remove three water molecules, and the efficient catalysts discovered for this activity are solid acid catalysts. Numerous reports have been published on the topic, of which a few are listed here. A high yield of 94% HMF was obtained from fructose using sulfonated graphene oxide [70]. Additionally, $TiOm_2$/silica has been found to be effective for this conversion, achieving an efficiency of over 90% [71]. Wang *et al.* have also reported on the use of carbonaceous material after sulfonic acid treatment for effective conversion of fructose (96%) to produce HMF (93.4%) [23]. The commercially available Amberlyst-15 has also been shown to work efficiently, achieving moderate to high HMF yields under moderate reaction conditions [72]. The only disadvantage of all the processes mentioned above is that the starting material used is fructose instead of glucose, which is easily available and has a low price [73]. Thus, there is a strong need to develop methodologies for producing HMF from glucose. The conversion of HMF from glucose involves the isomerization of glucose to fructose as the key step. To address the challenge of using the same catalyst for both steps, scientists have employed a biphasic system, as isomerization requires an aqueous solvent and dehydration demands the use of an organic solvent [74, 75]. Ionic liquids are also reported in the literature as an alternative to avoid the complexities of a biphasic system [76]. Brønsted acid sites-based ionic liquids for conversion of C-5 sugars to furfural *via* microwave-assisted reaction have been developed by Suarez *et al.* in 2012 [77].

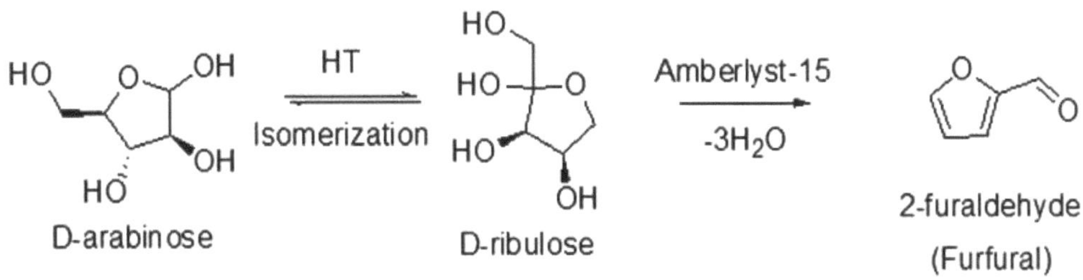

Scheme (2). Dehydration of arabinose to furfural.

Cascade Reactions

The next step, following the production of furfural and its derivatives, requires catalysts with greater versatility. The versatility can be achieved by employing multi-functional solid catalysts to achieve various reactions (hydrogenation, hydrogenolysis, dehydration, oxidation, *etc*). The purpose of these reactions is to produce value-added commodities from furan derivatives (Fig. **2**).

The ring-opening oxidation of furfural/HMF to succinic acid, formic acid, and levulinic acid has gained tremendous attention, owing to its application in plasticizers, surface coatings, pharmaceuticals, *etc* [78, 79]. Numerous heterogeneous catalysts have been developed or are being developed for the oxidative conversion to maleic acid, succinic acid, and formic acid. The reaction was initially studied by Yin *et al.* in 2011 involving an aqueous organic biphasic system [80]. Within a short time, in 2012, Choudhary *et al.* developed the green oxidation method for the conversion of furfural in water using Amberlyst-15 as a catalyst and H_2O_2 as a green oxidizing agent [81]. In 2018, metal-free graphene oxide was employed as an efficient catalyst for the transformation of furfural to succinic acid with high yield [82]. Lewis acidic sites of Sn-beta zeolites have also been observed to show great catalytic activity for conversion [83].

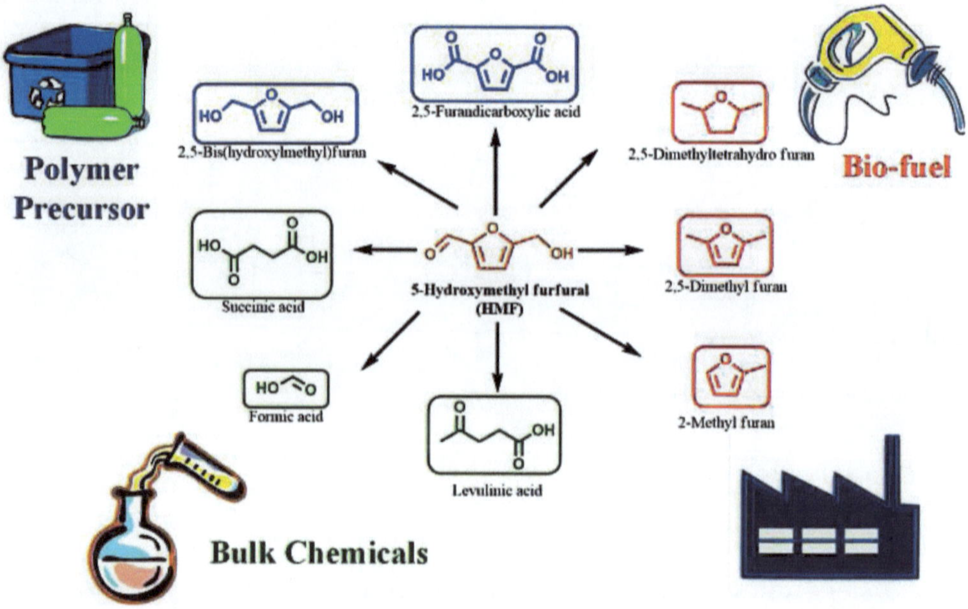

Fig. (2). Value-added commodities derived from the platform molecule 5-hydroxymethyl furfural.

Similar to this, the conversion of HMF to levulinic acid using solid catalysts is also a very important reaction, converting biomass to value-added chemicals [84]. A variety of solid acid catalysts based on carbon materials, metal oxides, polymer-supported acidic catalysts, zeolites, and heteropoly acids, have been developed. A new approach has suggested the optimization of the Brønsted/Lewis acid ratio to have a great impact on maximizing the yield of levulinic acid. Higher Brønsted acidic sites compared to Lewis ones have been found to result in a high yield of levulinic acid, accounting for 64.2% [85]. Table **3** lists a few solid acid catalysts for the effective production of levulinic acid from biomass.

Table 3. Synthesis of levulinic acid from biomass using different solid acid catalysts.

S. No.	Starting Material	Solid Acid Catalyst	LA Yield	References
1.	Glucose	Cr/HZSM-5 (0.75 g)	64.4	[86]
2.	Glucose	Fe-NbP (50 mg)	64.2	[87]
3.	Fructose	H-Resin (30 wt%)	74.6	[88]
4.	Fructose	S-beta zeolite (3 wt%)	43.5	[89]
5.	Cellulose	Ga_2O_3-UiO-66 (2.5%)	32	[90]
6.	Rice husk	Mn_3O_4/ZSM-5, H_3PO_4 (10%), H_2O_2 (2%)	39.8	[91]

Another very important reaction is the hydrogenation of furan derivatives to 2,5-dimethylfuran, 2-methylfuran, and 2,5-dimethyl tetrahydrofuran. A variety of precious metal (Au, Pd, Pt)-based and non-noble metal (Cu, Co, Bi, Fe, Ru, Rh)-based catalysts have been used to date for the hydrogenation reaction [92]. Though precious metals have shown tremendous activity for the said reaction, the focus is now on non-nobel catalysts as the latest developments are being made in this sector only. Among these, Cu-based solid catalysts have been reported in a huge number of publications due to their low price, high activity, easy accessibility, and good performance in hydrogenation. For example, Cu supported on acidic as well as basic supports has been well-studied by Wang *et al.*; they used Cu/ZSM-5, Cu/Al_2O_3, Cu/SiO_2, Cu/ZnO, and Cu/MgO. The studies have suggested the presence of Cu on basic support to provide much better results for hydrogenation with great selectivity, whereas Cu acidic supports have resulted in various side products [93]. Another study proposed that to achieve a high yield with high selectivity from the hydrogenation process, it is necessary to have both Cu^0 and Cu^+ active sites [94]. Further studies have been performed using Cu-based bimetallic catalysts where Co-Cu/Al_2O_3 has been employed to study the hydrogenation reaction, transforming furfural into methylfuran [95]. In addition, Co-Ni bimetallic systems have also received great attention for their effectiveness in producing tetrahydrofuran and 2,5-dimethylfuran from furan derivatives [92].

CONCLUSION

There is no doubt that, in alignment with the Sustainable Development Goals (SDGs) and the pursuit of a sustainable world, the use of renewable resources combined with solid catalysts is considered a key approach. In this context, various scientists have developed and are currently developing various methodologies to achieve maximum efficiency in terms of yield of value-added products with the highest selectivity following green chemistry principles. The products focused in this area include 2,5-dimethylfuran as a fuel precursor; hexanediol, pentanediol, levulinic acid, and formic acid as polymer precursors;

alcohols as biofuels; other products for the synthesis of bioplastics, *etc.* In addition, solid catalysts comprising metal catalysts, metal-supported catalysts, metal oxides, heteropoly acids, and sulfonated polymers are considered as potential catalysts. To design a reaction for converting biomass into the desired product, it is essential to focus on the type of reaction it will undergo. Based on this, an appropriate solid catalyst can be selected. However, it needs to be determined whether the reaction demands deoxygenation, ring-opening hydrogenation, ring-opening hydrogenolysis, oxidation, hydrolysis, or hydrogenation process. To convert cellulose and hemicellulose into C5 and C6 sugars, cation exchange resins and solid Brønsted acidic sites are considered effective. For further transformation of sugars to furan derivatives, metal oxides and monometallic and bimetallic catalysts supported on solid bases are considered efficient. The ring-opening hydrogenolysis process also demands the use of solid Brønsted acidic sites. The oxidative reaction of HMF resulting in 2,5-furandicarboxylic acid (FDCA) is particularly achieved by monometallic or bimetallic catalysts with a solid base used as the support.

REFERENCES

[1] Available from: http://ec.europa.eu/energy/res/legislation/doc/biofuels/en_final.pdf

[2] Available from: http://usgovinfo.about.com/b/2007/01/23/bush-delivers-his-seventh-state-of-the-union-address.htm

[3] Ragauskas, A.J.; Williams, C.K.; Davison, B.H.; Britovsek, G.; Cairney, J.; Eckert, C.A.; Frederick, W.J., Jr; Hallett, J.P.; Leak, D.J.; Liotta, C.L.; Mielenz, J.R.; Murphy, R.; Templer, R.; Tschaplinski, T. The path forward for biofuels and biomaterials. *Science,* **2006**, *311*(5760), 484-489. [http://dx.doi.org/10.1126/science.1114736] [PMID: 16439654]

[4] De, S.; Burange, A.S.; Luque, R. Conversion of biomass-derived feedstocks into value-added chemicals over single-atom catalysts. *Green Chem.,* **2022**, *24*(6), 2267-2286. [http://dx.doi.org/10.1039/D1GC04285H]

[5] Luque, R. Catalytic chemical processes for biomass conversion: Prospects for future biorefineries. *Pure Appl. Chem.,* **2014**, *86*(5), 843-857. [http://dx.doi.org/10.1515/pac-2013-0913]

[6] Tuteja, J. Heterogeneous catalysis for sustainable energy. *SynOpen,* **2023**, *7*(3), 353-361. [http://dx.doi.org/10.1055/a-2129-9076]

[7] Tuteja, J.; Nishimura, S.; Ebitani, K. One-pot synthesis of furans from various saccharides using a combination of solid acid and base catalysts. *Bull. Chem. Soc. Jpn.,* **2012**, *85*(3), 275-281. [http://dx.doi.org/10.1246/bcsj.20110287]

[8] Taarning, E.; Osmundsen, C.M.; Yang, X.; Voss, B.; Andersen, S.I.; Christensen, C.H. Zeolite-catalyzed biomass conversion to fuels and chemicals. *Energy Environ. Sci.,* **2011**, *4*(3), 793-804. [http://dx.doi.org/10.1039/C004518G]

[9] Boerjan, W.; Ralph, J.; Baucher, M. Lignin biosynthesis. *Annu. Rev. Plant Biol.,* **2003**, *54*(1), 519-546. [http://dx.doi.org/10.1146/annurev.arplant.54.031902.134938] [PMID: 14503002]

[10] Shuai, L.; Pan, X.J. Biohydrocarbons: Next generation liquid fuels from lignocellulosic biomass. *Research Progress in Paper Industry and Biorefinery: 4th ISETPP, Guangzhou, China*; South China University of Technology Press: Guangzhou, **2010**, 1–3, pp. 1293-1297.

[11] Iriondo, A.; Agirre, I.; Viar, N.; Requies, J. Value-added bio-chemicals commodities from catalytic conversion of biomass derived furan-compounds. *Catalysts,* **2020**, *10*(8), 895. [http://dx.doi.org/10.3390/catal10080895]

[12] Hara, M.; Nakajima, K.; Kamata, K. Recent progress in the development of solid catalysts for biomass conversion into high value-added chemicals *Sci. Technol. Adv. Mater., 16*, 034903. [http://dx.doi.org/10.1088/1468-6996/16/3/034903]

[13] Corma, A.; Iborra, S.; Velty, A. Chemical routes for the transformation of biomass into chemicals. *Chem. Rev.,* **2007**, *107*(6), 2411-2502. [http://dx.doi.org/10.1021/cr050989d] [PMID: 17535020]

[14] van Putten, R.J.; van der Waal, J.C.; de Jong, E.; Rasrendra, C.B.; Heeres, H.J.; de Vries, J.G. Hydroxymethylfurfural, a versatile platform chemical made from renewable resources. *Chem. Rev.,* **2013**, *113*(3), 1499-1597. [http://dx.doi.org/10.1021/cr300182k] [PMID: 23394139]

[15] Anastas, P.T.; Warner, J.C. *Green Chemistry: Theory and Practice*; Oxford University Press: New York, **1998**.

[16] Meurig Thomas, J.; Raja, R. Designing catalysts for clean technology, green chemistry, and sustainable development. *Annu. Rev. Mater. Res.,* **2005**, *35*(1), 315-350. [http://dx.doi.org/10.1146/annurev.matsci.35.102003.140852]

[17] Horváth, I.T. Green Chemistry. *Acc. Chem. Res.,* **2002**, *35*(9), 685-694. [http://dx.doi.org/10.1021/ar020160a] [PMID: 12234197]

[18] Tundo, P.; Anastas, P.; Black, D.S.; Breen, J.; Collins, T.J.; Memoli, S.; Miyamoto, J.; Polyakoff, M.; Tumas, W. Synthetic pathways and processes in green chemistry. Introductory overview. *Pure Appl. Chem.,* **2000**, *72*(7), 1207-1228. [special issue]. [http://dx.doi.org/10.1351/pac200072071207]

[19] Sheldon, R.A. Atom efficiency and catalysis in organic synthesis. *Pure Appl. Chem.,* **2000**, *72*(7), 1233-1246. [http://dx.doi.org/10.1351/pac200072071233]

[20] Pines, H.; Haag, W.O. Preparation of alkyl azides from hydrazine derivatives. *J. Org. Chem.,* **1958**, *23*(11), 328-329. [http://dx.doi.org/10.1021/jo01096a627]

[21] Zhang, G.; Hattori, H.; Tanabe, K. Aldol addition of acetone, catalyzed by solid base catalysts: magnesium oxide, calcium oxide, strontium oxide, barium oxide, lanthanum (III) oxide and zirconium oxide. *Appl. Catal.,* **1988**, *36*, 189-197. [http://dx.doi.org/10.1016/S0166-9834(00)80114-1]

[22] Hattori, H. Catalysis by basic metal oxides. *Mater. Chem. Phys.,* **1988**, *18*(5-6), 533-552. [http://dx.doi.org/10.1016/0254-0584(88)90021-1]

[23] Vaccari, A. Preparation and catalytic properties of cationic and anionic clays. *Catal. Today,* **1998**, *41*(1-3), 53-71. [http://dx.doi.org/10.1016/S0920-5861(98)00038-8]

[24] Allada, R.; Navrotsky, A.; Berbeco, H.T.; Casey, W.H. Thermochemistry and aqueous solubilities of hydrotalcite-like solids. *Science,* **2002**, *296*(5568), 721-723. [http://dx.doi.org/10.1126/science.1069797] [PMID: 11976450]

[25] Sels, B.F.; De Vos, D.E.; Jacobs, P.A. Hydrotalcite-like anionic clays in catalytic organic reactions. *Catal. Rev., Sci. Eng.,* **2001**, *43*(4), 443-488. [http://dx.doi.org/10.1081/CR-120001809]

[26] Thomas, S.J.M. Solid acid catalysts. *Sci. Am.,* **1992**, *266*(4), 112-118. [http://dx.doi.org/10.1038/scientificamerican0492-112]

[27] Tanabe, K. *Solid Acids and Bases: Their Catalytic Properties*; Academic Press: New York, **1970**. [http://dx.doi.org/10.1016/B978-0-12-683250-1.50005-5]

[28] Tanabe, K. Industrial application of solid acid-base catalysts *Appl. Catal, A,* , *181*(2), 399-434.

[29] Védrine, J. Heterogeneous catalysis on metal oxides. *Catalysts,* **2017**, *7*(11), 341. [http://dx.doi.org/10.3390/catal7110341]

[30] Busca, G. Bases and basic materials in industrial and environmental chemistry: a review of commercial processes. *Ind. Eng. Chem. Res.,* **2009**, *48*(14), 6486-6511. [http://dx.doi.org/10.1021/ie801878d]

[31] Figueras, F. Base catalysis in the synthesis of fine chemicals. *Top. Catal.,* **2004**, *29*(3/4), 189-196. [http://dx.doi.org/10.1023/B:TOCA.0000029802.75155.5a]

[32] Hattori, H. Solid base catalysts: generation of basic sites and application to organic synthesis *Appl. Catal, A,* **2001**, *222*(1–2), 247-259.

[33] Hattori, H. Solid base catalysts: generation, characterization, and catalytic behavior of basic sites. *J. Jpn. Petrol. Inst.,* **2004**, *47*(2), 67-81. [http://dx.doi.org/10.1627/jpi.47.67]

[34] Ono, Y. Solid base catalysts for the synthesis of fine chemicals. *J. Catal.,* **2003**, *216*(1-2), 406-415. [http://dx.doi.org/10.1016/S0021-9517(02)00120-3]

[35] Weitkamp, J.; Hunger, M.; Rymsa, U. Base catalysis on microporous and mesoporous materials: recent progress and perspectives. *Microporous Mesoporous Mater.,* **2001**, *48*(1-3), 255-270. [http://dx.doi.org/10.1016/S1387-1811(01)00366-3]

[36] Hattori, H. Solid base catalysts: fundamentals and their applications in organic reactions *Appl. Catal, A,* **2015**, *504*, 103-109.

[37] Heeres, H.; Handana, R.; Chunai, D.; Borromeus Rasrendra, C.; Girisuta, B.; Jan Heeres, H. Combined dehydration/(transfer)-hydrogenation of C6-sugars (D-glucose and D-fructose) to γ-valerolactone using ruthenium catalysts. *Green Chem.,* **2009**, *11*(8), 1247-1255. [http://dx.doi.org/10.1039/b904693c]

[38] Deng, L.; Li, J.; Lai, D.M.; Fu, Y.; Guo, Q.X. Catalytic conversion of biomass-derived carbohydrates into gamma-valerolactone without using an external H_2 supply. *Angew. Chem. Int. Ed.,* **2009**, *48*(35), 6529-6532. [http://dx.doi.org/10.1002/anie.200902281] [PMID: 19630045]

[39] Tuteja, J.; Nishimura, S.; Ebitani, K. Base-free chemoselective transfer hydrogenation of nitroarenes to anilines with formic acid as hydrogen source by a reusable heterogeneous Pd/ZrP catalyst. *RSC Advances,* **2014**, *4*(72), 38241-38249. [http://dx.doi.org/10.1039/C4RA06174H]

[40] Toledano, A.; Serrano, L.; Pineda, A.; Romero, A.A.; Luque, R.; Labidi, J. Microwave-assisted depolymerisation of organosolv lignin *via* mild hydrogen-free hydrogenolysis: Catalyst screening. *Appl. Catal. B,* **2014**, *145*, 43-55. [http://dx.doi.org/10.1016/j.apcatb.2012.10.015]

[41] Toledano, A.; Serrano, L.; Labidi, J.; Pineda, A.; Balu, A.M.; Luque, R. Heterogeneously catalysed mild hydrogenolytic depolymerisation of lignin under microwave irradiation with hydrogen-donating solvents. *ChemCatChem,* **2013**, *5*(4), 977-985. [http://dx.doi.org/10.1002/cctc.201200616]

[42] Toledano, A.; Serrano, L.; Balu, A.M.; Luque, R.; Pineda, A.; Labidi, J. Fractionation of organosolv lignin from olive tree clippings and its valorization to simple phenolic compounds. *ChemSusChem,* **2013**, *6*(3), 529-536. [http://dx.doi.org/10.1002/cssc.201200755] [PMID: 23404837]

[43] Pineda, A.; Balu, A.M.; Campelo, J.M.; Romero, A.A.; Carmona, D.; Balas, F.; Santamaria, J.; Luque,

R. A dry milling approach for the synthesis of highly active nanoparticles supported on porous materials. *ChemSusChem,* **2011**, *4*(11), 1561-1565.
[http://dx.doi.org/10.1002/cssc.201100265] [PMID: 22191094]

[44] Loges, B.; Boddien, A.; Junge, H.; Beller, M. Controlled generation of hydrogen from formic acid amine adducts at room temperature and application in H_2/O_2 fuel cells. *Angew. Chem. Int. Ed.,* **2008**, *47*(21), 3962-3965.
[http://dx.doi.org/10.1002/anie.200705972] [PMID: 18457345]

[45] Boddien, A.; Loges, B.; Junge, H.; Beller, M. Hydrogen generation at ambient conditions: application in fuel cells. *ChemSusChem,* **2008**, *1*(8-9), 751-758.
[http://dx.doi.org/10.1002/cssc.200800093] [PMID: 18686291]

[46] Geboers, J.A.; Van de Vyver, S.; Ooms, R.; Op de Beeck, B.; Jacobs, P.A.; Sels, B.F. Chemocatalytic conversion of cellulose: opportunities, advances and pitfalls. *Catal. Sci. Technol.,* **2011**, *1*(5), 714-726.
[http://dx.doi.org/10.1039/c1cy00093d]

[47] Suganuma, S.; Nakajima, K.; Kitano, M.; Yamaguchi, D.; Kato, H.; Hayashi, S.; Hara, M. Hydrolysis of cellulose by amorphous carbon bearing SO_3H, COOH, and OH groups. *J. Am. Chem. Soc.,* **2008**, *130*(38), 12787-12793.
[http://dx.doi.org/10.1021/ja803983h] [PMID: 18759399]

[48] Cai, H.; Li, C.; Wang, A.; Xu, G.; Zhang, T. Zeolite-promoted hydrolysis of cellulose in ionic liquid, insight into the mutual behavior of zeolite, cellulose and ionic liquid. *Appl. Catal. B,* **2012**, *123-124*, 333-338.
[http://dx.doi.org/10.1016/j.apcatb.2012.04.041]

[49] Li, X.; Jiang, Y.; Wang, L.; Meng, L.; Wang, W.; Mu, X. Effective low-temperature hydrolysis of cellulose catalyzed by concentrated $H_3PW_{12}O_{40}$ under microwave irradiation. *RSC Advances,* **2012**, *2*(17), 6921-6925.
[http://dx.doi.org/10.1039/c2ra21022c]

[50] Onda, A.; Ochi, T.; Yanagisawa, K. Selective hydrolysis of cellulose into glucose over solid acid catalysts. *Green Chem.,* **2008**, *10*(10), 1033-1037.
[http://dx.doi.org/10.1039/b808471h]

[51] Akiyama, G.; Matsuda, R.; Sato, H.; Takata, M.; Kitagawa, S. Cellulose hydrolysis by a new porous coordination polymer decorated with sulfonic acid functional groups. *Adv. Mater.,* **2011**, *23*(29), 3294-3297.
[http://dx.doi.org/10.1002/adma.201101356] [PMID: 21661069]

[52] Kitano, M.; Yamaguchi, D.; Suganuma, S.; Nakajima, K.; Kato, H.; Hayashi, S.; Hara, M. Adsorption-enhanced hydrolysis of β-1,4-glucan on graphene-based amorphous carbon bearing SO_3H, COOH, and OH groups. *Langmuir,* **2009**, *25*(9), 5068-5075.
[http://dx.doi.org/10.1021/la8040506] [PMID: 19397353]

[53] Rinaldi, R.; Palkovits, R.; Schüth, F. Depolymerization of cellulose using solid catalysts in ionic liquids. *Angew. Chem. Int. Ed.,* **2008**, *47*(42), 8047-8050.
[http://dx.doi.org/10.1002/anie.200802879] [PMID: 18814164]

[54] Kobayashi, H.; Komanoya, T.; Hara, K.; Fukuoka, A. Water-tolerant mesoporous-carbon-supported ruthenium catalysts for the hydrolysis of cellulose to glucose. *ChemSusChem,* **2010**, *3*(4), 440-443.
[http://dx.doi.org/10.1002/cssc.200900296] [PMID: 20198680]

[55] Lai, D.; Deng, L.; Li, J.; Liao, B.; Guo, Q.; Fu, Y. Hydrolysis of cellulose into glucose by magnetic solid acid. *ChemSusChem,* **2011**, *4*(1), 55-58.
[http://dx.doi.org/10.1002/cssc.201000300] [PMID: 21226211]

[56] Takagaki, A.; Nishimura, M.; Nishimura, S.; Ebitani, K. Hydrolysis of sugars using magnetic silica nanoparticles with sulfonic acid groups. *Chem. Lett.,* **2011**, *40*(10), 1195-1197.
[http://dx.doi.org/10.1246/cl.2011.1195]

[57] Zhang, C.; Wang, H.; Liu, F.; Wang, L.; He, H. Magnetic core–shell Fe_3O_4@C-SO_3H nanoparticle catalyst for hydrolysis of cellulose. *Cellulose,* **2013**, *20*(1), 127-134. [http://dx.doi.org/10.1007/s10570-012-9839-5]

[58] Van de Vyver, S.; Peng, L.; Geboers, J.; Schepers, H.; de Clippel, F.; Gommes, C.J.; Goderis, B.; Jacobs, P.A.; Sels, B.F. Sulfonated silica/carbon nanocomposites as novel catalysts for hydrolysis of cellulose to glucose. *Green Chem.,* **2010**, *12*(9), 1560-1563. [http://dx.doi.org/10.1039/c0gc00235f]

[59] Suganuma, S.; Nakajima, K.; Kitano, M.; Yamaguchi, D.; Kato, H.; Hayashi, S.; Hara, M. Hydrolysis of cellulose by amorphous carbon bearing SO_3H, COOH, and OH groups. *J. Am. Chem. Soc.,* **2008**, *130*(38), 12787-12793. [http://dx.doi.org/10.1021/ja803983h] [PMID: 18759399]

[60] Rinaldi, R.; Palkovits, R.; Schüth, F. Depolymerization of cellulose using solid catalysts in ionic liquids. *Angew. Chem. Int. Ed.,* **2008**, *47*(42), 8047-8050. [http://dx.doi.org/10.1002/anie.200802879] [PMID: 18814164]

[61] Harmer, M.A.; Farneth, W.E.; Sun, Q. High Surface area nafion resin/silica nanocomposites: A new class of solid acid catalyst. *J. Am. Chem. Soc.,* **1996**, *118*(33), 7708-7715. [http://dx.doi.org/10.1021/ja9541950]

[62] Fraga, A.C.; Quitete, C.P.B.; Ximenes, V.L.; Sousa-Aguiar, E.F.; Fonseca, I.M.; Rego, A.M.B. Biomass derived solid acids as effective hydrolysis catalysts. *J. Mol. Catal. Chem.,* **2016**, *422*, 248-257. [http://dx.doi.org/10.1016/j.molcata.2015.12.005]

[63] To, A.T.; Chung, P.W.; Katz, A. Weak-acid sites catalyze the hydrolysis of crystalline cellulose to glucose in water: importance of post-synthetic functionalization of the carbon surface. *Angew. Chem. Int. Ed.,* **2015**, *54*(38), 11050-11053. [http://dx.doi.org/10.1002/anie.201504865] [PMID: 26276901]

[64] Foo, G.S.; Van Pelt, A.H.; Krötschel, D.; Sauk, B.F.; Rogers, A.K.; Jolly, C.R.; Yung, M.M.; Sievers, C. Hydrolysis of cellobiose over selective and stable sulfonated activated carbon catalysts. *ACS Sustain. Chem.& Eng.,* **2015**, *3*(9), 1934-1942. [http://dx.doi.org/10.1021/acssuschemeng.5b00530]

[65] Rastogi, M.; Shrivastava, S. Recent advances in second generation bioethanol production: An insight to pretreatment, saccharification and fermentation processes. *Renew. Sustain. Energy Rev.,* **2017**, *80*, 330-340. [http://dx.doi.org/10.1016/j.rser.2017.05.225]

[66] Onda, A.; Ochi, T.; Yanagisawa, K. Selective hydrolysis of cellulose into glucose over solid acid catalysts. *Green Chem.,* **2008**, *10*(10), 1033-1037. [http://dx.doi.org/10.1039/b808471h]

[67] Kobayashi, H.; Yabushita, M.; Komanoya, T.; Hara, K.; Fujita, I.; Fukuoka, A. High-yielding one-pot synthesis of glucose from cellulose using simple activated carbons and trace hydrochloric acid. *ACS Catal.,* **2013**, *3*(4), 581-587. [http://dx.doi.org/10.1021/cs300845f]

[68] Bozell, J.J.; Petersen, G.R. Technology development for the production of biobased products from biorefinery carbohydrates—the US Department of Energy's "Top 10" revisited. *Green Chem.,* **2010**, *12*(4), 539-554. [http://dx.doi.org/10.1039/b922014c]

[69] Hou, Q.; Li, W.; Ju, M.; Liu, L.; Chen, Y.; Yang, Q. One-pot synthesis of sulfonated graphene oxide for efficient conversion of fructose into HMF. *RSC Advances,* **2016**, *6*(106), 104016-104024. [http://dx.doi.org/10.1039/C6RA23420H]

[70] Songo, M.M.; Moutloali, R.; Ray, S.S. Development of TiO_2-carbon composite acid catalyst for

dehydration of fructose to 5-hydroxymethylfurfural. *Catalysts,* **2019**, *9*(2), 126-141.
[http://dx.doi.org/10.3390/catal9020126]

[71] Marullo, S.; Rizzo, C.; D'Anna, F. Activity of a heterogeneous catalyst in deep eutectic solvents: The case of carbohydrate conversion into 5-hydroxymethylfurfural. *ACS Sustain. Chem.& Eng.,* **2019**, *7*(15), 13359-13368.
[http://dx.doi.org/10.1021/acssuschemeng.9b02605]

[72] Boisen, A.; Christensen, T.B.; Fu, W.; Gorbanev, Y.Y.; Hansen, T.S.; Jensen, J.S.; Klitgaard, S.K.; Pedersen, S.; Riisager, A.; Ståhlberg, T.; Woodley, J.M. Process integration for the conversion of glucose to 2,5-furandicarboxylic acid. *Chem. Eng. Res. Des.,* **2009**, *87*(9), 1318-1327.
[http://dx.doi.org/10.1016/j.cherd.2009.06.010]

[73] Román-Leshkov, Y.; Chheda, J.N.; Dumesic, J.A. Phase modifiers promote efficient production of hydroxymethylfurfural from fructose. *Science,* **2006**, *312*(5782), 1933-1937.
[http://dx.doi.org/10.1126/science.1126337] [PMID: 16809536]

[74] Nikolla, E.; Román-Leshkov, Y.; Moliner, M.; Davis, M.E. One-Pot Synthesis of 5-(hydroxy-methyl)furfural from carbohydrates using tin-beta zeolite. *ACS Catal.,* **2011**, *1*(4), 408-410.
[http://dx.doi.org/10.1021/cs2000544]

[75] Zunita, M.; Yuan, D.M.; Syafi' Laksono, A. Glucose conversion into hydroxymethylfurfural *via* ionic liquid-based processes. *Chem. Eng. J. Adv.,* **2022**, *11*, 100307-100321.
[http://dx.doi.org/10.1016/j.ceja.2022.100307]

[76] Serrano-Ruiz, J.C.; Campelo, J.M.; Francavilla, M.; Romero, A.A.; Luque, R.; Menéndez-Vázquez, C.; García, A.B.; García-Suárez, E.J. Efficient microwave-assisted production of furfural from C5 sugars in aqueous media catalysed by Brönsted acidic ionic liquids. *Catal. Sci. Technol.,* **2012**, *2*(9), 1828-1832.
[http://dx.doi.org/10.1039/c2cy20217d]

[77] Li, X.; Ho, B.; Lim, D.S.W.; Zhang, Y. Highly efficient formic acid-mediated oxidation of renewable furfural to maleic acid with H_2O_2. *Green Chem.,* **2017**, *19*(4), 914-918.
[http://dx.doi.org/10.1039/C6GC03020C]

[78] Alonso-Fagúndez, N.; Granados, M.L.; Mariscal, R.; Ojeda, M. Selective conversion of furfural to maleic anhydride and furan with VO(x)/Al($_2$)O($_3$) catalysts. *ChemSusChem,* **2012**, *5*(10), 1984-1990.
[http://dx.doi.org/10.1002/cssc.201200167] [PMID: 22847991]

[79] Guo, H.; Yin, G. Catalytic aerobic oxidation of renewable furfural with phosphomolybdic acid catalyst: an alternative route to maleic acid. *J. Phys. Chem. C,* **2011**, *115*(35), 17516-17522.
[http://dx.doi.org/10.1021/jp2054712]

[80] Choudhary, H.; Nishimura, S.; Ebitani, K. Highly efficient aqueous oxidation of furfural to succinic acid using reusable heterogeneous acid catalyst with hydrogen peroxide. *Chem. Lett.,* **2012**, *41*(4), 409-411.
[http://dx.doi.org/10.1246/cl.2012.409]

[81] Zhu, W.; Tao, F.; Chen, S.; Li, M.; Yang, Y.; Lv, G. Efficient oxidative transformation of furfural into succinic acid over acidic metal-free graphene oxide. *ACS Sustain. Chem.& Eng.,* **2019**, *7*(1), 296-305.
[http://dx.doi.org/10.1021/acssuschemeng.8b03373]

[82] Palai, Y.N.; Shrotri, A.; Fukuoka, A. Selective oxidation of furfural to succinic acid over lewis acidic sn-beta. *ACS Catal.,* **2022**, *12*(6), 3534-3542.
[http://dx.doi.org/10.1021/acscatal.1c05348]

[83] Signoretto, M.; Taghavi, S.; Ghedini, E.; Menegazzo, F. Catalytic production of Levulinic Acid (LA) from actual biomass. *Molecules,* **2019**, *24*(15), 2760-2780.
[http://dx.doi.org/10.3390/molecules24152760] [PMID: 31366018]

[84] Liu, Y.; Li, H.; He, J.; Zhao, W.; Yang, T.; Yang, S. Catalytic conversion of carbohydrates to levulinic acid with mesoporous niobium-containing oxides. *Catal. Commun.,* **2017**, *93*, 20-24.

[http://dx.doi.org/10.1016/j.catcom.2017.01.023]

[85] Wei, W.; Wu, S. Experimental and kinetic study of glucose conversion to levulinic acid in aqueous medium over Cr/HZSM-5 catalyst. *Fuel,* **2018**, *225*, 311-321. [http://dx.doi.org/10.1016/j.fuel.2018.03.120]

[86] Liu, Y.; Li, H.; He, J.; Zhao, W.; Yang, T.; Yang, S. Catalytic conversion of carbohydrates to levulinic acid with mesoporous niobium-containing oxides. *Catal. Commun.,* **2017**, *93*, 20-24. [http://dx.doi.org/10.1016/j.catcom.2017.01.023]

[87] Pyo, S.H.; Glaser, S.J.; Rehnberg, N.; Hatti-Kaul, R. Clean production of levulinic acid from fructose and glucose in salt water by heterogeneous catalytic dehydration. *ACS Omega,* **2020**, *5*(24), 14275-14282. [http://dx.doi.org/10.1021/acsomega.9b04406] [PMID: 32596564]

[88] Bisen, S.K.; Niphadkar, P.S.; Nandanwar, S.U.; Simakova, I.; Bokade, V.V. Milder operating parameters for one-step conversion of fructose to levulinic acid over sulfonated H-β zeolite in aqueous media. *Environ. Prog. Sustain. Energy,* **2021**, *40*(1), e13530. [http://dx.doi.org/10.1002/ep.13530]

[89] Wang, K.; Liu, Y.; Wu, W.; Chen, Y.; Fang, L.; Li, W.; Ji, H. Production of levulinic acid *via* cellulose conversion over metal oxide-loaded mof catalysts in aqueous medium. *Catal. Lett.,* **2020**, *150*(2), 322-331. [http://dx.doi.org/10.1007/s10562-019-03023-y]

[90] Pratama, A.P.; Rahayu, D.U.C.; Krisnandi, Y.K. Levulinic acid production from delignified rice husk waste over manganese catalysts: Heterogeneous *versus* homogeneous. *Catalysts,* **2020**, *10*(3), 327. [http://dx.doi.org/10.3390/catal10030327]

[91] Wang, Y.; Zhao, D.; Rodríguez-Padrón, D.; Len, C. Recent Advances in catalytic hydrogenation of furfural. *Catalysts,* **2019**, *9*(10), 796. [http://dx.doi.org/10.3390/catal9100796]

[92] Wang, Y.; Zhu, W.; Sang, S.; Gao, L.; Xiao, G. Supported Cu catalysts for the hydrogenation of furfural in aqueous phase: effect of support. *Asia-Pac. J. Chem. Eng.,* **2017**, *12*(3), 422-431. [http://dx.doi.org/10.1002/apj.2085]

[93] Zhang, H.; Lei, Y.; Kropf, A.J.; Zhang, G.; Elam, J.W.; Miller, J.T.; Sollberger, F.; Ribeiro, F.; Akatay, M.C.; Stach, E.A.; Dumesic, J.A.; Marshall, C.L. Enhancing the stability of copper chromite catalysts for the selective hydrogenation of furfural using ALD overcoating. *J. Catal.,* **2014**, *317*, 284-292. [http://dx.doi.org/10.1016/j.jcat.2014.07.007]

[94] Srivastava, S.; Jadeja, G.C.; Parikh, J. Copper-cobalt catalyzed liquid phase hydrogenation of furfural to 2-methylfuran: An optimization, kinetics and reaction mechanism study. *Chem. Eng. Res. Des.,* **2018**, *132*, 313-324. [http://dx.doi.org/10.1016/j.cherd.2018.01.031]

CHAPTER 4

Oxidation of Biomass to Value-Added Chemicals Using Heterogeneous Solid Base Catalysts: A Recent Advancement

Kajal Pawar[1] and **Pavan More**[1,*]

[1] *Institute of Chemical Technology, Department of Chemistry, Matunga, Mumbai-400019, India*

Abstract: Biomass is considered one of the alternative resources with the greatest potential to compete both as fuels and as chemical intermediates. Biomass-derived molecules like glucose, xylose, HMF, and levulinic acid can be converted to value-added chemicals *via* dehydration, oxidation, isomerization, reforming, and aldol condensation processes using heterogeneous catalysts. The drawbacks of homogeneous catalysts are their low solvent solubility, susceptibility to breakdown under oxidation conditions, and requirement for product separation. Biomass contains a high amount of water, and the formation of water as a byproduct during transformation is a significant obstacle to reaction. Therefore, aqueous phase reaction using heterogeneous catalysts is of major interest. There have been reports of studies employing expensive noble transition metals as catalysts. Non-noble metal oxides are more widely available, less expensive, and have greater thermal stability and poisoning resistance than noble metals. In this chapter, we discuss some noble and non-noble metal heterogeneous catalysts. Heterogeneous metal catalyst involves single metal, bimetallic catalysts, metal oxide, spinel, perovskites, *etc*.

Keywords: Aqueous phase, Biomass, Heterogeneous catalyst, Intermediate, Noble metals.

INTRODUCTION

Biomass, as a sustainable and renewable resource, has gained significant attention in recent years as an alternative feedstock for the production of value-added products [1]. One of the key challenges in biomass valorization is the efficient conversion of biomass-derived compounds into high-value chemicals and fuels. Catalytic processes are crucial in this conversion, enabling selective transformations of biomass components into desired products [2].

* **Corresponding author Pavan More:** Institute of Chemical Technology, Department of Chemistry, Matunga, Mumbai-400019, India; E-mails: pm.more@ictmumbai.edu.in; chy21kb.pawar@pg.ictmumbai.edu.in

Over the past ten years, the public and commercial sectors have significantly supported scientific research and process evolution for synthesizing chemicals from sugars derived from lignocellulosic biomass [3]. New methods for synthetic biology and metabolic engineering, as well as new chemical and biological pathways and homogeneous and heterogeneous catalysts, have been developed as a result of this progress, allowing the conversion of waste products like glycerol, lignin, and sugars into useful chemicals and hydrocarbon fuels [4]. Inexpensive feeds and low-quality heat drive large-scale C-C coupling, hydrogenation, and deoxygenation processes. However, chemical catalysis requires clean feed streams to prevent catalyst deactivation [5]. Corn-derived sugar, sucrose (sugar cane), lignocellulosic sugars (glucose, xylose), and glycerol are examples of renewable resources that can be used to produce industrial chemicals and fuels [6]. Carbohydrates from C3 to C6 are available from these sources as basic ingredients. Biodiesel may also be made from vegetable and palm oils [7]. C–C coupling is well-established through methods like Methanol-To-Gasoline (MTG) and Fischer–Tropsch, and large-scale heterogeneous hydrogenation, deoxygenation, and decarboxylation reactors are widely utilized in the chemical and petroleum-based fuel sectors [8]. Theoretical modeling has been used to study the reaction mechanism of molecules formed from biomass on the surface of metal particles. The development of multifunctional catalysts that enable the intensification of processes has led to significant improvements in efficiency and product yields [9]. Catalyst modification considers the effects of water, porosity, hydrophilic, acidic properties of supports, and catalyst deactivation caused by organic pollutants found in bio-sourced raw materials [10].

Two key processes for the biomass valorization process are esterification reactions and catalytic selective oxidation. Alcohols and other oxygen-containing molecules generated from biomass can be selectively oxidized to produce useful compounds, including aldehydes, ketones, carboxylic acids, and esters [11]. However, esterification makes it easier to produce esters from acids and alcohols obtained from biomass. These esters can be used in a variety of industrial applications, such as flavouring agents, solvents, and precursors for biodiesel [12]. Demand for sustainable and renewable alternatives to fossil fuels has increased due to the global depletion of fossil fuels. These alternatives are used to produce energy, chemicals, and fuels [13]. Although solar, wind, hydropower, and geothermal energy are examples of renewable energy sources, biomass is the only readily accessible renewable source of carbon needed for producing fuels and chemicals [14]. Currently, the majority of industrial processes rely on fossil fuels like petroleum and other fossil resources. However, there is more than enough biomass available right now to meet the needs of the modern chemical industry in terms of raw materials [15].

Review articles on bio-sourced chemicals have been published in large quantities. A wide variety of biomass feedstocks and reaction types are covered in these articles. They focus on specific feedstocks like cellulose, hemicelluloses and pentoses, glycerol, 5-hydroxymethylfurfural, carbohydrates, triglycerides, lignin, and lignocellulose [16]. A few review articles also discuss certain types of reactions, including oxidation, telomerization, hydrogenolysis/dehydroxylation, and metathesis [17]. Catalysts tailored to the biomass-to-chemical value chain have received relatively less attention. Besides these pure chemicals, consideration will also be given to metal-catalyzed processes that yield a mixture of chemicals that can be utilized to produce high-tonnage end products, such as plasticizers, paints, foams, resins, surfactants, and paper additives [18].

This chapter focuses on the application of both noble and non-noble metal catalysts to the selective oxidation and esterification of chemicals produced from biomass. It discusses the fundamentals of catalyst design, reaction mechanisms, and process optimization for the effective conversion of biomass to value-added products.

Catalyst Design and Synthesis

The conversion of biomass into value-added chemicals is a sustainable solution to environmental and resource challenges, with heterogeneous solid base catalysts playing a key role in oxidation processes. These catalysts are valued for their efficiency, reusability, and selectivity, making their design crucial for optimizing reactions. Factors, such as active site selection, support materials, and properties like porosity and thermal stability, significantly impact the performance of catalysts [19]. Enhancements in catalytic efficiency are achieved through the incorporation of alkali metals, rare-earth elements, or metal oxides. Advanced synthesis techniques, such as sol-gel and hydrothermal methods, enable precise control over catalyst structure. Innovations like bifunctional active sites and nanostructures further enhance reaction rates and selectivity for products like lactic acid, vanillin, and HMF. Developing cost-effective and eco-friendly catalysts is essential for advancing biomass oxidation on an industrial scale [20]. Gold (Au), platinum (Pt), and palladium (Pd) are noble metals known for their excellent conversion efficiency and selectivity across various oxidation processes. These metals can activate oxygen-containing functional groups in biomass-derived molecules, converting them into useful products [21]. Transition metals, such as nickel (Ni), cobalt (Co), and iron (Fe), have been widely explored as non-noble metal catalysts for biomass valorization due to their cost-effectiveness and abundance compared to noble metals [22, 23]. The preparation of these catalysts involves diverse methods, including deposition-precipitation, impregnation, and sol-gel techniques [24]. To enhance catalytic performance, these materials are often supported on porous substrates like alumina, silica, or zeolites, which improve surface area and catalyst dispersion, thereby boosting their efficiency [25].

Selective Oxidation Reaction of Biomass

The selective oxidation process is crucial for the chemical industry and organic synthesis. When alcohol is selectively oxidized, carbonyl molecules, such as aldehydes, ketones, and esters, are produced [26]. However, this process produces hazardous and poisonous substances. It has been observed that these limitations are overcome by the heterogeneous and homogeneous catalytic oxidation methods [27]. Catalysts, such as MnO_2, hypervalent iodine reagents, and chromium (VI) salts, have long been utilized in the oxidation of alcohols. However, it utilizes costly, ecologically dangerous chemicals that are not appropriate for large-scale operations [28]. The drawbacks of homogeneous catalysts are their low solvent solubility, susceptibility to breakdown under oxidation conditions, and requirement for product separation [29]. As a result, more beneficial techniques utilizing recyclable catalysts are crucial for both the environment and the economy [30]. The advantages of using heterogeneous metal oxide catalysts include their recyclability, high thermal stability, ease of synthesis at a low cost, chemical stabilities, and the absence of metallic species contamination in the products [31]. This chapter discusses noble and non-noble metal catalysts supported by a review of the literature. Molecular oxygen and heterogeneous catalysis used in oxidation processes provide a sustainable substitute for conventional hazardous chemical oxidants [32]. Similarly, esterification reactions, which involve the condensation of biomass-derived acids with alcohols, can also benefit from these green catalytic methods to produce ester products.

METAL CATALYSTS

Noble Metal Catalysts

Selective oxidation reactions of biomass-derived compounds, such as alcohols and acids, require catalysts capable of activating specific functional groups while minimizing overoxidation and side reactions [33]. The first catalysts suggested for oxidation were platinum, palladium, gold, rhodium, and iridium, ruthenium. Noble metal catalysts exhibit very high activity for oxidation processes, particularly at low temperatures. The high activity and selectivity of supported and mixed oxide-doped noble metal catalysts are especially popular due to their enhanced activity and selectivity in these reactions [34]. Noble metal catalysts, particularly supported Pt and Pd catalysts, have shown excellent activity and selectivity in these reactions [35]. For example, Pt-based catalysts have been used to selectively oxidize sugars and alcohols produced from biomass to useful intermediates like glycolic acid and glyceric acid [36]. Pd catalysts have demonstrated high selectivity in the oxidation of biomass-derived aldehydes to carboxylic acids, which are important building blocks for polymers and pharmaceuticals [37]. Catalysts for esterification should promote the formation of ester bonds while minimizing side reactions, such as dehydration and polymerization [38]. For example, the esterification of fatty acids with alcohols using Pt catalysts supported on acidic supports has resulted in biodiesel, a sustainable substitute for diesel fuel made from petroleum [39]. Alcohols and carboxylic acids produced from biomass have been esterified using Au-based catalysts to create flavored esters [40].

For oxidation and esterification processes, noble metals, such as Pt, Rh, and Pd supported on alumina, have shown effective activity and selectivity [41]. To create value-added products using precious metal catalysts, including platinum, gold, and palladium, have been widely used for the selective oxidation of chemicals produced from biomass. These catalysts possess unique properties that make them highly effective in promoting specific chemical transformations while minimizing undesired side reactions [42]. Following are some examples of noble metal catalysts used for the selective oxidation of biomass to value-added products:

Platinum (Pt) Catalysts

Alcohols and sugars produced from biomass have been selectively oxidized and esterified to useful intermediates using Pt catalysts supported on a variety of materials, including carbon, alumina, or zeolites [43]. For instance, Pt catalysts have been employed for the oxidation of glycerol to glyceric acid, a valuable platform chemical with applications in the pharmaceutical and food industries. M. Faroppa *et al.* [44] synthesized PtSn0.4/C catalyst by impregnation to oxidize glycerol to dihydroxyacetone (DHA), with a 97% improved conversion selectivity. The amount of tin deposited was verified by TEM (Fig. **1**) and XPS (Fig. **2**) Scheme **1**.

Additionally, Pt catalysts have shown promise in the oxidation of lignocellulosic biomass-derived compounds, such as sugars and polyols, to produce valuable building blocks [45]. Platinum catalysts supported on acidic supports have been utilized for the esterification of biomass-derived carboxylic acids with alcohols Scheme **2**. The acidic support facilitates protonation of the carboxylic acid, making it more reactive towards nucleophilic attack by the alcohol [46]. Pt catalysts can promote the formation of ester bonds while minimizing side reactions, leading to high selectivity in the esterification process [47]. Yang Tang *et al.* [48] synthesized 5%Pt/$Al_2(SiO_3)_3$ catalyst using the wet impregnation method, which exhibited higher activity for butyl aldehyde to butyl acetate ester (93% selectivity) formation. Its greater surface area, wider pore diameter, smaller metal particle size, and more active acid sites are probably the reasons for this result. 5%Pt/$Al_2(SiO_3)_3$ has a surface area of 160 m^2/g. A greater surface area can result from a larger pore volume, and this surface area may encourage a greater dispersion of Pt to enhance activity.

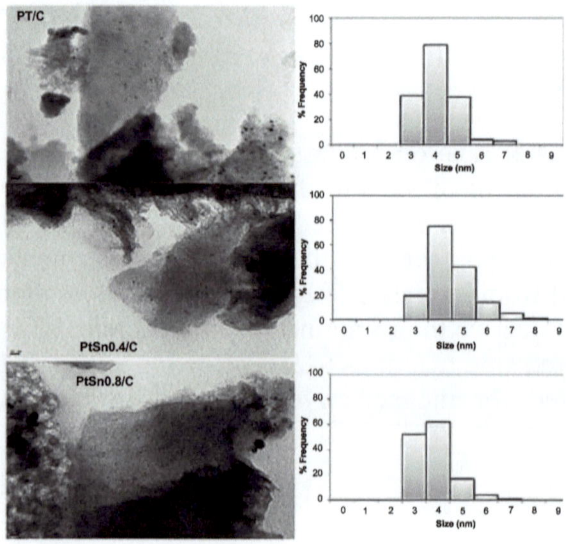

Fig. (1). TEM images and particle size of Pt/C, PtSn0.4/C, and PtSn0.8/C. (Reprinted from M. Faroppa *et al.* [44] with permission from MDPI, Copyright 2023).

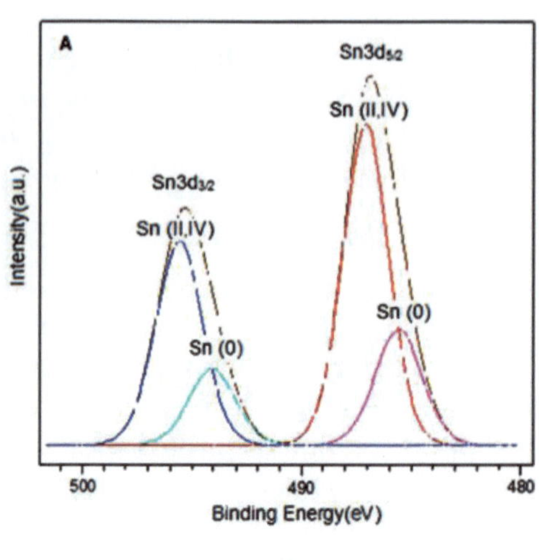

Fig. (2). Sn3d XPS spectra of PtSn0.4/C catalyst (Reprinted with permission from [44]).

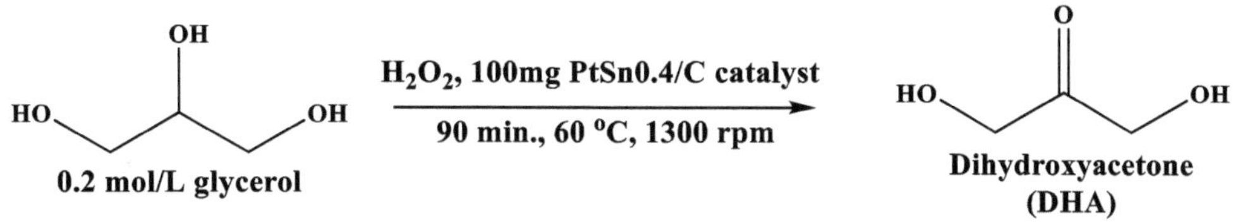

Scheme (1). Selective oxidation of glycerol to dihydroxyacetone reaction.

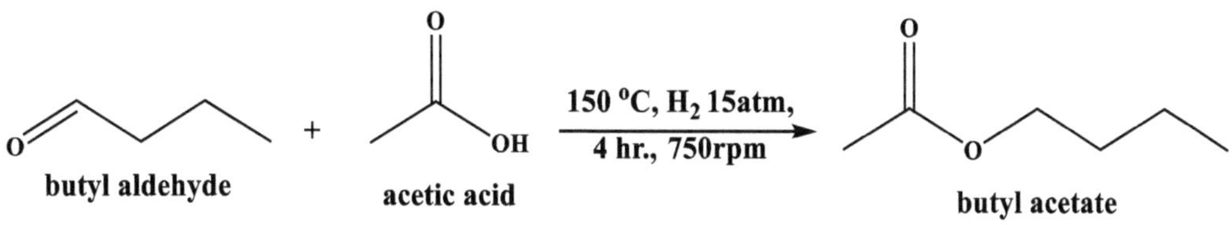

Scheme (2). Esterification reaction of butyl aldehyde to butyl acetate.

Palladium (Pd) Catalysts

The selective oxidation of aldehydes, alcohols, and acids produced from biomass to useful compounds, such as carboxylic acids, ketones, and esters, has been extensively studied using Pd catalysts [49]. The oxidation of biomass-derived alcohols to ketones and esters, as well as biomass-derived aldehydes to carboxylic acids, has demonstrated remarkable activity and selectivity when using supported Pd catalysts. These are crucial intermediates in the production of polymers, perfumes, and medications. The catalytic behaviors of Pd supported on NaX zeolite catalysts with different mean Pd sizes were investigated by Li *et al.* [50]. Among the catalysts used in the study, Pd with a mean size of 2.8 nm showed the maximum BzOH conversion, while catalysts with bigger and smaller Pd particles demonstrated lower conversions (Table **1**). The mean size of the Pd nanoparticles was found to be significantly influenced by the interaction between the Pd precursor and the support.

Table 1. Conversion rates and TOF of Pd/NaX catalysts (Pd content, 1.35 wt%) with different mean sizes of Pd particle.

Mean size Pd (nm)	Initial conversion rate (mmol h^{-1})	Intrinsic TOF (h^{-1})
2.6	7.6	1720
2.8	11.7	2970
2.9	10.7	2730
5.4	3.5	1550

Palladium catalysts supported by silica-alumina (SiO_2-Al_2O_3) were produced and investigated by Chen *et al.* [51] for selective oxidation of alcohols. The mean size of Pd particles strongly impacted the oxidation of BzOH. When the mean size of Pd particles was adjusted from 2.2 to 10 nm, the intrinsic TOF per surface Pd atom for BzOH conversion peaked at a medium size of 3.6–4.3 nm, suggesting that the reaction was structure-sensitive (Table **2**).

Table 2. Conversion of solvent-free aerobic oxidation of benzyl alcohol

Catalysts	Pd content [wt%]	Mean size of Pd [nm]	Initial conversion rate [$mol.g^{-1}s^{-1}$]
Pd/SiO_2-hexanol	0.55	5.7	0.26
Pd/SiO_2-Al_2O_3-hexanol (2/1)	0.52	3.2	2.12
Pd/SiO_2-Al_2O_3-H_2 (2/1)	0.52	4.3	3.22
Pd/SiO_2-H_2	0.55	10	0.094

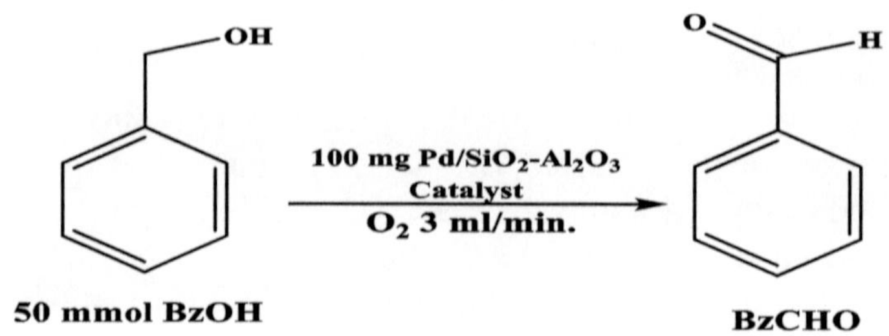

Scheme (3). The selective oxidation reaction of BzOH to BzCHO.

Gold (Au) Catalysts

Au catalysts supported on metal oxides or carbonaceous materials have demonstrated excellent catalytic activity in the selective oxidation of biomass-derived compounds under mild reaction conditions [52]. Au catalysts have been employed for the oxidation of lignin-derived phenolic compounds to valuable aromatic aldehydes, which are used as intermediates in the synthesis of fine chemicals and pharmaceuticals [53]. Gold catalysts have also been investigated for esterification reactions involving biomass-derived compounds. Au catalysts supported on appropriate supports, such as metal oxides or carbon materials, can exhibit catalytic activity in esterification reactions [54]. The precise mechanism by which gold catalyzes esterification is still under investigation, but it likely involves activation of the carbonyl group in the carboxylic acid and subsequent nucleophilic attack by the alcohol.

Maela Manzoli *et al.* [55] examined an effective Au/CeO_2 catalytic system that could function without a base, which has a detrimental impact on the affordability of the 2-furfuryl alcohol esterification reaction. The study investigated the impact of both gold particle size and the CeO2 support on catalytic activity in the esterification reaction. The higher activity of the most effective catalyst calcined at 500 °C can be attributed to the presence of more coordinated Ce^{4+} sites on the surface and a decrease in carbonate species concentrations. The possible mechanism of furfuryl alcohol oxidative esterification is shown in Fig. (**3**). The ability of Ceria to supply activated oxygen leads to the formation of oxygen vacancies, with its capacity to release oxygen increasing as surface cleanliness improves. It has been suggested that gold sites may be responsible for reverse oxygen overflow and the activation of methanol molecules near the edge of the nanoparticles.

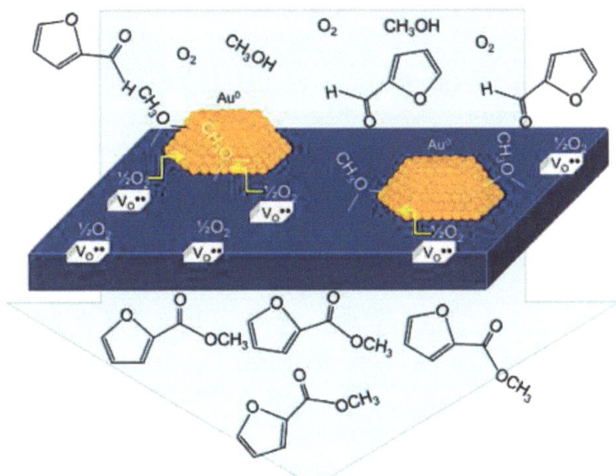

Fig. (3). Proposed mechanism of Ce and Au nanoparticles in the 2-Furfuryl alcohol oxidative esterification reaction.

In summary, Pt, Pd, and Au are examples of noble metal catalysts that are required for the selective oxidation of chemicals produced from biomass to create products with added value. These catalysts offer high activity, selectivity, and stability, making them promising candidates for various biomass valorization processes. Although noble metal catalysts have higher selectivity and catalytic activity than non-noble metal catalysts, their application in biomass esterification processes may be constrained by cost. However, ongoing research aims to develop more efficient and cost-effective noble metal catalysts or to optimize reaction conditions to minimize catalyst usage while maintaining high catalytic performance. Additionally, the choice of catalyst depends on factors, such as the specific biomass feedstock, desired product, and process economics.

Non-Noble Metal Catalyst

Non-noble metal catalysts have also shown promise in selective oxidation reactions. Considering its well-known products and production pathways, a typical reaction for the selective oxidation of alcohols is the oxidation of BzOH [56]. The selective oxidation of alcohols to aldehydes is the main focus of organic reactions. Since aromatic aldehydes are crucial for creating flavors and scents, much study has been carried out to ensure their efficient manufacturing. Nickel-based catalysts, supported on mesoporous materials, have been utilized for the oxidation of biomass-derived alcohols to aldehydes and ketones with good selectivity [57]. Iron-based catalysts have been investigated for the oxidation of lignin-derived phenolic compounds to aromatic aldehydes, which are valuable intermediates for the synthesis of fine chemicals [58]. Non-noble metal catalysts have been extensively researched and utilized for the selective oxidation of biomass-derived compounds to produce various value-added products. These catalysts offer advantages, such as cost-effectiveness, abundance, and often comparable catalytic performance to noble metal catalysts [59].

Researchers are currently exploring the esterification of molecules produced from biomass through the use of heterogeneous non-noble metal catalysts in the biomass valorization field [60]. Ester products, which have a variety of industrial uses, such as solvents, flavouring agents, and biodiesel precursors, are formed through the condensation of acids produced from biomass with alcohols in esterification reactions [61]. Heterogeneous catalysts offer advantages, such as ease of separation, reusability, and, often, tunable acidity, making them attractive for esterification processes [62]. For esterification processes, non-noble metal catalysts, such as solid acid catalysts like sulfated zirconia and hetero polyacids, have also been thoroughly researched. Advantages of these catalysts include adjustable acidity and stability in reaction conditions [63]. For instance, sulfated zirconia catalysts have been utilized for the esterification of levulinic acid, a platform chemical derived from biomass, with various alcohols to produce fuel additives and flavoring agents. Metal oxide catalysts, such as titanium dioxide (TiO_2) and zirconium oxide (ZrO_2), have been investigated for esterification reactions [64]. These catalysts can offer Lewis acid sites that promote the activation of biomass-derived acids and alcohols. Additionally, mixed metal oxide catalysts, such as cerium-zirconium oxides and niobium-tungsten oxides, have shown enhanced catalytic performance in esterification reactions due to synergistic effects between different metal components [65].

Mitran *et al.* [66] studied the selective oxidation of n-C_4H_9-OH to butyraldehyde using $MnCo_2O_4$ spinel as a catalyst, which resulted in 50% n-butanol conversion. Three different bands appeared in the FT-IR spectra at 655, 551, and 500 cm^{-1}, representing Co^{2+}-O, Co^{3+}-O, and $Mn^{3+}(Mn^{4+})$-O, respectively. For the selective oxidation reaction, the active sites Co^{3+} and Mn^{3+}/Mn^{4+} were mainly located in octahedral rather than tetrahedral sites. The average crystallite size of $MnCo_2O_4$ was found to be 20.4 nm. The average pore diameter was 14.1 nm, and the specific surface area was 71 m^2g^{-1}. The active sites of the catalyst in a selective oxidation reaction are schematically represented in Fig. (**4**). It has been observed that Co^{3+} and Mn^{4+} in octahedral locations were more active than in tetrahedral locations in oxidation, involving the Mars–van Krevelen mechanism. As the temperature rose to 200 °C, the yield of n-butyraldehyde increased. Afterward, it decreased due to further oxidation of butyraldehyde. The ratios of Co/Mn and Co^{3+}/Co^{2+} were found to be directly correlated with the yield of n-butyraldehyde. The presence of oxygen vacancies enhanced lattice oxygen mobility, leading to an increased quantity of active oxygen species and improved catalytic activity.

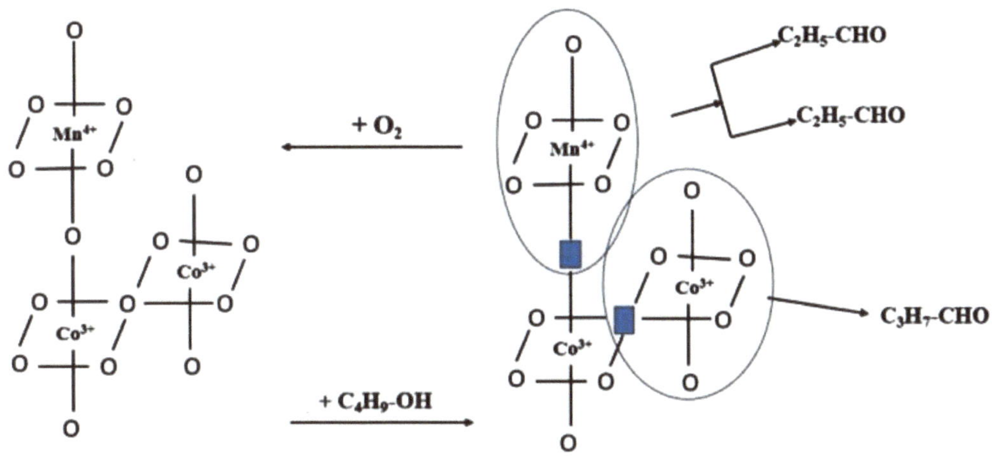

Fig. (4). Schematic illustration of the active sites in $MnCo_2O_4$ spinel oxide.

Kumar *et al.* [67] synthesized a nano Co-Sr aluminate catalyst using the modified sol-gel approach (size 7–27 nm) that oxidized BzOH to BzCHO with 75% yield and 99% selectivity. After strontium was added to the Co-aluminate matrix, the BET surface area and the average pore diameter were 97.0 m^2/g and 4.28 nm, respectively. Marek Pawel Checinski *et al.* [68] investigated the effectiveness of Ru catalysts, enhanced with Mn and Ce oxides, in facilitating the oxidation of 2-octanol and other alcohols into ketones and aldehydes. Yingjie Li *et al.* [69] synthesized a Zn/Co-ZIF, Co@NC catalyst. Under optimal reaction conditions, the catalysts facilitated the oxidation of 5-hydroxymethylfurfural (HMF) to yield 5-formyl-2-furancarboxylic acid (FFCA) with a selectivity of 95% and 2,5-furandicarboxylic acid (FDCA) with a selectivity of 98%. In the initial stage, a diol intermediate was formed on the catalyst surface as hydroxyl groups interacted with the aldehyde side chain of HMF. This intermediate underwent oxidative dehydrogenation to generate 5-hydroxymethyl-2-furancarboxylic acid (HMFCA). Subsequently, on the Co-N structure, molecular oxygen (O_2) was converted into superoxide, enabling the elimination of the β-C-H bond in the alcohol side chain of HMFCA, resulting in aldehyde formation. The oxidation of FFCA to FDCA followed a pathway similar to the first phase, with O_2 acting as an electron scavenger to facilitate proton removal from intermediates. The energy barrier for this step was 0.81 eV when Co@NC was used as the catalyst and 1.84 eV when NC was used. The higher energy barrier associated with NC prevented the complete oxidation of FFCA to FDCA, allowing selective oxidation of HMF to FFCA (95% selectivity using NC) or to FDCA (98% selectivity using Co@NC). G. Dhinagaran *et al.* [70] synthesized an SBA-15-supported CuO catalyst for the selective oxidation of veratryl alcohol to veratraldehyde. The liquid-phase oxidation of alcohols was investigated using catalytically activated tert-butyl hydroperoxide (TBHP) under a heterogeneous catalytic system. CuO/SBA-15 was employed as the catalyst, with activated TBHP achieving efficient oxidation of alcohol types at 80 °C. Notably, CuO(10 wt.%)/SBA-15 demonstrated a conversion rate of 82.5% and 100% selectivity compared to other catalysts, which could be attributed to its smaller particle size, as confirmed by TEM analysis. Khan M. *et al.* [71] studied the electrooxidation of low molecular weight ethanol and methanol to valve-added acetate and formate using CuO nanosheets on a copper foam electrode with ~100% selectivity.

These examples indicate how non-noble metal catalysts can be used in a variety of selective oxidation processes to increase the value of biomass. Continued research in catalyst development and optimization is essential for improving the efficiency and selectivity of these processes, ultimately enabling the sustainable conversion of biomass into valuable products.

CONCLUSION

The catalytic selective oxidation and esterification of biomass-derived chemicals with noble and non-noble metal catalysts provide promising pathways for converting biomass into useful products. The rational design of catalysts, understanding of reaction mechanisms, and optimizing reaction conditions are critical for achieving high selectivity, activity, and stability in these processes. Continued research into catalyst development and process optimization will help to progress biomass valorization technology and the transition to a sustainable bio-based economy.

REFERENCES

[1] Afraz, M.; Muhammad, F.; Nisar, J.; Shah, A.; Munir, S.; Ali, G.; Ahmad, A. Production of value added products from biomass waste by pyrolysis: An updated review. *Waste Management Bulletin,* **2024**, *1*(4), 30-40.
[http://dx.doi.org/10.1016/j.wmb.2023.08.004]

[2] Deng, W.; Feng, Y.; Fu, J.; Guo, H.; Guo, Y.; Han, B.; Jiang, Z.; Kong, L.; Li, C.; Liu, H.; Nguyen, P.T.T.; Ren, P.; Wang, F.; Wang, S.; Wang, Y.; Wang, Y.; Wong, S.S.; Yan, K.; Yan, N.; Yang, X.; Zhang, Y.; Zhang, Z.; Zeng, X.; Zhou, H. Catalytic conversion of lignocellulosic biomass into chemicals and fuels. *Green Energy & Environment,* **2023**, *8*(1), 10-114.
[http://dx.doi.org/10.1016/j.gee.2022.07.003]

[3] Mujtaba, M.; Fernandes Fraceto, L.; Fazeli, M.; Mukherjee, S.; Savassa, S.M.; Araujo de Medeiros, G.; do Espírito Santo Pereira, A.; Mancini, S.D.; Lipponen, J.; Vilaplana, F. Lignocellulosic biomass from agricultural waste to the circular economy: a review with focus on biofuels, biocomposites and bioplastics. *J. Clean. Prod.,* **2023**, *402*, 136815.
[http://dx.doi.org/10.1016/j.jclepro.2023.136815]

[4] Okoro, V.; Azimov, U.; Munoz, J. Recent advances in production of bioenergy carrying molecules, microbial fuels, and fuel design - A review. *Fuel,* **2022**, *316*, 123330.
[http://dx.doi.org/10.1016/j.fuel.2022.123330]

[5] Latif, M.N.; Wan Isahak, W.N.R.; Samsuri, A.; Hasan, S.Z.; Manan, W.N.; Yaakob, Z. Recent advances in the technologies and catalytic processes of ethanol production. *Catalysts,* **2023**, *13*(7), 1093.
[http://dx.doi.org/10.3390/catal13071093]

[6] Brown, J.; Lindstrom, J.K.; Ghosh, A.; Rollag, S.A.; Brown, R.C. Production of sugars from lignocellulosic biomass *via* biochemical and thermochemical routes. *Front. Energy Res.,* **2024**, *12*, 1347373.
[http://dx.doi.org/10.3389/fenrg.2024.1347373]

[7] Singh, S.P.; Singh, D. Biodiesel production through the use of different sources and characterization of oils and their esters as the substitute of diesel: A review. *Renew. Sustain. Energy Rev.,* **2010**, *14*(1), 200-216.
[http://dx.doi.org/10.1016/j.rser.2009.07.017]

[8] Wheeldon, I.; Christopher, P.; Blanch, H. Integration of heterogeneous and biochemical catalysis for production of fuels and chemicals from biomass. *Curr. Opin. Biotechnol.,* **2017**, *45*, 127-135.
[http://dx.doi.org/10.1016/j.copbio.2017.02.019] [PMID: 28365403]

[9] Babu, S.; Jojo, L.; James, A.; Melethil, K.; Thomas, B. Metal-based catalysis for biomass and renewables valorization- current status. *Tetrahedron Green Chem,* **2023**, *2*, 100018.
[http://dx.doi.org/10.1016/j.tgchem.2023.100018]

[10] Wei, M.; Kuang, Y.; Duan, Z.; Li, H. The crucial role of catalyst wettability for hydrogenation of biomass and carbon dioxide over heterogeneous catalysts. *Cell Reports Physical Science,* **2023**, *4*(5), 101340.
[http://dx.doi.org/10.1016/j.xcrp.2023.101340]

[11] Esser, T.; Wassenberg, A.; Voß, D.; Albert, J. Selective catalytic oxidation of humins to carboxylic acids using the $H_4[PVMo_{11}O_{40}]$ Keggin-type polyoxometalate enhanced by alcohol doping and solubilizer. *React. Chem. Eng.,* **2024**, *9*(7), 1666-1684.
[http://dx.doi.org/10.1039/D3RE00672G]

[12] Lodhi, A.; Maheria, K.C. Zeolite-catalysed esterification of biomass-derived acids into high-value esters products: Towards sustainable chemistry. *Catal. Commun.,* **2024**, *187*, 106883.
[http://dx.doi.org/10.1016/j.catcom.2024.106883]

[13] Megía, P.J.; Vizcaíno, A.J.; Calles, J.A.; Carrero, A. Hydrogen production technologies: From fossil fuels toward renewable sources. A mini review. *Energy Fuels,* **2021**, *35*(20), 16403-16415.

[http://dx.doi.org/10.1021/acs.energyfuels.1c02501]

[14] Ang, T.; Salem, M.; Kamarol, M.; Das, H.S.; Nazari, M.A.; Prabaharan, N. A comprehensive study of renewable energy sources: Classifications, challenges, and suggestions *Energy Strategy Rev.,* **2022**, *43*, 100939.
[http://dx.doi.org/10.1016/j.esr.2022.100939]

[15] Gabrielli, P.; Rosa, L.; Gazzani, M.; Meys, R.; Bardow, A.; Mazzotti, M.; Sansavini, G. Net-zero emissions chemical industry in a world of limited resources. *One Earth,* **2023**, *6*(6), 682-704.
[http://dx.doi.org/10.1016/j.oneear.2023.05.006]

[16] Kohli, K.; Prajapati, R.; Sharma, B.K. Bio-based chemicals from renewable biomass for integrated biorefineries. *Energies,* **2019**, *12*(2), 233.
[http://dx.doi.org/10.3390/en12020233]

[17] Zhang, C. Recent progress of hydrogenation and hydrogenolysis catalysts derived from layered double hydroxides *Catalysts,* **2022**, *12*(1484), 1-31. Available from: https://doi.org/

[18] Khopade, K.V.; Chikkali, S.H.; Barsu, N. Metal-catalyzed plastic depolymerization. *Cell Reports Physical Science,* **2023**, *4*(5), 101341.
[http://dx.doi.org/10.1016/j.xcrp.2023.101341]

[19] Kabe, T.; Ishihara, A.; Qian, Eika; Sutrisna, I.P.; Kabe, Y. Studies in Surface Science and Catalysis. Coal and Coal-related Compounds. *Structure Reactivity and Catalytic Reactions,* **2024.**

[20] Basahel, S.N.; Medkhali, A.H.A.; Mokhtar, M.; Narasimharao, K. Noble metal (Pd, Pt and Rh) incorporated LaFeO3 perovskite oxides for catalytic oxidative cracking of n-propane. *Catal. Today,* **2022**, *397-399*, 81-93.
[http://dx.doi.org/10.1016/j.cattod.2021.11.032]

[21] Tang, T.; Wang, Y.; Dong, W.; Liu, C.; Xu, C. Reusable and active Pt@Co-NC catalysts for oxidation of glycerol. *Renew. Energy,* **2020**, *153*, 472-479.
[http://dx.doi.org/10.1016/j.renene.2020.02.029]

[22] Yao, C.; Fan, H.; Adogwa, A.; Xiong, H.; Yang, M.; Liu, F.; Chen, Z.; Lou, Y. Recent advances in carbon dioxide selective hydrogenation and biomass valorization *via* single-atom catalysts. *Resources Chemicals and Materials,* **2023**, *2*(3), 189-207.
[http://dx.doi.org/10.1016/j.recm.2023.05.003]

[23] Li, X.P.; Huang, C.; Han, W.K.; Ouyang, T.; Liu, Z.Q. Transition metal-based electrocatalysts for overall water splitting. *Chin. Chem. Lett.,* **2021**, *32*(9), 2597-2616.
[http://dx.doi.org/10.1016/j.cclet.2021.01.047]

[24] Hutchingsa, G. J.; Vedrineb, J. C. Basic principles in applied catalysis. *Springer Series in Chemical Physics,* , *75*, 215-257.

[25] Shinde, P.S.; Suryawanshi, P.S.; Patil, K.K.; Belekar, V.M.; Sankpal, S.A.; Delekar, S.D.; Jadhav, S.A. A brief overview of recent progress in porous silica as catalyst supports. *Journal of Composites Science,* **2021**, *5*(3), 75.
[http://dx.doi.org/10.3390/jcs5030075]

[26] Yasu-eda, T.; Kitamura, S.; Ikenaga, N.; Miyake, T.; Suzuki, T. Selective oxidation of alcohols with molecular oxygen over Ru/CaO–ZrO_2 catalyst. *J. Mol. Catal. Chem.,* **2010**, *323*(1-2), 7-15.
[http://dx.doi.org/10.1016/j.molcata.2010.03.018]

[27] Hama Aziz, K.H.; Mustafa, F.S. Advanced oxidation processes for the decontamination of heavy metal complexes in aquatic systems: A review. *Case Studies in Chemical and Environmental Engineering,* **2024**, *9*, 100567.
[http://dx.doi.org/10.1016/j.cscee.2023.100567]

[28] Zhu, Y.; Xu, J.; Lu, M. Oxidation of primary and secondary alcohols to the corresponding carbonyl compounds with molecular oxygen using 1,1-diphenyl-2-picrylhydrazyl and WO_3/Al_2O_3 as catalysts. *Catal. Commun.,* **2014**, *48*, 78-84.

[http://dx.doi.org/10.1016/j.catcom.2014.01.025]

[29] Miceli, M.; Frontera, P.; Macario, A.; Malara, A. Recovery/reuse of heterogeneous supported spent catalysts. *Catalysts,* **2021**, *11*(5), 591.
[http://dx.doi.org/10.3390/catal11050591]

[30] Ludwig, J.R.; Schindler, C.S. Catalyst: Sustainable catalysis. *Chem,* **2017**, *2*(3), 313-316.
[http://dx.doi.org/10.1016/j.chempr.2017.02.014]

[31] Gadore, V.; Mishra, S.R.; Yadav, N.; Yadav, G.; Ahmaruzzaman, M. Metal oxide-based heterogeneous catalysts for biodiesel production. *Next Sustainability,* **2023**, *2*, 100012.
[http://dx.doi.org/10.1016/j.nxsust.2023.100012]

[32] Cavani, F.; Teles, J.H. Sustainability in catalytic oxidation: an alternative approach or a structural evolution? *ChemSusChem,* **2009**, *2*(6), 508-534.
[http://dx.doi.org/10.1002/cssc.200900020] [PMID: 19536755]

[33] Thiruvengetam, P.; Chand, D.K. Controlled and predictably selective oxidation of activated and unactivated C(sp^3)–H bonds catalyzed by a molybdenum-based metallomicellar catalyst in water. *J. Org. Chem.,* **2022**, *87*(6), 4061-4077.
[http://dx.doi.org/10.1021/acs.joc.1c02855] [PMID: 35098716]

[34] Mariño, F.; Descorme, C.; Duprez, D. Noble metal catalysts for the preferential oxidation of carbon monoxide in the presence of hydrogen (PROX). *Appl. Catal. B,* **2004**, *54*(1), 59-66.
[http://dx.doi.org/10.1016/j.apcatb.2004.06.008]

[35] Han, Y.; Xu, H.; Su, Y.; Xu, Z.; Wang, K.; Wang, W. Noble metal (Pt, Au@Pd) nanoparticles supported on metal organic framework (MOF-74) nanoshuttles as high-selectivity CO_2 conversion catalysts. *J. Catal.,* **2019**, *370*, 70-78.
[http://dx.doi.org/10.1016/j.jcat.2018.12.005]

[36] Liu, Y.; Zhou, S.; Wang, X.; Qin, J.; Hu, C.; Li, J. Coproduction of glyceric acid and glycolic acid from biomass-based sugars over a Ru/Co_3O_4 catalyst. *ACS Catal.,* **2024**, *14*(10), 7609-7623.
[http://dx.doi.org/10.1021/acscatal.4c00782]

[37] Megías-Sayago, C.; Navarro-Jaén, S.; Castillo, R.; Ivanova, S. Recent advances in selective oxidation of biomass-derived platform chemicals over gold catalysts. *Curr. Opin. Green Sustain. Chem.,* **2020**, *21*, 50-55.
[http://dx.doi.org/10.1016/j.cogsc.2019.12.001]

[38] Khan, Z.; Javed, F.; Shamair, Z.; Hafeez, A.; Fazal, T.; Aslam, A.; Zimmerman, W.B.; Rehman, F. Current developments in esterification reaction: A review on process and parameters. *J. Ind. Eng. Chem.,* **2021**, *103*, 80-101.
[http://dx.doi.org/10.1016/j.jiec.2021.07.018]

[39] Barrón Cruz, A.E.; Melo Banda, J.A.; Mendoza, H.; Ramos-Galvan, C.E.; Meraz Melo, M.A.; Esquivel, D. Pt and Ni supported catalysts on SBA-15 and SBA-16 for the synthesis of biodiesel. *Catal. Today,* **2011**, *166*(1), 111-115.
[http://dx.doi.org/10.1016/j.cattod.2011.01.008]

[40] Manzoli, M.; Menegazzo, F.; Signoretto, M.; Marchese, D. Biomass derived chemicals: Furfural oxidative esterification to methyl-2-furoate over gold catalysts. *Catalysts,* **2016**, *6*(7), 107.
[http://dx.doi.org/10.3390/catal6070107]

[41] Chidunchi, I.; Kulikov, M.; Safarov, R.; Kopishev, E. Extraction of platinum group metals from catalytic converters. *Heliyon,* **2024**, *10*(3), e25283.
[http://dx.doi.org/10.1016/j.heliyon.2024.e25283] [PMID: 38327460]

[42] Xu, H.; Li, X.; Hu, W.; Yu, Z.; Zhou, H.; Zhu, Y.; Lu, L.; Si, C. Research progress of highly efficient noble metal catalysts for the oxidation of 5-Hydroxymethylfurfural. *ChemSusChem,* **2022**, *15*(13), e202200352.
[http://dx.doi.org/10.1002/cssc.202200352] [PMID: 35575041]

[43] Nishimura, S.; Ebitani, K. Selective oxidation of biomass-derived alcohols with supported metal catalysts. *J. Jpn. Petrol. Inst.,* **2017**, *60*(2), 72-84.
[http://dx.doi.org/10.1627/jpi.60.72]

[44] Faroppa, M.L.; Chiosso, M.E.; Musci, J.J.; Ocsachoque, M.A.; Merlo, A.B.; Casella, M.L. Oxidation of glycerol in base-free aqueous solution using carbon-supported Pt and PtSn catalyst. *Catalysts,* **2023**, *13*(7), 1071.
[http://dx.doi.org/10.3390/catal13071071]

[45] Kobayashi, H.; Fukuoka, A. Synthesis and utilisation of sugar compounds derived from lignocellulosic biomass. *Green Chem.,* **2013**, *15*(7), 1740-1763.
[http://dx.doi.org/10.1039/c3gc00060e]

[46] Qu, R.; Junge, K.; Beller, M. Hydrogenation of carboxylic acids, esters, and related compounds over heterogeneous catalysts: A step toward sustainable and carbon-neutral processes. *Chem. Rev.,* **2023**, *123*(3), 1103-1165.
[http://dx.doi.org/10.1021/acs.chemrev.2c00550] [PMID: 36602203]

[47] Belousov, A.S.; Esipovich, A.L.; Kanakov, E.A.; Otopkova, K.V. Recent advances in sustainable production and catalytic transformations of fatty acid methyl esters. *Sustain. Energy Fuels,* **2021**, *5*(18), 4512-4545.
[http://dx.doi.org/10.1039/D1SE00830G]

[48] Tang, Y.; Yu, W.; Mo, L.; Lou, H.; Zheng, X. One-step hydrogenation esterification of Aldehyde and acid to ester over bifunctional Pt catalysts: A model reaction as a novel route for catalytic upgrading of fast pyrolysis bio-oil. *Energy Fuels,* **2008**, *22*(5), 3484-3488.
[http://dx.doi.org/10.1021/ef800148q]

[49] Chan-Thaw, C.E.; Savara, A.; Villa, A. Selective benzyl alcohol oxidation over Pd catalysts. *Catalysts,* **2018**, *8*(10), 431.
[http://dx.doi.org/10.3390/catal8100431]

[50] Li, F.; Zhang, Q.; Wang, Y. Size dependence in solvent-free aerobic oxidation of alcohols catalyzed by zeolite-supported palladium nanoparticles. *Appl. Catal. A Gen.,* **2008**, *334*(1-2), 217-226.
[http://dx.doi.org/10.1016/j.apcata.2007.10.008]

[51] Chen, J.; Zhang, Q.; Wang, Y.; Wan, H. Size-dependent catalytic activity of supported palladium nanoparticles for aerobic oxidation of alcohols. *Adv. Synth. Catal.,* **2008**, *350*(3), 453-464.
[http://dx.doi.org/10.1002/adsc.200700350]

[52] Gil, S.; Muñoz, L.; Sánchez-Silva, L.; Romero, A.; Valverde, J.L. Synthesis and characterization of Au supported on carbonaceous material-based catalysts for the selective oxidation of glycerol. *Chem. Eng. J.,* **2011**, *172*(1), 418-429.
[http://dx.doi.org/10.1016/j.cej.2011.05.108]

[53] Zhu, Y.; Li, Z.; Chen, J. Applications of lignin-derived catalysts for green synthesis. *Green Energy & Environment,* **2019**, *4*(3), 210-244.
[http://dx.doi.org/10.1016/j.gee.2019.01.003]

[54] Menegazzo, F.; Signoretto, M.; Pinna, F.; Manzoli, M.; Aina, V.; Cerrato, G.; Boccuzzi, F. Oxidative esterification of renewable furfural on gold-based catalysts: Which is the best support? *J. Catal.,* **2014**, *309*, 241-247.
[http://dx.doi.org/10.1016/j.jcat.2013.10.005]

[55] Manzoli, M.; Menegazzo, F.; Signoretto, M.; Cruciani, G.; Pinna, F. Effects of synthetic parameters on the catalytic performance of Au/CeO_2 for furfural oxidative esterification. *J. Catal.,* **2015**, *330*, 465-473.
[http://dx.doi.org/10.1016/j.jcat.2015.07.030]

[56] Savara, A.; Chan-Thaw, C.E.; Rossetti, I.; Villa, A.; Prati, L. Benzyl alcohol oxidation on carbon-supported Pd nanoparticles: elucidating the reaction mechanism. *ChemCatChem,* **2014**, *6*(12), 3464-

3473.
[http://dx.doi.org/10.1002/cctc.201402552]

[57] Sahoo, B.; Surkus, A-E.; Pohl, M-M.; Radnik, J.; Schneider, M.; Bachmann, S.; Scalone, M.; Junge, K.; Beller, M. Dr. Surkus, A. E.; Dr. Pohl, M. M.; Dr. Radnik, J.; Dr. Schneider, M.; Dr. Bachmann, S.; Dr. Scalone, M.; Dr. Junge, K.; Prof. Dr. Beller, M. A Biomass-derived non-noble cobalt catalyst for selective hydrodehalogenation of alkyl and (hetero)aryl halides. *Angew. Chem.,* **2017**, *129*(37), 11394-11399.
[http://dx.doi.org/10.1002/ange.201702478]

[58] Tarabanko, V.; Tarabanko, N. Catalytic oxidation of lignins into the aromatic aldehydes: general process trends and development prospects. *Int. J. Mol. Sci.,* **2017**, *18*(11), 2421.
[http://dx.doi.org/10.3390/ijms18112421] [PMID: 29140301]

[59] Cui, Y.; He, S.; Yang, J.; Gao, R.; Hu, K.; Chen, X.; Xu, L.; Deng, C.; Lin, C.; Peng, S.; Zhang, C. Research progress of non-noble metal catalysts for carbon dioxide methanation. *Molecules,* **2024**, *29*(2), 374.
[http://dx.doi.org/10.3390/molecules29020374] [PMID: 38257287]

[60] Dwivedi, A. D.; Priya, B.; Chinthala, R.; Pandey, D. S.; Singh, S. K. Valorization of biomass-derived furans over molecular catalysts *Tetrahedron Green Chem.,* **2023**, *1*, 100008.
[http://dx.doi.org/10.1016/j.tgchem.2023.100008]

[61] Lodhi, A.; Maheria, K.C. Zeolite-catalysed esterification of biomass-derived acids into high-value esters products: Towards sustainable chemistry. *Catal. Commun.,* **2024**, *187*, 106883.
[http://dx.doi.org/10.1016/j.catcom.2024.106883]

[62] Orege, J.I.; Oderinde, O.; Kifle, G.A.; Ibikunle, A.A.; Raheem, S.A.; Ejeromedoghene, O.; Okeke, E.S.; Olukowi, O.M.; Orege, O.B.; Fagbohun, E.O.; Ogundipe, T.O.; Avor, E.P.; Ajayi, O.O.; Daramola, M.O. Recent advances in heterogeneous catalysis for green biodiesel production by transesterification. *Energy Convers. Manage.,* **2022**, *258*, 115406.
[http://dx.doi.org/10.1016/j.enconman.2022.115406]

[63] Saravanan, K.; Tyagi, B.; Bajaj, H.C. Sulfated zirconia: an efficient solid acid catalyst for esterification of myristic acid with short chain alcohols. *Catal. Sci. Technol.,* **2012**, *2*(12), 2512-2520.
[http://dx.doi.org/10.1039/c2cy20462b]

[64] Kuwahara, Y.; Fujitani, T.; Yamashita, H. Esterification of levulinic acid with ethanol over sulfated mesoporous zirconosilicates: Influences of the preparation conditions on the structural properties and catalytic performances. *Catal. Today,* **2014**, *237*, 18-28.
[http://dx.doi.org/10.1016/j.cattod.2013.11.008]

[65] Babu, S.; Jojo, L.; James, A.; Melethil, K.; Thomas, B. Metal-based catalysis for biomass and renewables valorization- current status. *Tetrahedron Green Chem,* **2023**, *2*, 100018.
[http://dx.doi.org/10.1016/j.tgchem.2023.100018]

[66] Mitran, G.; Chen, S.; Seo, D.K. Selective oxidation of *n* -buthanol to butyraldehyde over $MnCo_2O_4$ spinel oxides. *RSC Advances,* **2020**, *10*(42), 25125-25135.
[http://dx.doi.org/10.1039/D0RA04738D] [PMID: 35517452]

[67] Kumar, R.T.; Sagaya Selvam, N.C.; Adinaveen, T.; Kennedy, L.J.; Vijaya, J.J. Strontium(II)-added CoAl2O4 nanocatalysts for the selective oxidation of alcohols. *React. Kinet. Mech. Catal.,* **2012**, *106*(2), 379-394.
[http://dx.doi.org/10.1007/s11144-012-0428-2]

[68] Chęciński, M.P.; Brückner, A.; Radnik, J.; Köckritz, A. Oxidation of alcohols using RuMnCe catalysts. *Appl. Catal. A Gen.,* **2009**, *366*(1), 212-219.
[http://dx.doi.org/10.1016/j.apcata.2009.07.007]

[69] Li, Y.; Wang, Y.; Gao, H.; Zhang, Z.; Chen, L.; Liao, X. Tunable products of 5-hydroxymethylfurfural oxidation by Zn/Co-ZIF derived catalysts. *Appl. Catal. A Gen.,* **2024**, *678*, 119712.
[http://dx.doi.org/10.1016/j.apcata.2024.119712]

[70] Dhinagaran, G.; Harichandran, G.; Suvaitha, S.P.; Venkatachalam, K. Catalytic activity of SBA-15 supported CuO for selective oxidation of veratryl alcohol to veratraldehyde. *Molecular Catalysis,* **2022**, *528*, 112454.
[http://dx.doi.org/10.1016/j.mcat.2022.112454]

[71] Khan, M.; Hameed, A.; Samad, A.; Mushiana, T.; Abdullah, M.I.; Akhtar, A.; Ashraf, R.S.; Zhang, N.; Pollet, B.G.; Schwingenschlögl, U.; Ma, M. *In situ* grown oxygen-vacancy-rich copper oxide nanosheets on a copper foam electrode afford the selective oxidation of alcohols to value-added chemicals. *Commun. Chem.,* **2022**, *5*(1), 109.
[http://dx.doi.org/10.1038/s42004-022-00708-1] [PMID: 36697633]

CHAPTER 5

Solid Base Catalysts in Multicomponent Reactions: A Green Pathway to Heterocyclic Compounds

Divyani P. Patel[1] and **Satish Kumar Singh**[1,*]

[1] *Department of Chemistry, Government College Daman (Affiliated to Veer Narmad South Gujarat University, Surat), Daman, U. T. of Dadra & Nagar Haveli and Daman & Diu, India*

Abstract: Solid base catalysis has emerged as a promising strategy for driving Multicomponent Reactions (MCRs) in organic synthesis. MCRs, involving more than two starting reagents, occur under specific reaction conditions, yielding a single product embodying key features of the starting materials. These reactions play a crucial role in modern drug discovery, offering a rich origin of molecular variation and enabling quick, automatic, and scalable production of organic molecules. Solid base catalysts can facilitate various reactions, including condensation, rearrangement, and hydrogenation, making them valuable across different industries. This chapter provides a comprehensive overview of recent trends in solid base-catalyzed MCRs for eco-compatible heterocyclic synthesis, emphasizing their potential to advance environmentally conscious and highly effective synthetic methodologies in organic chemistry.

Keywords: Catalysis, Heterocycle, Multicomponent reactions, Solid base catalyst.

INTRODUCTION

The history of solid base catalysts is intertwined with the broader development of heterogeneous catalysis and the quest for more efficient and sustainable chemical processes.[1-3] While the concept of solid base catalysis has ancient roots, its systematic exploration and application in modern chemistry began to gain momentum in the 20th century [4]. Early investigations into solid base catalysis primarily focused on metal oxides and hydroxides, which exhibited basic properties due to the presence of surface oxygen ions. One of the pioneering studies in this field dates back to the 1920s when researchers explored the catalytic properties of metal oxides such as MgO (Magnesium Oxide) and CaO (Calcium Oxide) [5, 6]. Through the basic sites of metal oxides on their surface,

* **Corresponding author Satish Kumar Singh:** Department of Chemistry, Government College Daman (Affiliated to Veer Narmad South Gujarat University, Surat), Daman, U. T. of Dadra & Nagar Haveli and Daman & Diu, India; E-mail: drsatish.singh@ddd.gov.in

Ravi Tomar, Vivek Mishra, Kapil Arya & Garima Pandey (Eds.)

various organic reactions, including aldol condensation, Michael addition, and Knoevenagel condensation, were promoted [7, 8].

A new class of solid base catalysts emerged in the 1950s and 1960s due to the advent of zeolites [9-11]. Zeolites are crystalline aluminosilicate distinctive materials with nanoporous structures and high surface areas. The presence of basic sites within the zeolite framework, typically associated with bridging oxygen atoms, endowed them with solid base catalytic properties [12-14]. Zeolites proved to be efficient catalysts for an extensive scope of reactions, counting isomerization, alkylation, and cracking, in both petrochemical and fine chemical industries [15, 16]. In the following decades, researchers continued to explore and expand the repertoire of solid base catalysts. Metal oxides, like hydrotalcite (layered double hydroxides) and mixed metal oxides, gained attention for their tunable basicity and catalytic activity in various organic transformations [17]. Mesoporous resources, like mesoporous silica and metal oxide, also emerged as promising solid base catalysts due to their large surface areas and interstitial volumes, which facilitated efficient mass transfer and catalytic activities [18-23].

The development of nanoporous materials and advances in nanotechnology further propelled the field of solid-base catalysis [24]. Nanostructured catalysts, including metal nanoparticles supported on solid bases, exhibited enhanced catalytic performance and more selectivity when compared to their larger equivalents [25]. The design and synthesis of tailored solid base catalysts with specific pore structures, surface chemistries, and functionalities became a research focus for optimizing catalytic efficiency and selectivity in various chemical processes [26, 27].

In recent years, the application of solid base catalysts has extended beyond traditional organic synthesis to areas such as biomass conversion, environmental remediation, and renewable energy [28-33]. Solid base catalysts perform a key role in promoting sustainable and environmentally friendly procedures by enabling the efficient utilization of renewable feedstocks, reducing energy consumption, and minimizing waste generation [34-36].

Recognition of certain materials as solid base catalysts is based on the following key factors [4, 37-40].

1. **Surface characterization:** Various analytical methods, including changes in the color of acid-base indicators, acidic molecule adsorption, and spectroscopic techniques (such as UV, IR, XPS, ESR, *etc.*), detect the occurrence of basic sites on the surfaces of these substances.
2. **Correlation with catalytic activity:** The effectiveness of these materials in catalysis shows a clear connection to the quantity and potency of basic sites

they possess. Moreover, the active sites can be negatively impacted by acidic molecules, including hydrochloric acid, water, and carbon dioxide, rendering them less effective.

3. **Comparable activity to homogeneous catalysts:** These materials demonstrate catalytic behaviors akin to those observed in homogeneous basic catalysts involved in "base-catalyzed reactions" within homogeneous reactions. The proposed mechanistic pathway for reactions taking place on their surfaces closely mimics those found in homogeneous basic solutions.
4. **Insight into mechanism:** Research into reaction mechanisms, yield distributions, and surface species spectral analysis suggests the participation of anionic intermediates in these reactions.

The list of some common solid-base catalysts used in organic synthesis is given below:

1. Alkali metal hydroxides (*e.g.*, KOH, NaOH)
2. Carbonates of alkali metals (*e.g.*, K_2CO_3, Na_2CO_3)
3. Hydrides of alkali metals (*e.g.* KH, NaH)
4. Alkali metal alkoxides (*e.g.*, KO^tBu, NaOEt)
5. Alumina (aluminum oxide)
6. Silica gel
7. Zeolites (*e.g.*, zeolites (alkali ion-added): Cs_2O/Cs–X, Zeolites (alkali ion-exchanged): Na–X, Cs–X, imide and nitride of lanthanide on zeolite)
8. Ion exchange resins
9. Metallic oxides (*e.g.*, MgO, CaO, and BaO).
10. Clay minerals (*e.g.*, hydrotalcite, chrysolite, sepiolite)
11. Non-oxide (*e.g.*, KF impregnated on alumina, metal oxynitrides: VALPON ALPON)
12. Nanoparticles, *etc*

These solid base catalysts are often employed in various organic transformations, including multicomponent reactions, due to their ability to facilitate reactions efficiently under mild conditions [41-45].

Sr No.	Solid Base Catalysts	Liquid Base Catalysts
1	Neutralizing the reaction mixture is unnecessary, reducing waste generation.	In liquid base-catalyzed reactions, the necessity to neutralize the base results in waste production.
2	High-temperature reactions are possible without limitations.	Liquid base-catalyzed reactions are limited to temperatures below boiling point due to solvent use.

(Table) cont.....

Sr No.	Solid Base Catalysts	Liquid Base Catalysts
3	The employment of flow reactor systems enhances process productivity.	-
4	Catalyst separation from products is straightforward, simplifying the recycling of spent catalysts.	-
5	The reactor system does not necessitate base-tolerant materials.	-
6	Bifunctional catalysis is possible due to the simultaneous presence of both basic and acidic functionalities on solid bases.	It is impossible to achieve bifunctional catalysis in the liquid phase.
7	Certain reactions unique to solid-base catalysts include the heterolytic splitting of hydrogen into hydrogen ions (H^+ and H^-).	-

In the field of organic chemistry, the formation of heterocyclic compounds catalyzes innovation and exploration. These intricate molecular structures form the basis of countless natural products, pharmaceuticals, and materials, driving progress across a spectrum of fields from medicine to materials science [46-50]. However, conventional synthetic methods often come at an environmental cost, relying on harsh reagents, generating waste, and depleting finite resources. In response to escalating environmental concerns, the pursuit of eco-compatible heterocyclic synthesis emerges as imperative [51-53]. This transformative shift seeks to the demand for novel molecular designs with the need to minimize environmental impact, ushering in a new era where green chemistry principles shape the landscape of heterocyclic chemistry [52, 54-56].

Multicomponent Reactions (MCRs) represent a powerful strategy in organic synthesis, offering efficiency and diversity in the construction of complex molecular architectures [57]. Recently, there has been an increasing fascination with employing MCRs in synthesizing heterocyclic compounds, driven by their significance in medicinal chemistry, materials science, and various interdisciplinary domains [58, 59]. One of the significant benefits of MCRs lies in their capacity to assemble multiple reactants in a single reaction vessel, leading to the simultaneous formation of multiple bonds and functional groups [60]. This atom-economic approach minimizes waste and simplifies synthetic pathways, making MCRs particularly attractive from the green chemistry perspective [61]. This versatility is particularly valuable in drug discovery and development, where access to structurally diverse compound libraries is essential for identifying lead compounds with desired biological activities [62,63]. Furthermore, MCRs frequently occur under gentle conditions for reaction, therefore avoiding the requirement for harsh reagents and high temperatures. They are abundantly present in a variety of natural substances, including vitamins, hormones, antibiotics, alkaloids, herbicides, and colors [64, 65]. MCRs in heterocyclic chemistry offer a potent strategy for the rapid and efficient synthesis of diverse heterocyclic compounds. In contrast to traditional stepwise synthesis methods, which require multiple reaction steps and intermediate purification, MCRs allow for the simultaneous assembly of multiple components into complex heterocyclic frameworks in a single operation [66-69].

Moreover, MCRs often exhibit high levels of functional group acceptance, allowing for combining various functional groups into final heterocyclic compounds [57]. This flexibility enhances the synthetic utility of MCRs and enables the rapid exploration of chemical space [70]. From the venerable Passerini and Ugi reactions to more recent developments such as the Biginelli and Hantzsch reactions, a plethora of MCRs were developed for the formation of several heterocyclic moieties [71-77]. These reactions have found applications in material science and drug discovery, among other fields, underscoring their broad utility and significance [73, 78-81].

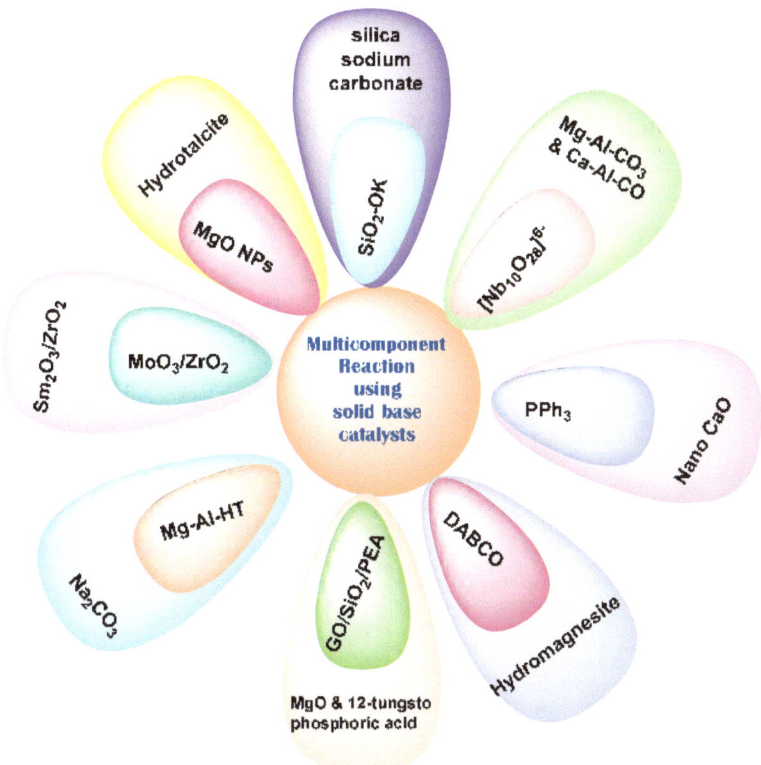

Fig. (1). Versatilities of solid base catalysts in MCRs.

Solid base catalysis plays a crucial role in facilitating multicomponent reactions by providing an active and selective catalytic surface for the transformation of reactants into desired products [82]. It was in 1958 when Pines and Haag began a string of experiments into solid base catalysts [44]. Applications of solid base catalysts in multicomponent reactions have been explored across a variety of synthetic transformations, with the synthesis of heterocycles, carbocycles, and functionalized organic molecules [82]. Catalysts, such as metal oxides, zeolites, hydrotalcite, *etc.*, have demonstrated remarkable efficacy in promoting these transformations under mild reaction conditions (Fig. **1**) [83-86].

Substituted Steroidal Pyridines

Ansari and coworkers utilized the potential of magnesium oxide nanoparticles as a versatile and recyclable catalyst in synthesizing a range of highly substituted steroidal pyridines (Scheme **1**) [89]. This inventive method entailed one-pot MCRs under microwave irradiation. The process commenced with the combination of steroidal ketones, aromatic aldehydes, $CH_2(CN)_2$/ $CH_2(CN)(CO_2Et)$, and NH_4OAc in ethanol. Magnesium oxide nanoparticles acted as catalysts for the steroidal ketones, ensuring successful synthesis. Moreover, the other team achieved successful synthesis utilizing pregnenolone acetate. While various catalysts such as ZnO, CaO, K_2CO_3, and Al_2O_3 were explored, the most promising results consistently emerged with the presence of MgO nanoparticles.

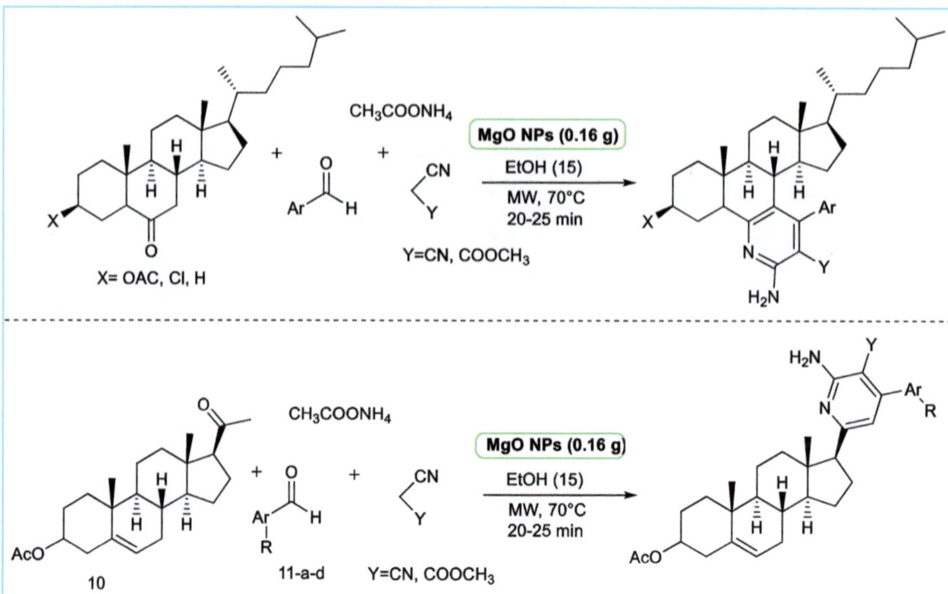

Scheme (1). Steroidal pyridine synthesis.

The optimized reaction conditions determined ethanol to be the most appropriate solvent. MgO nanoparticles (NPs) markedly decreased the reaction time, attaining product yields of up to 89% in 20 minutes under microwave irradiation. Conversely, traditional heating necessitated 6 to 24 hours, producing yields ranging from 25% to 79%. The substrate range includes steroidal ketones and diverse substituted aromatic aldehydes, including benzaldehyde, 3-nitrobenzaldehyde, and 4-chlorobenzaldehyde, in addition to active methylene compounds such as malononitrile and methyl cyanoacetate. This demonstrates the adaptability of the strategy. Substituents on the aromatic aldehydes had a negligible effect on the yields, with 3-nitrobenzaldehyde demonstrating the most reactivity. Moreover, steroidal ketones with less steric hindrance, such as pregnenolone acetate, exhibited accelerated reaction kinetics and superior yields relative to their more sterically hindered equivalents.

The product is formed through the first synthesis of imine (**I**), which occurs when a steroidal ketone reacts with ammonium acetate. Simultaneously, arylidene (**II**) is produced *via* the conventional Knoevenagel condensation of aromatic aldehydes with $CH_2(CN)_2$ or $CH_2(CN)(CO_2Et)$. Imine (**I**) undergoes a Michael addition to the activated double bond of the intermediate (**II**), followed by intramolecular cyclization, oxidation, and aromatization to give the target products (Scheme **2**). MgO nanoparticles have been acknowledged for their exceptional efficacy as a catalyst exhibiting heterogeneous basic characteristics in facilitating the Michael addition and Knoevenagel condensation reactions. These nanoparticles possess magnesium ions as Lewis acid and oxide ions as Lewis base sites, as well as different cationic and anionic vacancies inside their lattice structure. During the reaction, these sites are extensively exposed, enabling the stabilization of reactants on them and facilitating the attack of reactants, resulting in the generation of the end product at a large scale.

Synthesis of Hantzsch Dihydropyridine (DHP)

Antonyraj successfully conducted a one-pot MCR of Hantzsch dihydropyridine by blending benzaldehyde, EAA, and ammonium acetate at ambient temperature. This research utilized eco-friendly hydrotalcite (HT) and hydrotalcite-like (HT-like) substances, synthesized *in situ*, as solid-base catalysts, which were also deemed medically suitable (Scheme **3**) [90]. Various catalysts, like $NiAl_3$-HT, $MgAl_2$-CHT, $MgAl_4$-RHT, and $CoAl_3$-HT, were tested, but $MgAl_2$-HT was discovered to be the greatest effective for benzaldehyde production, yielding 73% with NH_4OAc as the nitrogen source in 6.5 hours, using CH_3CN as the solvent. Various catalysts, like $NiAl_3$-HT, $MgAl_2$-CHT, and $MgAl_4$-RHT, $CoAl_3$-HT, were tested but, $MgAl_2$-HT was discovered to be of utmost effectiveness for benzaldehyde production, yielding 73% with NH_4OAc as the nitrogen source in

6.5 hours, using CH_3CN as the solvent. This could be attributed to the effective collaboration between the acid-base sites in this catalyst, specifically the higher concentration of Lewis acid sites (Al^{3+}) and stronger Bronsted basic. Experiments investigating changes in solvents and nitrogen sources have demonstrated that CH_3CN is the most effective solvent and NH_4OAc is a suitable nitrogen source for achieving excellent yields of dihydropyrimidine. An ion exchange took place when NH_4OAc was utilised as the nitrogen supply in the reaction, rendering the catalyst non-reusable. The molten $MgAl_2$-HT ($MgAl_2$-CHT) showed minimal conversion, highlighting the necessity of structural hydroxyl groups within a lattice-like HT.

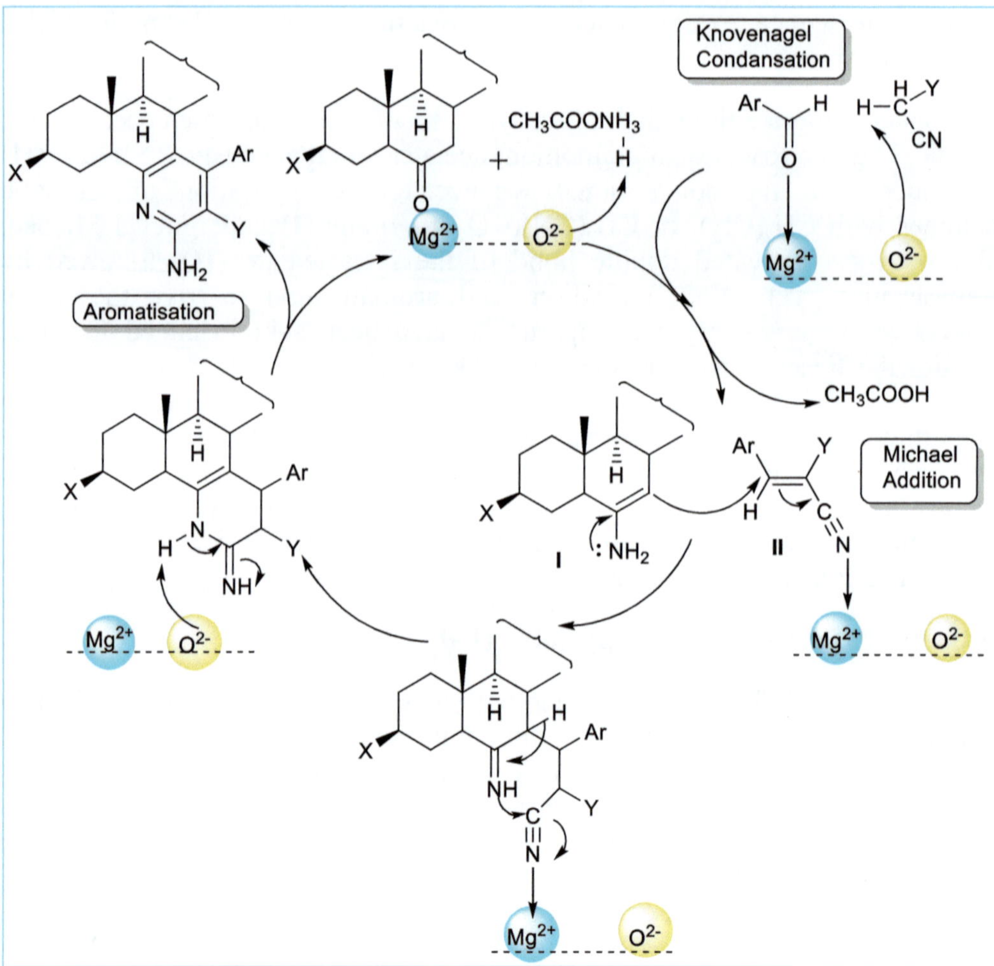

Scheme (2). Proposed reaction of steroidal pyridines.

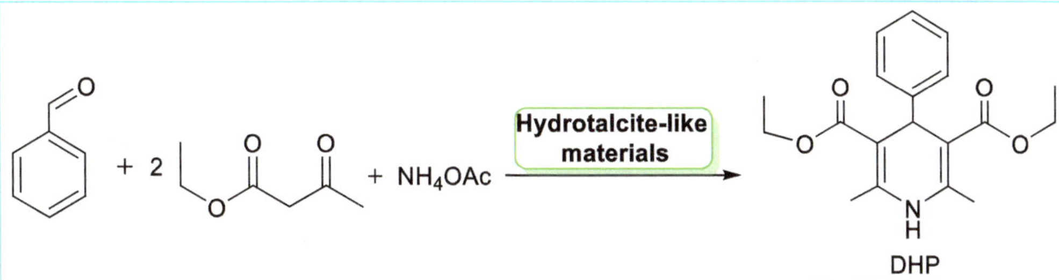

Scheme (3). Synthesis of Hantzsch dihydropyridines.

The reaction exhibited a broad substrate scope, tolerating a variety of aldehydes, including aromatic, cyclic, and aliphatic types, with minimal influence on yield, although electron-withdrawing nitro groups slightly reduced reactivity. Catalyst concentration was optimized, with 2.9 wt.% ideal for achieving maximum yields. This study establishes a sustainable and efficient pathway for synthesizing bioactive DHPs, with yields ranging from 18% to 73% depending on reaction conditions and substrate variations.

A likely chemical pathway for the production of DHP using MgAl hydrotalcite is shown in (Scheme **4**). The active-Me group of EAA undergoes proton abstraction at a basic site, followed by *N*-addition to a protonated carbonyl group. Subsequent dehydration leads to the formation of ethyl 3-aminobut-2-enoate (**I**). Following this, the aforementioned molecule undergoes a reaction with a separate molecule of EAA, leading to the formation of an imine referred to as ethyl 3-(4 ethoxy---oxobutan-2-ylidenamino)but-2-enoate (**III**) *via* a condensation process. The imine will undergo base-catalyzed tautomerization to produce an enamine (**II**). The reaction between this enamine intermediate and benzaldehyde results in the production of DHP (**IV**).

2-Amino-4*H*-benzo[*b*]pyrans synthesis

Chandralekha and coworkers devised a new and effective method for synthesizing 2-amino-4*H*-benzo[*b*]pyran derivatives. This approach entails a one-step, atom-efficient, multicomponent reaction involving aromatic aldehydes, dimedone, and malononitrile under solvent-free conditions at 80°C, resulting in excellent yields of the final products. The catalysis of the reaction is facilitated by SiO_2-OK, a solid basic catalyst capable of being regenerated for multiple cycles. The reaction was implemented by the group-assistant-purification chemistry protocol, thereby eliminating the need for traditional chromatography and recrystallization purification methods (Scheme **5**) [91]. An investigation involving benzaldehyde, dimedone, and malononitrile was conducted using a catalytic amount of SiO_2-OK. At room temperature, the reaction rate was observed to be sluggish, yet it demonstrated an increase at elevated temperatures. The maximum reaction rate was achieved at 80°C, with no further rise observed with higher temperatures. Triethylamine, *L*-proline, piperidine, and silica gel catalysts yield a product of moderate quality, whereas SiO_2-OK catalyst produces an excellent product.

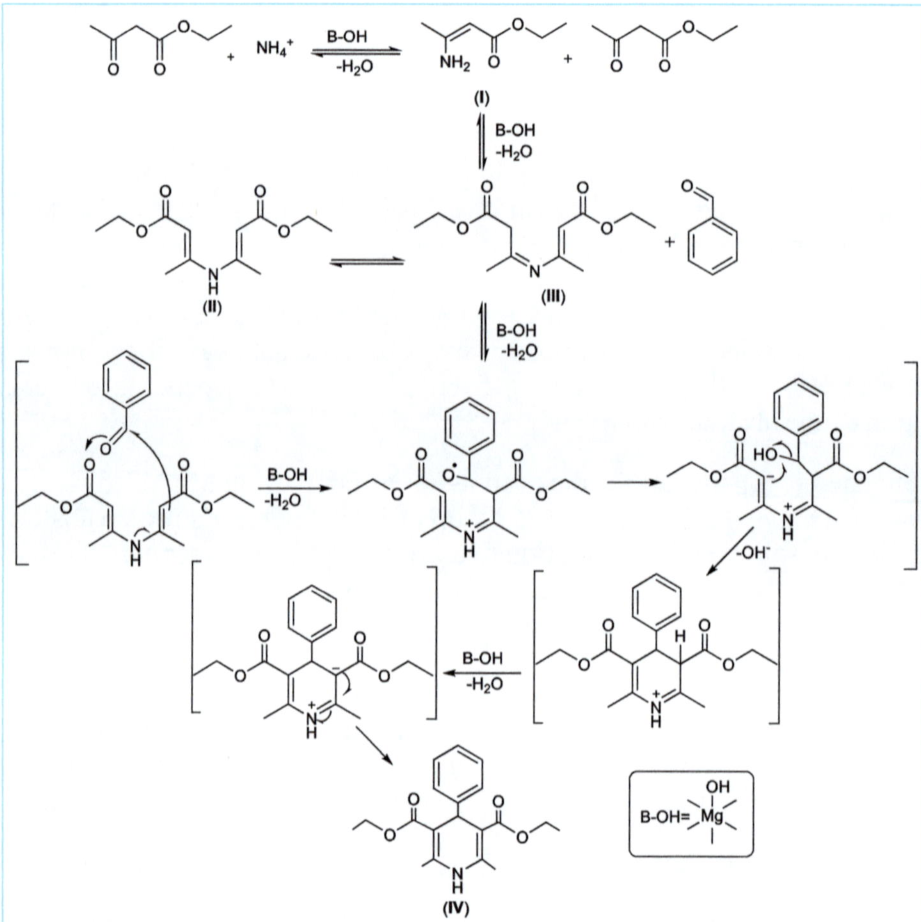

Scheme (4). Probable mechanisms of DHP.

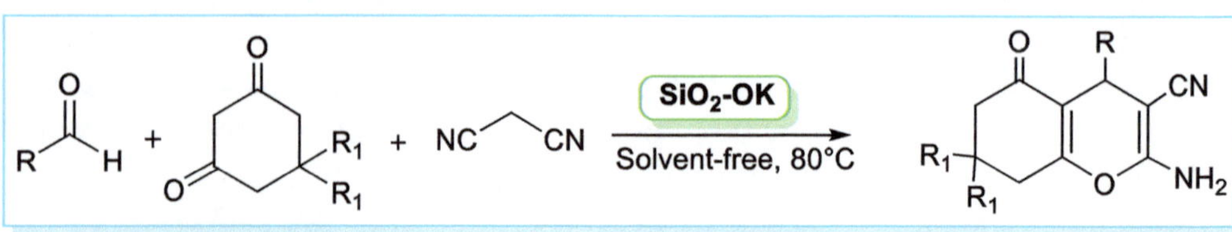

Scheme (5). 2-Amino-4*H*-benzo[*b*]pyrans synthesis.

The substrate scope of the reaction was extensively explored, demonstrating remarkable versatility and functional group tolerance. Aromatic aldehydes bearing electron-donating groups, such as methyl, methoxy, and dimethoxy, and electron-withdrawing groups, including nitro, fluoro, chloro, bromo, and formyl functionalities, were successfully converted into the desired products with high efficiency. Furthermore, ortho, meta, and para substitution patterns were well tolerated, with negligible steric or electronic effects on yields, indicating the robustness of the method. Even challenging substrates, such as furanyl and allyl-substituted aldehydes, produced the corresponding products in high yields (91–94%).

The mechanism commences with the initial step of Knoevenagel condensation involving an aromatic aldehyde and malononitrile catalyzed by SiO_2-OK (Scheme **6**). This yields arylidenemalononitrile, accompanied by the elimination of an H_2O molecule. Subsequently, the enolizable dimedone undergoes nucleophilic (Michael) addition to arylidenemalononitrile, followed by intramolecular cyclization and tautomerization, ultimately leading to the formation final product.

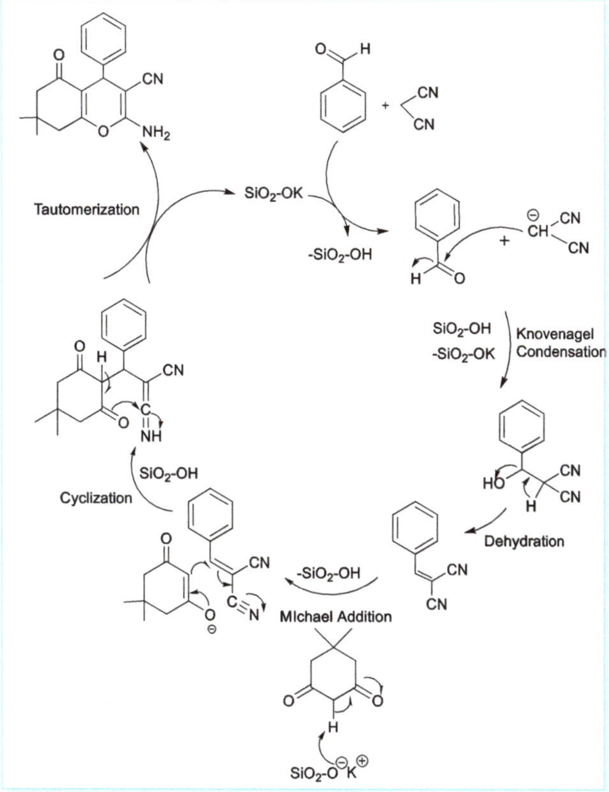

Scheme (6). Proposed mechanism of 2-amino-4*H*-benzo[*b*]pyrans.

Synthesis of acyl-substituted bis(pyrazolyl)methanes

Eskandari and coworkers synthesized an effective and environmentally conscious protocol for synthesizing novel substituted acyl-substituted bis(pyrazolyl) methanes. This is accomplished by a multicomponent condensation reaction employing 3-methyl-1-phenyl-2-pyrazolin-5-one, arylglyoxal derivatives, and silica sodium carbonate (SSC) as a reusable eco-friendly catalyst (Scheme 7) [92]. An investigation was conducted to determine the impact of a catalyst on the model reaction of compound 1-(4-nitrophenyl)-2,2-*bis*(5-hydroxy-3-meth-l-1-phenyl-1*H*-pyrazol-4-yl)ethan-1-one. No significant product was produced without a catalyst. When silica was used by itself, a product was obtained with a low yield. Nevertheless, increasing the proportion of the catalyst's utilization resulted in increased production. The reaction involving furopyrazole derivatives was not observed in good results.

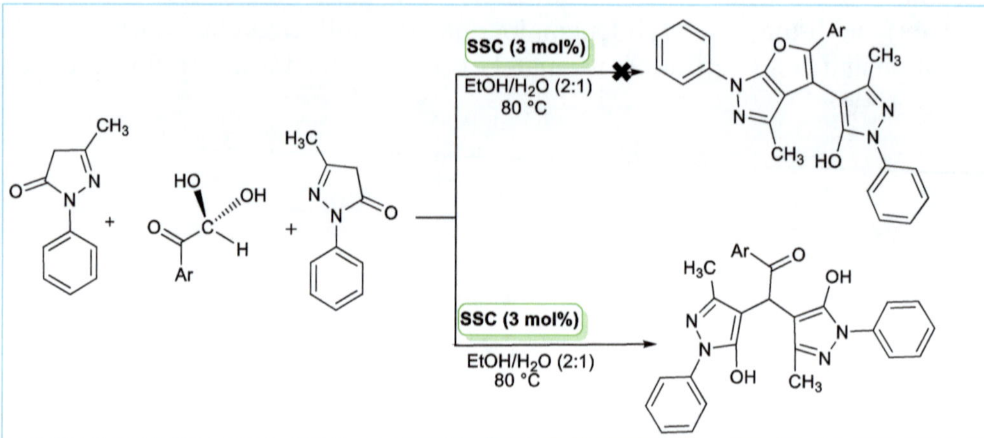

Scheme (7). Synthesis of acyl-substituted bis(pyrazolyl)methanes.

The study explored a broad substrate scope, demonstrating tolerance for various arylglyoxals bearing electron-donating and electron-withdrawing groups, such as chloro, bromo, nitro, methoxy, and methyl, as well as naphthyl derivatives. Substituted arylglyoxals like 4-chlorophenyl (89% in 45 minutes), 4-bromophenyl (86% in 50 minutes), 4-nitrophenyl (90% in 40 minutes), and 4-methoxyphenyl (84% in 50 minutes) yielded the target products efficiently. Moreover, naphthyl derivatives, such as 1-naphthyl and 2-naphthyl, also performed well, with yields of 88% and 90%, respectively, achieved in 60 minutes. The optimized reaction conditions (3 mol% SSC, aqueous medium, 80°C) favored the selective formation of acyl-substituted bis(pyrazolyl)methanes, avoiding potential side products like furopyrazoles.

The proposed synthesis of bis(pyrazolyl)methane derivatives is depicted in Initially, the compound 3-methyl-1 phenyl-2-pyrazolin-5-one (**1**) undergoes tautomerization to form compound **2**. The nucleophile, activated by SSC, undergoes a chemo-selective attack on the hemiacetal carbon of aryl glyoxals, leading to the creation of an intermediate **4** through the elimination of water. In the subsequent stage, compound **2**, which is activated by SSC, undergoes a chemo-selective attack on the hemiacetal carbon of compound **4**. This reaction yields intermediate compound **5** by eliminating water. Through an enol-keto tautomerization, **5** ultimately yields the final compound **6**.

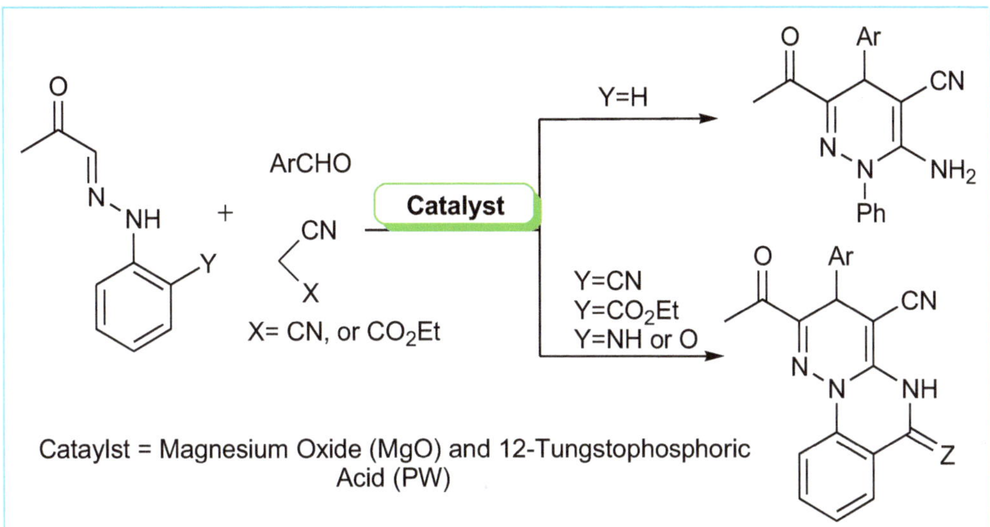

Scheme (8). Proposed mechanisms of acyl-substituted bis(pyrazolyl)methane derivatives.

Synthesis of 4*H*-pyrans

Gutierrez and colleagues focus on exploring novel applications of polyoxometalates, specifically the decaniobate ions $[Nb_{10}O_{28}]^{6-}$. Polyoxometalates (POMs) are clusters of metal atoms and oxygen atoms exhibiting a negative charge. These clusters are formed by octahedral MO_6 units, where M denotes the metal atom, and the units share their edges or corners. The extensive utility of polyoxometalates in chemistry and the medicinal sector has spurred investigations into lesser-known polyoxometalates. One instance of this is the decaniobate ion $[Nb_{10}O_{28}]^{6-}$, with its most notable compound being $TMA_6[Nb_{10}O_{28}]$ (TMA^+: tetramethylammonium ion). This compound remains stable within a pH range of 5-8 and is not soluble in solvents that are not water-based. The synthesis of 4*H*-pyrans involved the utilization of malononitrile, an aromatic aldehyde, and a 1,3-diketo molecule. The decaniobates can serve as effective solid catalysts in MCR under solvent-free microwave irradiations, leading to high yields of 4*H*-pyrans.

This reaction can be performed using various aldehydes, which may contain both electron-withdrawing groups (such as nitro or chloro groups) and electron-donating groups (like methyl or ethyl groups). These variations typically yield good to excellent results. Additionally, the substrates included malononitrile and *β*-dicarbonyl compounds, such as ethyl acetoacetate and methyl acetoacetate, which are commonly used in the synthesis of 4*H*-pyrans. The synthesis of $(TMA)_6[Nb_{10}O_{28}].6H_2O$ was carried out using a straightforward methodology (HPNb), which was adjusted to reduce the formation of hexaniobate impurities (HPNb-HF) (Scheme **9**) [93].

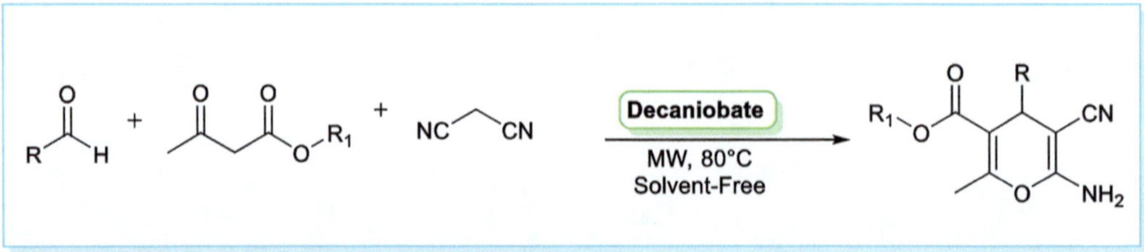

Scheme (9). Synthesis of 6-amino-4*H*-pyran derivative.

Synthesis of 3,4-dihydropyrimidinones

Lal and coworkers conducted a study on the synthesis of 3,4-dihydropyrimidinones/thiones. They utilized a multicomponent synthesis approach, including EAA, aromatic aldehydes, along with urea/thiourea. The reaction was facilitated by the Mg-Al–CO_3 and Ca–Al–CO_3 hydrotalcite, which served as a solid catalyst that can be reused. The reaction proceeds in a high-temperature, solvent-free environment both with and without microwave irradiation (Scheme **10**) [94]. In this approach, several catalysts, such as metal hydroxides, alkali-treated alumina, calcined hydrotalcite, rehydrated hydrotalcite, and metal oxides, were employed to facilitate the aforementioned reaction. It was determined that Mg–Al–CO_3 and Ca–Al–CO_3 exhibited the highest catalytic activity. The present method is efficient, and the product yield is excellent.

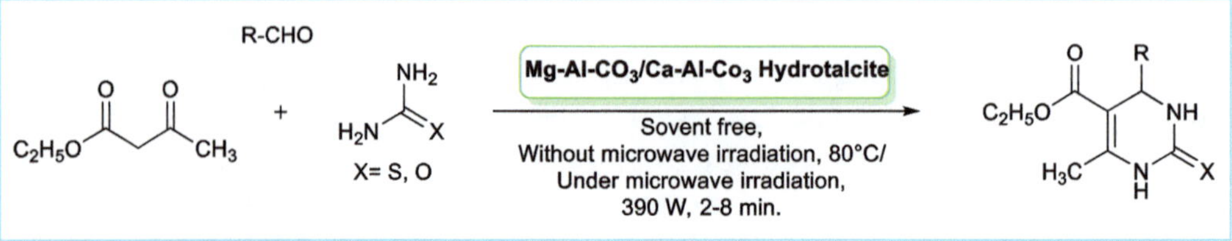

Scheme (10). Synthesis of dihydropyrimidinones.

Mg–Al–CO_3 was synthesized using a hydrothermal reaction conducted at a temperature of 175 °C for 6 hrs, under intrinsic pressure, in an aqueous solution. This method was employed to produce particles that are both tiny in size and possess a large surface area. In a standard procedure, a mixture of Mg and Al hydroxides in a metallic ratio of 3:1 was combined with the appropriate ratio of sodium bicarbonate. The pH level was carefully controlled at 8.5. Following a 12-hour aging period, a white precipitate was separated by filtration, washed, and then dried. The Ca–Al–CO_3 compound was created through the amalgamation of salt solutions containing calcium ions and aluminum ions in a 3:1 molar ratio. The pH value of the solution was elevated to 8.0 through the addition of 25% concentrated ammonium hydroxide solution. The viscous white mixture was subjected to 175 °C and high pressure in an autoclave. Upon completion, the resulting solid was subjected to a 12-hour aging period, subsequently undergoing filtration, rinsing with deionized water, and ultimately drying.

The proposed mechanism for 3,4-dihydropyrimidinones is illustrated in (Scheme **11**). Three probable pathways for synthesizing 3,4-dihydropyrimidinones are the Knoevenagel process, iminium mechanism, and enamine mechanism. Initially, the compound ethyl acetoacetate underwent a reaction with benzaldehyde, leading to the creation of a Knoevenagel condensate. This condensate was then subjected to a reaction with urea, using a solvent-free system and employing Mg-Al-CO_3 and Ca-Al-CO_3 hydrotalcite catalysts. Unfortunately, the intended outcome was not obtained.

Additionally, benzaldehyde reacted with urea to form a Schiff base/iminium ion. This compound was subsequently reacted with EAA under solvent-free conditions, utilizing Mg-Al-CO_3 and Ca-Al-CO_3 hydrotalcite as catalysts, resulting in the formation of 3,4-dihydropyrimidinones. Ultimately, EAA reacted with urea, leading to the creation of an enamine product. Subsequent treatment of this product with benzaldehyde in the presence of hydrotalcite did not yield the anticipated product.

Synthesis of 3,4-dihydropyrimidin-2-(1*H*)-ones/thiones/imines

The 3,4-dihydropyrimidin-2(1*H*)-one/thione/imine derivatives were synthesized by Mansoor and groups using a modified Biginelli catalysed by PPh_3 as Lewis-base; however, the reaction for the furopyrazole derivatives was not detected (Scheme **12**) [95]. The optimization of the reaction was carried out in several solvents, including ethyl alcohol, CH_3CN, methylbenzene, methylene chloride, and cyclohexane at 100 °C under solvent-free conditions. The reaction is optimized with 10 mol% triphenylphosphine, leading to a substantial product yield.

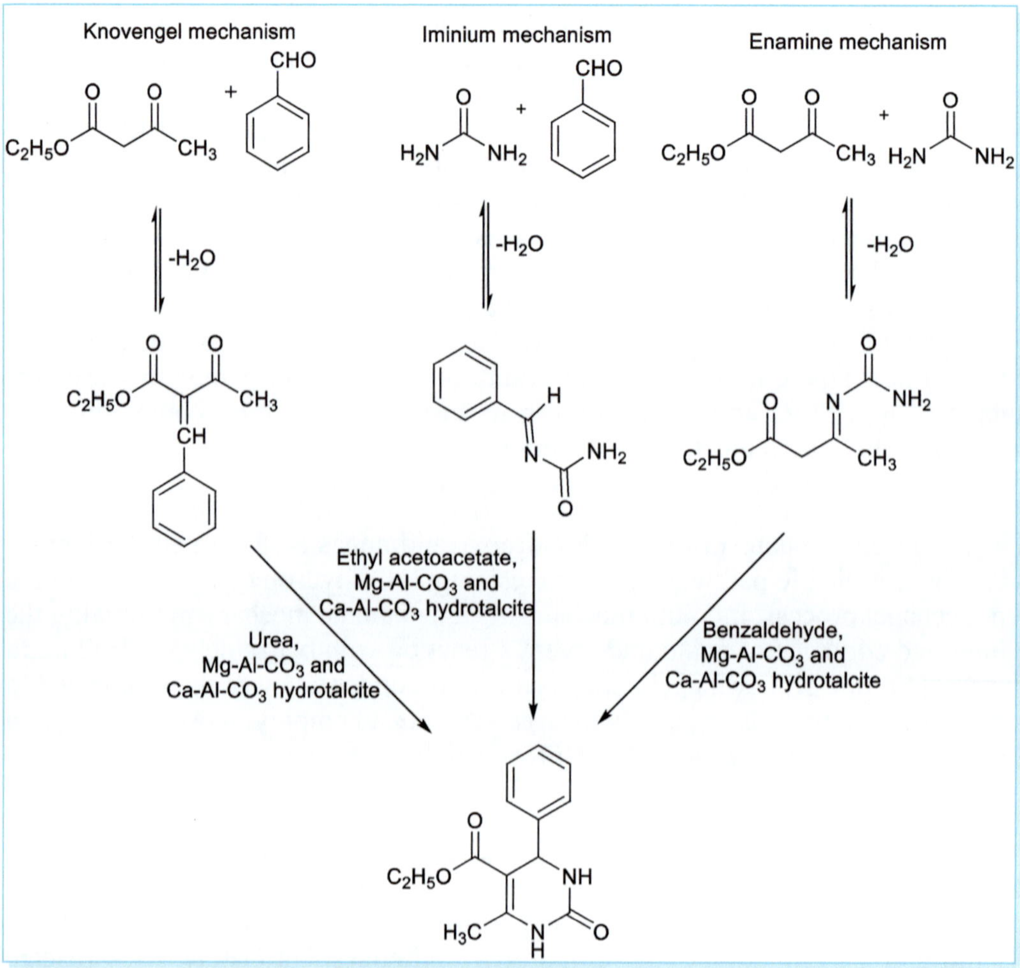

Scheme (11). Dihydropyrimidinone synthesis.

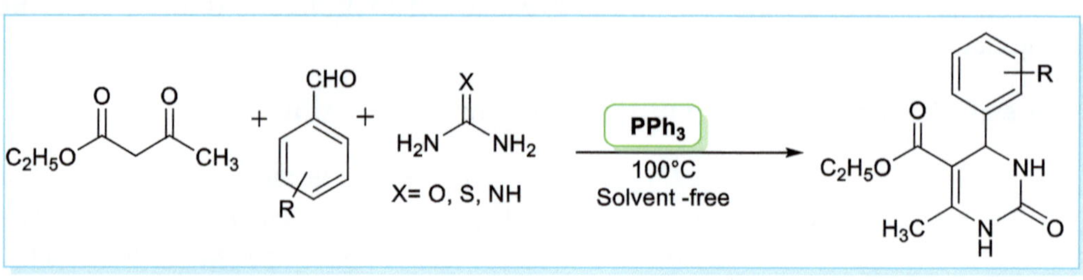

Scheme (12). Synthesis of 3,4-dihydropyrimidin-2-(1*H*)-ones/thiones/imines.

To investigate the reaction parameters, various derivatives of 3,4-dihydropyrimidin-2(1*H*)-one, 3,4-dihydropyrimidin-2(1*H*)-thione, and 3,4-dihydropyrimidin-2(1*H*)-imine were synthesized using a diverse array of substrates. These included aromatic aldehydes, ethyl acetoacetate, and either urea/thiourea or guanidine, facilitated by PPh_3 as the catalytic agent. The benzaldehyde derivatives employed featured a range of substitutions on the aromatic ring, including 4-methyl, halogens at various positions (n = differing placements), 2-hydroxy, *n*-methoxy, and diverse nitro positions, along with *N, N*-dimethylamino groups. The reactions with urea, thiourea, and guanidine resulted in yields of 80-90%.

A triphenylphosphine-catalyzed mechanism for the Biginelli reaction is established through a three-step process (Scheme **13**). The initial stage involves the synthesis of the acylimine intermediate through the reaction of an aldehyde with urea. During the second step, triphenylphosphine reacts with the electrophilic carbon of the aldehyde. To initiate the formation of the β-ketoester enolate, the aldehyde is initially coordinated with PPh_3, enabling the subsequent deprotonation of the *β*-ketoester. Subsequently, in the third step, the enol reacts with the acylimine, yielding an intermediate. This intermediate undergoes cyclization, succeeded by dehydration, ultimately leading to the generation of the corresponding dihydropyrimidines.

Synthesis of Highly Substituted Pyridines

Mehrabi and his team have developed a highly effective one-pot multicomponent reaction approach utilizing aldehydes, thiols, and malononitrile in the presence of CaO nanoparticles (Scheme **14**) [96]. This approach facilitates the synthesis of highly substituted pyridines, recognized as crucial medicinal structures, *via* the creation of C-C and C-hetero bonds. The technique offers a fresh and enhanced route for producing 2-amino-4-aryl 3,5-dicyano-6-sulfanylpyridines, showcasing high yields and shortened reaction times. The cyclization reaction involving aldehydes, thiols, and malononitrile was explored under various nanocatalysts such as CuO, Mn_3O_4, MgO, NiO, SiO_2, and CaO. Among these catalysts, CaO nanoparticles emerged as the highly effective catalyst for this cyclization compared to others. Numerous studies have highlighted the remarkable catalytic performance of CaO nanoparticles, attributed to their surface properties, which include acidity facilitated by Ca^{2+} ions and basicity provided by O^{2-} ions. The interaction between these ions and reactants facilitates the acceleration of reaction conditions, underscoring the catalytic ability of CaO nanoparticles.

In the substrate scope of the reaction, a range of aldehydes and thiols were tested under optimized conditions using CaO nanoparticles as the catalyst. Aldehydes with both electron-withdrawing and electron-donating groups, such as nitro (NO_2), chlorine (Cl), bromine (Br), methyl (Me), and hydroxyl (OH), reacted efficiently with malononitrile, resulting in the formation of the desired 2-amin-

-3,5-dicyano-6-sulfanyl pyridines. Aldehydes with electron-withdrawing groups, such as NO_2, Cl, and Br, yielded high amounts of pyridine derivatives in shorter reaction times, typically around 100 to 120 minutes. In contrast, aldehydes featuring electron-donating groups like methyl (Me) and methoxy (OMe) demonstrated slightly slower reactions but still produced good yields in the range of 80-90%.

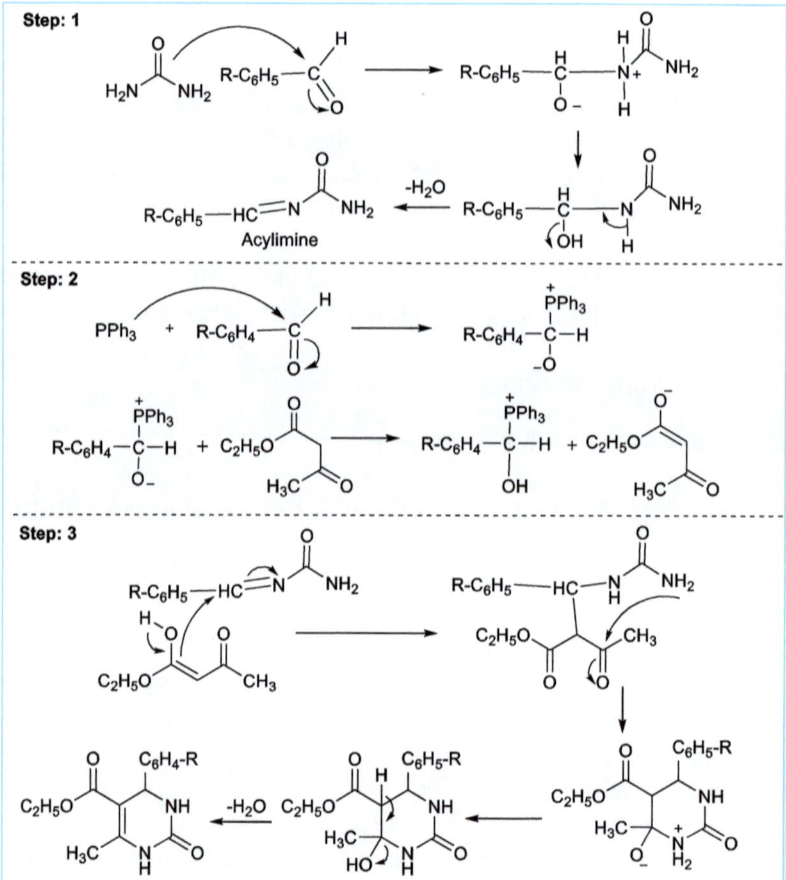

Scheme (13). Proposed mechanism for 3,4-dihydropyrimidine-2-(1*H*)-ones/thiones/imines.

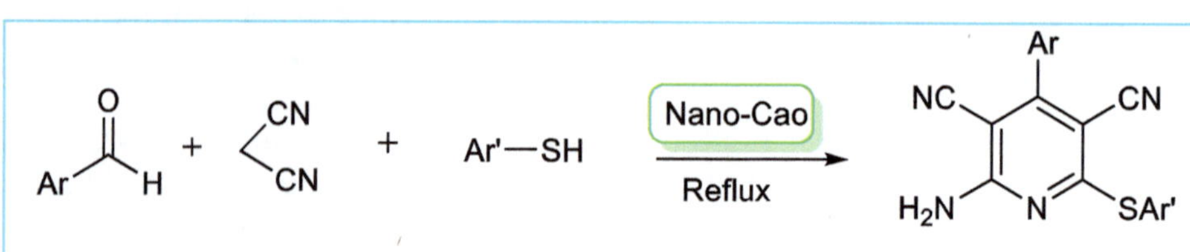

Scheme (14). Synthesis of highly substituted pyridines.

Sterically hindered aldehydes showed a decrease in reaction rate compared to their unstrained counterparts, yet they still produced the desired products with yields of approximately 70-80%, although reaction times extended to 130 to 150 minutes. Overall, the yields varied from 70-95%, highlighting the broad applicability of the reaction across a diverse array of aldehydes and thiols, with the highest yields generally observed for aldehydes containing electron-withdrawing substituents.

Synthesis of 1,4-dihydropyridines

Rajesh and his team explored the flexibility of hydromagnesite (HM) catalysts in the water-based multicomponent synthesis of 1,4-dihydropyridines. Leveraging the Hantzsch reaction, they accessed 1,4-dihydropyridines known for their diverse biological activities. By condensing benzaldehyde, 1,3-cyclohexanedione, EAA, NH_4OAc, and HM catalyst (30 mol%) in H_2O at 90 °C for 20 minutes, they achieved excellent yields of 1,4-dihydropyridine (Scheme **15**) [97].

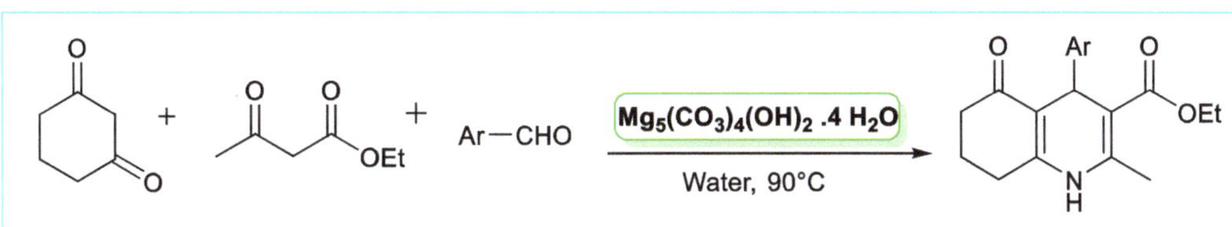

Scheme (15). Synthesis of 1,4-dihydropyridines.

Hydromagnesite (HM), a heterogeneous solid base catalyst with flower-like sheet morphology, has been explored for the synthesis of flavanones, flavonols, and 1,4-dihydropyridines in aqueous media. The catalyst's basic and hydrophilic surface groups, including OH and HCO_3, make it an effective candidate for these transformations. HM has shown excellent catalytic activity in the synthesis of flavanones through intramolecular oxa-Michael cyclization, with high yields (up to 94%) achieved under optimized conditions in water. The substrate scope for flavanone synthesis includes a variety of electron-rich and electron-deficient aldehydes, with yields ranging from 80-94%. For flavonol synthesis, HM catalyzed the formation of products with yields between 72-88%, demonstrating broad functional group tolerance. Moreover, HM was highly efficient in the synthesis of 1,4-dihydropyridines, offering yields of 70-98%, further establishing its versatility. In this one-pot synthesis, aromatic aldehydes featuring electron-donating groups such as -OMe and -Me (at *-o,-m,-p*) exhibited heightened reactivity, yielding excellent results within 20-35 minutes. Conversely, substrates bearing electron-deficient groups such as $-NO_2$ and -Cl (at *-o,-m,-p*) demonstrated comparatively moderate yields within 40–45 minutes.

Synthesis of bis-spiropyranopyrazole Derivatives

Safari and coworkers successfully synthesized novel bis-spiro pyranopyrazoles through a condensation reaction involving hydrazine hydrate, ß-ketoester derivatives, malononitrile, and bis-isatins in ethanol under reflux conditions with a catalytic amount of DABCO. Initially, they investigated the condensation of isatin and 1,6-dibromohexane to determine the optimal reaction medium for bis-isatin derivative synthesis. The model reaction was explored in various solvents (H_2O, PEG-400, EtOH, *t*-BuOH) and with different reagents (Na_2CO_3, Cs_2CO_3, K_2CO_3, CaH_2, NaOH, Et_3N, tetrahydro-1,4-oxazine, *L*-Proline, DABCO). The findings demonstrated that the maximum yield of the target bis-isatin compound was attained when using PEG-400 as the solvent with K_2CO_3 at room temperature. Subsequently, they synthesized a series of bis-isatin derivatives under optimized conditions, yielding good to excellent results. Next, they optimized reaction conditions for bis-spiropyranopyrazole synthesis by studying the reaction of 1-(--(2,3-dioxoindolin-1-yl)hexyl)indoline-2,3-dione, EAA, hydrazine hydrate, and malononitrile. Employing DABCO (6 mol%) in ethanol under reflux conditions has been identified as the optimal catalyst, showcasing superior efficacy compared to larger quantities of DABCO, which proved ineffective in enhancing the yield (Scheme **16**) [98].

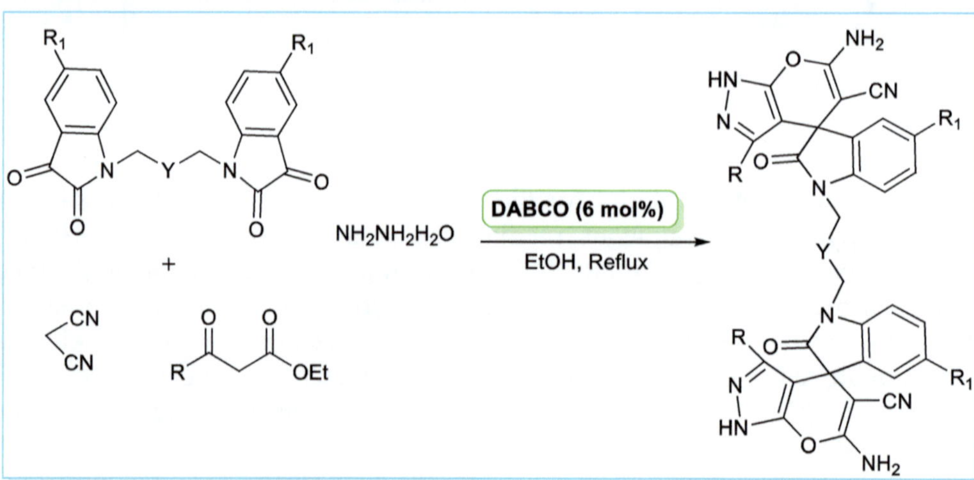

Scheme (16). Synthesis of bis-spiro pyrano pyrazole derivatives.

A variety of bis-isatin derivatives were synthesized under optimized conditions, achieving yields ranging from 80-95%. The scope of the reaction was further expanded to include bis-spiropyrano pyrazoles, synthesized by reacting bis-isatins with *β*-ketoesters, hydrazine hydrate, and malononitrile in the presence of

DABCO as a catalyst. Substituted bis-isatins, including those with electron-withdrawing groups such as bromine, showed faster reaction times, and products were obtained in 90-97% yields. The process demonstrated excellent reproducibility, with no loss of catalytic activity observed after five cycles. This approach provides a simple, effective, and sustainable method for synthesizing a wide range of bis-isatin and bis-spiropyranopyrazole derivatives.

The process commences with the condensation reaction with ß-ketoester **2** and hydrazine hydrate **3**, leading to the formation of pyrazolone **6**. In this particular mechanism, DABCO functions as a proficient catalyst for the production of olefin **A**. This olefin is generated spontaneously *in situ* employing the Knoevenagel condensation reaction involving carbonyl compounds **1** and the active methylene group of malononitrile **4**. Following this, the CH-activated compound **6** transforms into its corresponding enolate form, **B**, under the influence of DABCO. Enolate **B** then reacts with olefin **A** to give intermediate **C**, which subsequently undergoes proton transfer and tautomerization processes, ultimately yielding the final compound **5** (Scheme **17**).

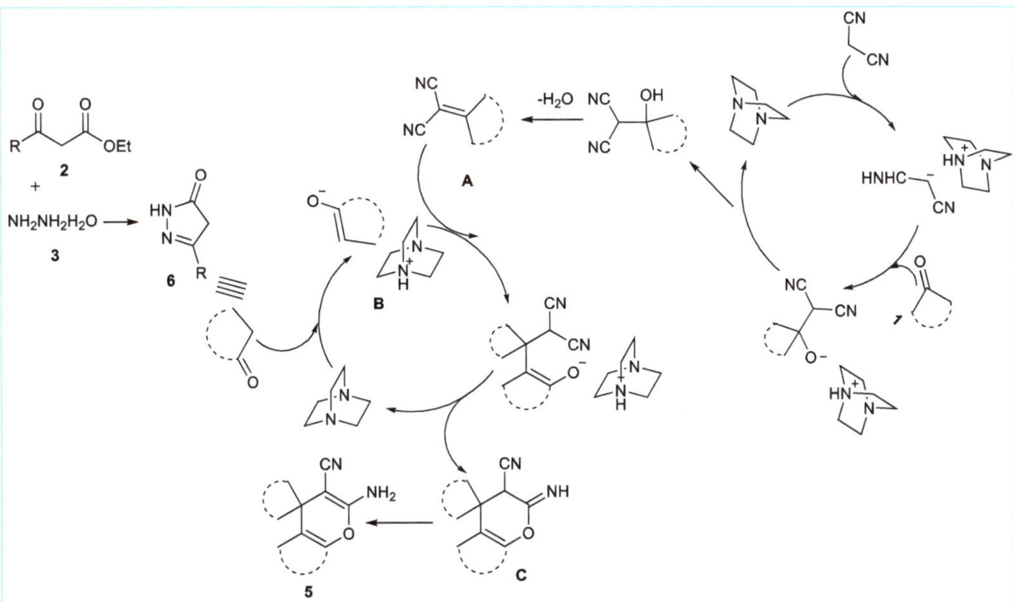

Scheme (17). Proposed mechanism bis-spiro pyrano pyrazole derivatives.

Pyridazine and Pyridazinoquinazoline Derivatives

Saibani and team have conversely synthesized a variety of 3-amino-2-5-dihydropyridazines and fused pyridazine quinazolines in three-component reactions by (arylhydrazono)-propan-2-ones, aldehydes, and malononitrile or ethyl 2-cyanoacetate (Scheme **18**) [99]. The work was directed with highly efficient heterogeneous catalysts, *i.e.*, MgO and 12-tungstophosphoric acid (PW). In the first stage, they attempt the reaction with two kinds of magnesium crystals: MgO and high-surface-area MgO (HSA-MgO). The latter was more amorphous and active than CM-MgO. In the second phase, 12-tungstophosphoric acid (PW) and Et_3N were used as a heterogeneous catalyst with a catalytic quantity, which showed good yields for all compounds.

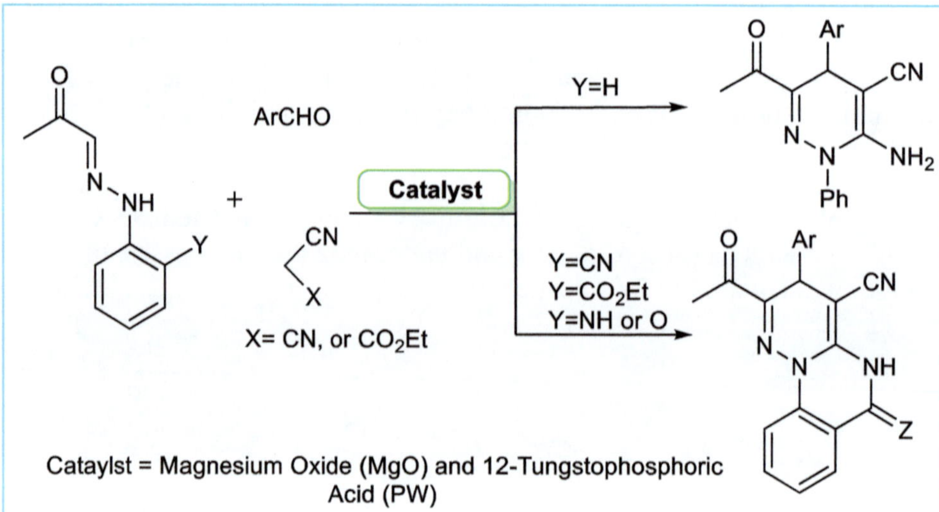

Scheme (18). Pyridazine and pyridazinoquinazoline derivatives.

The substrate scope of the three-component reaction exhibited good to excellent yields across a range of aldehydes, arylhydrazones, and cyano compounds. For instance, phenylhydrazone produced a yield of 95% with the catalyst PW, while 4-chlorophenylhydrazone achieved 94% under the same conditions. Arylhydrazones containing electron-withdrawing groups, such as 4-nitrophenyl hydrazone, yielded 65-80%, with PW delivering the best results. Ethyl cyanoacetate derivatives showed yields ranging from 65-92%. Additionally, 4-methoxyphenyl hydrazone yielded 50-82%, and pyridazine quinazoline derivatives were obtained in 65-78% yields. These findings underscore the versatility of the method, emphasizing the importance of catalyst selection and solvent conditions in optimizing yields and reaction times.

The mechanism involving (arylhydrazono)-propan-2-ones, aldehydes, and either malononitrile or EAA under MgO, resulting in the synthesis of 3-amino-2-5-dihydropyridazine derivatives, as illustrated in (Scheme **19**).

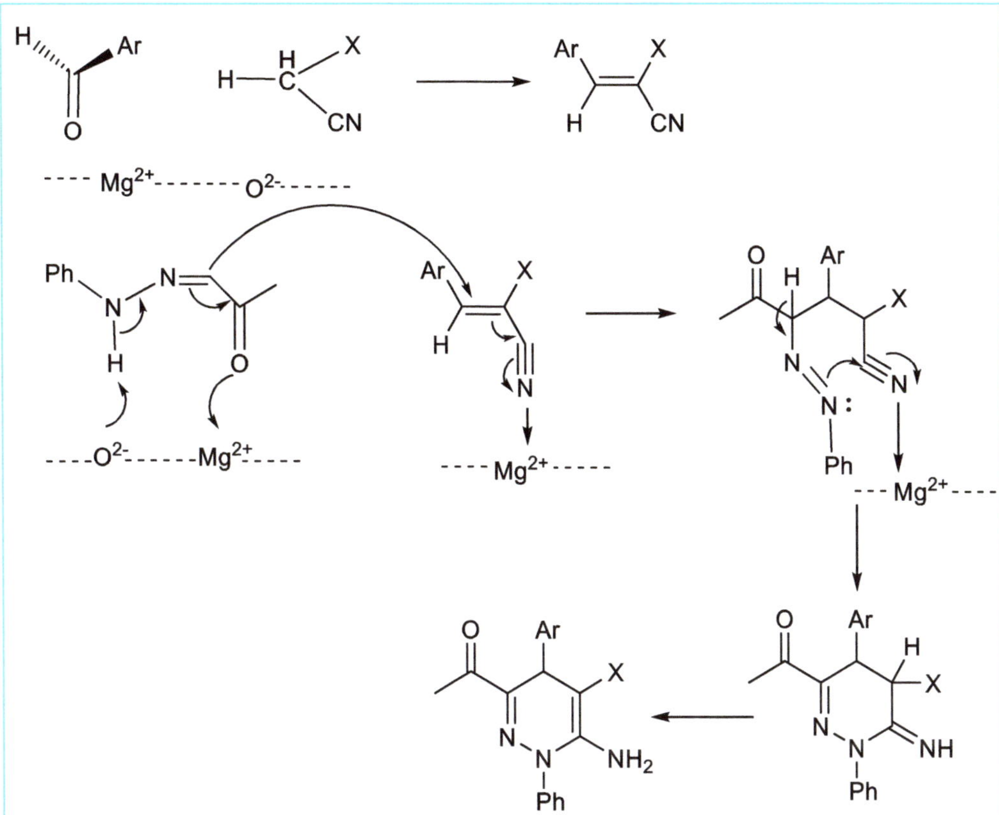

Scheme (19). Plausible mechanism.

Synthesis of Spirooxindole Derivatives

Toorbaf *et al.* investigated the highly effective and eco-friendly, one-pot MCRs of spirooxindole derivatives using graphene oxide functionalized with 2-(--piperazinyl) ethylamine (GO/SiO_2/PEA) as a solid base catalyst (Scheme **20**) [100]. They demonstrated that GO/SiO_2/PEA acts effective, recyclable, and solid base catalyst for synthesizing spirooxindole derivatives. The approach provides several benefits, such as high yields, environmentally benign reaction conditions, the capability for catalyst recycling, and a straightforward work-up protocol.

The reaction system exhibited remarkable versatility, being applicable to a wide variety of substrates. In the synthesis of spiro[indolin-3,4'-pyrano [2,3-*c*]pyrazole]-5'-carbonitrile derivatives, different isatins with varying electronic characteristics, both electron-rich and electron-deficient were successfully coupled with malononitrile under optimized reaction conditions. The yields for this synthesis ranged from moderate to excellent, with a consistent trend showing that electron-rich isatins produced higher yields compared to their electron-deficient counterparts. Notably, substituents at the 5-position of the isatin ring significantly influenced the reaction rate and final yield, with yields of 85% and 83% observed for nitro and chloro substituents, respectively. Furthermore, in the synthesis of 7-amino-2,4-dihydro[indoline-3,5-pyrano(2,3-*d*)pyrimidine]- 6-carbonitriles, a range of barbituric acid derivatives were effectively incorporated into the reaction, yielding products in good to excellent yields. Substituents on the barbituric acid ring, including fluoro, chloro, and methyl groups, had only minor effects on the overall yield, which remained above 80%. The reaction system also showed tolerance to functional groups such as -NO_2, -Cl, -Br, and -F, common in biologically active compounds, demonstrating the robustness and wide applicability of the catalyst.

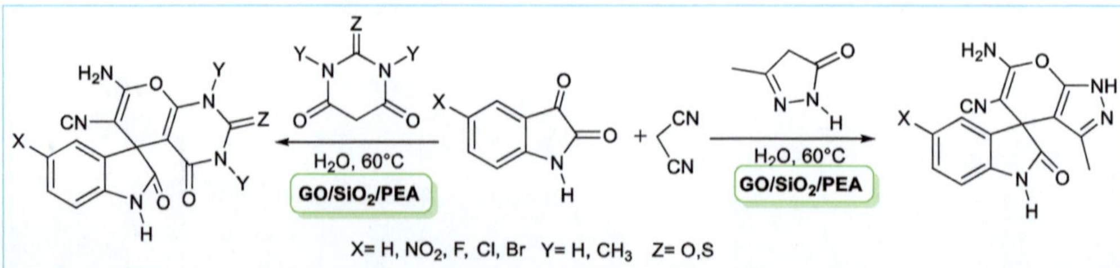

Scheme (20). Synthesis of spirooxindoles.

Using GO/SiO_2/PEA for the synthesis of spirooxindoles is also possible, and its mechanism is presented in (Scheme **21**). In this regard, hydrazine hydrate interacts with EAA, leading to the formation of 3-methyl-2-pyrazoline-5-one: then, in the presence of a catalyst, Knoevenagel condensation of malononitrile and isatin results in the generation of an intermediate. Next, Michael's addition of intermediate, followed by intra-molecular cyclization and tautomerization led to the spirooxindole formation.

Synthesis of Substituted Pyridines

Kale *et al.* conducted a one-pot, four-component synthesis of highly substituted pyridines through MCR involving aldehydes, malononitrile, and thiophenol, utilizing a solid base Mg-Al hydrotalcite catalyst. The catalytic performance is closely linked to the basicity of the catalyst, with the most favorable actions observed using the Mg/Al: 5 catalyst (Scheme **22**) [101]. They noted that aldehydes featuring electron-donating groups readily underwent the reaction to produce corresponding pyridine derivatives in moderate yield, whereas aldehydes containing electron-deficient groups like –NO_2 and Cl (-I effect) yielded the product efficiently.

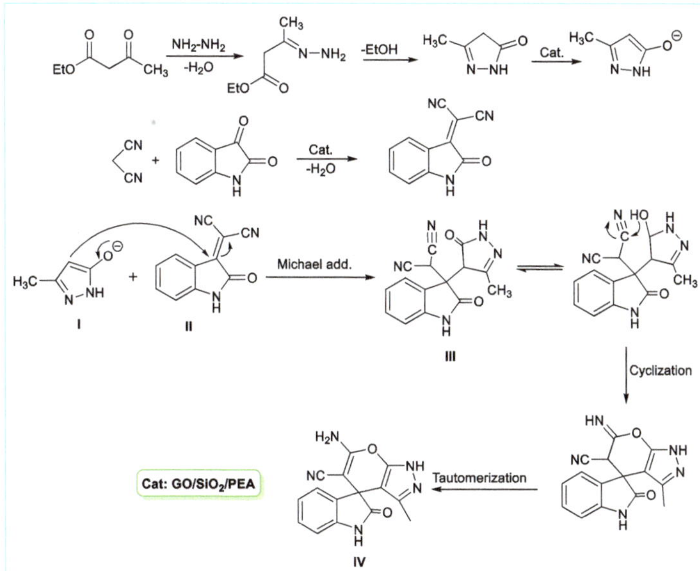

Scheme (21). Proposed mechanism of spirooxindoles.

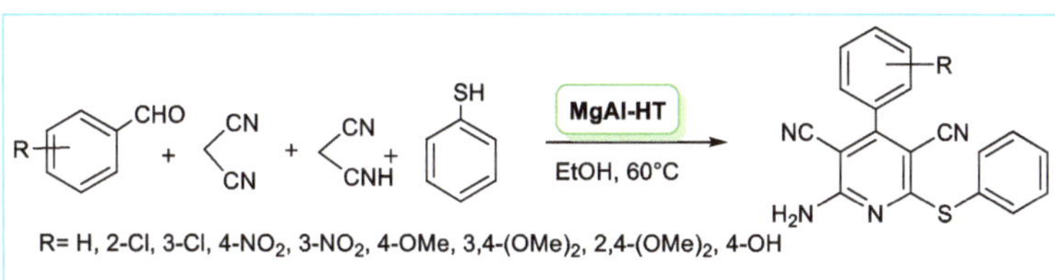

Scheme (22). Synthesis of highly substituted pyridines.

The substrate scope of the multicomponent reaction involving aldehydes, malononitrile, and thiophenol was systematically investigated to assess the general applicability of the reaction. A variety of aromatic aldehydes with different electronic properties were employed, including those with electron-donating groups (*e.g.*, methoxy and hydroxyl) and electron-withdrawing groups (*e.g.*, nitro and chloro). Aldehydes containing electron-withdrawing groups, such as 4-NO_2 and 3-NO_2, showed enhanced reactivity, resulting in higher yields (81-82%). This increased reactivity is attributed to the higher electrophilicity of the carbonyl carbon, which promotes nucleophilic attack. In contrast, electron-donating groups, like methoxy and hydroxyl, reduced the electrophilicity of the carbonyl carbon, leading to more moderate yields (64-69%). The position of substituents on the aromatic ring also affected the reaction efficiency. Generally, para-substituted aldehydes provided higher yields due to lower steric hindrance and greater accessibility of the carbonyl group. Aldehydes with multiple substituents, such as 3,4-$(OMe)_2$, yielded slightly lower results, likely due to steric effects. Halogenated aldehydes, such as 4-Cl and 3-Cl, exhibited good reactivity, with yields ranging from 69% to 72%. These findings demonstrate that the reaction applies to a wide variety of aldehydes, both electron-rich and electron-deficient, highlighting the versatility of the methodology. Additionally, the Mg-Al hydrotalcite catalyst proved highly effective across different substrates, facilitating the efficient formation of pentasubstituted pyridines under mild and environmentally benign conditions.

Synthesis of Pyrano [2,3-*h*]coumarins

Tanuraghaj and his co-worker synthesized novel pyrano [2,3-*h*]coumarin derivatives regioselectively in good yield. Specifically, the compounds were made *via* the reaction of 5,7-dihydroxy-4-methylcoumarin, aromatic aldehydes, and dialkyl acetylenedicarboxylates in the presence of Na_2CO_3 as a catalyst (Scheme **23**) [102]. First, the precursor 5,7-dihydroxy-4-methylcoumarin was synthesized by $ZrOCl_2/SiO_2$-catalyzed condensation controlled by phloroglucinol and EAA. Thereafter, a reaction involving 5,7-dihydroxy-4-methylcoumarin, benzaldehyde, and DMAD was selected as a model to optimize the reaction condition. Bases, solvents, reaction temperatures, and solvent usage were investigated. The optimal conditions have proven to be Na_2CO_3 in ethyl acetate at reflux as a solvent, providing exceptional yields of the desired pyrano [2,3-*h*]coumarinic product.

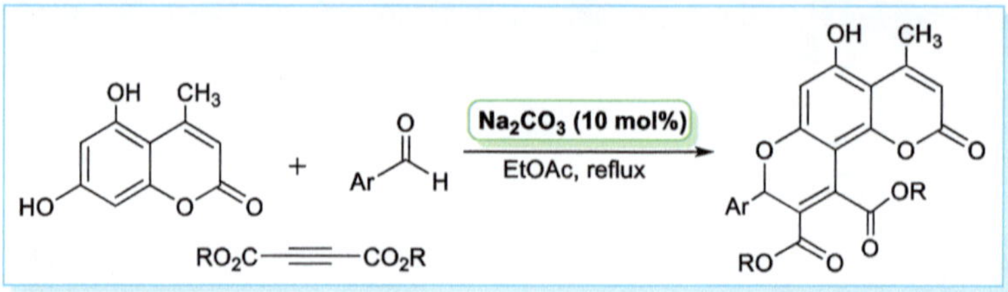

Scheme (23). Synthesis of pyrano [2,3-*h*] coumarins.

The reaction demonstrated an excellent substrate scope, as various aldehydes were well tolerated, including both electron-donating groups (such as methoxy and ethoxy) and electron-withdrawing groups (like chloro and nitro). Additionally, different dialkyl acetylenedicarboxylates, including dimethyl and diethyl, along with their substituted derivatives, also produced good to excellent yields of pyrano [2,3-*h*]coumarin products. For instance, the reaction of benzaldehyde with dimethyl acetylenedicarboxylate yielded pyrano [2,3-*h*]coumarin with an 85% yield, while using 4-chlorobenzaldehyde with dimethyl acetylenedicarboxylate resulted in an 88% yield. Other substituent combinations on the aldehyde, including methyl, ethoxy, and methoxy groups, led to yields ranging from 80% to 90%. Overall, this method is highly versatile, providing novel pyrano [2,3-*h*]coumarins in moderate to excellent yields.

An imaginable mechanism for the synthesis of pyrano [2,3-*h*] coumarins is presented in (Scheme **24**). The operation begins with the deprotonation of 5,7-dihydroxy-4–methylcoumarin with Na_2CO_3 to give the intermediate **1**, and then the Michael type of the intermediate **1** with dialkyl-acetylene dicarboxylates in C-8 to intermediates **2** and then **3**. After the aldehyde condensation reaction with **3** to form the adduct **4**, it will signal the formation of the end product by intramolecular ring closing and dehydration.

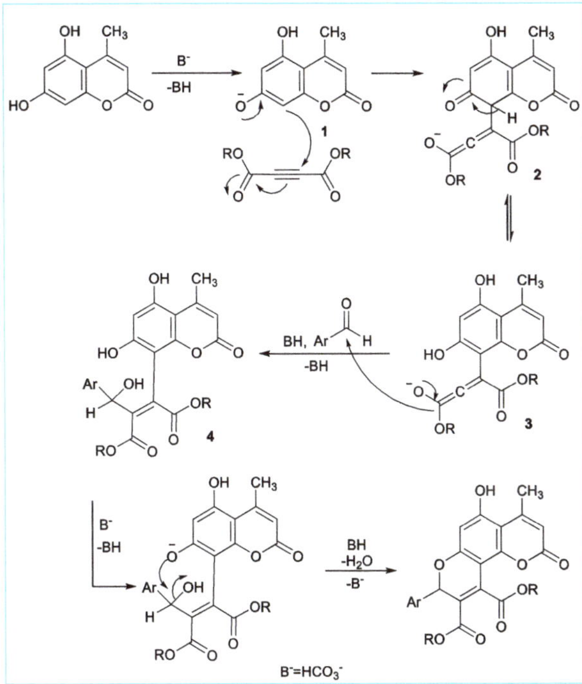

Scheme (24). Proposed mechanisms for pyrano [2,3-*h*] coumarins.

Synthesis of Functionalized 1,4-dihydropyridine Derivatives

Shabalala and coworkers have recently elaborated an effective single-step method for synthesizing new functionalized 1,4-dihydropyridines quickly at ambient temperature. This was conducted using a malononitrile, substituted aldehydes, dimethylacetylenedicarboxylate, and 4-fluoroaniline reaction in EtOH over Sm_2O_3/ZrO_2 (Scheme **25**) [103]. Remarkably, all reactions were carried out in under 20 minutes. Initial studies revealed the essential model reaction of 2-methoxybenzaldehyde, malononitrile, dimethylacetylenedicarboxylate, and 4-fluoroaniline using several factors. In the absence of solvent and catalyst, no reaction occurred. Furthermore, using ethanol as a solvent without a catalyst, neither room temperature nor reflux conditions induced any reaction even after 12 hours. Exploring various homogeneous catalysts, including triethylamine, piperidine, NaOH, $NaHCO_3$, and K_2CO_3 in ethanol, yielded moderate product yields after 6-10 hours. Pure oxides like ZrO_2, SiO_2, Fe_2O_3, and Al_2O_3, with their multifaceted properties, proved suitable for heterogeneous catalysis. When utilized with ethanol as solvent, these pure oxides achieved excellent yields at room temperature within 1.5-3.0 hours. Finally, they noted exceptional yields using ethanol in conjunction with mixed oxides, specifically Sm_2O_3/ZrO_2, accomplishing the reaction in just 20 minutes.

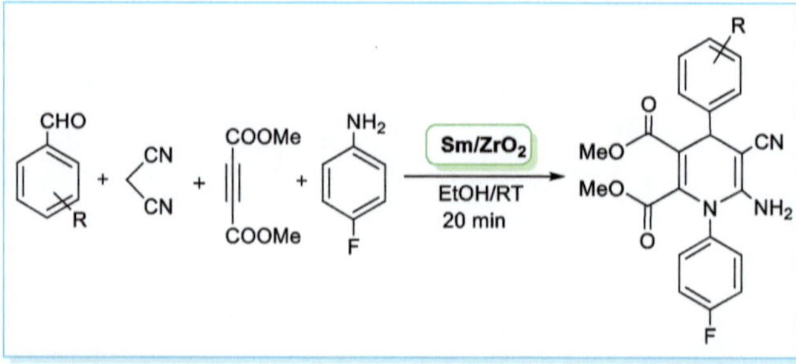

Scheme (25). Synthesis of 1,4-dihydropyridine.

The substrate scope was thoroughly investigated by varying the functional groups on the aryl aldehydes. Both electron-donating and electron-withdrawing groups were successfully incorporated at different positions on the aromatic ring (ortho, meta, and para). These substitutions had a minimal impact on the reaction yields and durations, with most functionalized 1,4-dihydropyridine derivatives being obtained in yields ranging from good to excellent (up to 96%). The reaction demonstrated high tolerance for a wide variety of aldehydes, including those with bulky substituents and diverse electronic properties. Additionally, aldehydes containing different functional groups, such as methoxy, chloro, and bromo, also yielded excellent results, further confirming the broad applicability of the method. Overall, the substrate scope illustrates that the reaction is versatile and can accommodate a wide range of functional groups without compromising yield or reaction time.

The reaction process can be outlined step by step, starting with the occurrence of a Knoevenagel condensation. Here, upon the fast condensation of malononitrile with arylaldehyde, a 2-arylidenemalononitrile is produced. Meanwhile, the 1,3-dipolar compound **6** is created immediately following the reaction between 4-fluoroaniline and dimethylacetylenedicarboxylate through the active sites of Sm_2O_3 and ZrO_2 of the catalyst surface. All of these processes occur simultaneously. Afterward, the Michael addition also took place on the catalyst surface and formed an intermediate compound. Further, an intramolecular cyclization was taken place to produce the final product which is the functionalized derivative of 1,4-dihydropyridine (Scheme **26**).

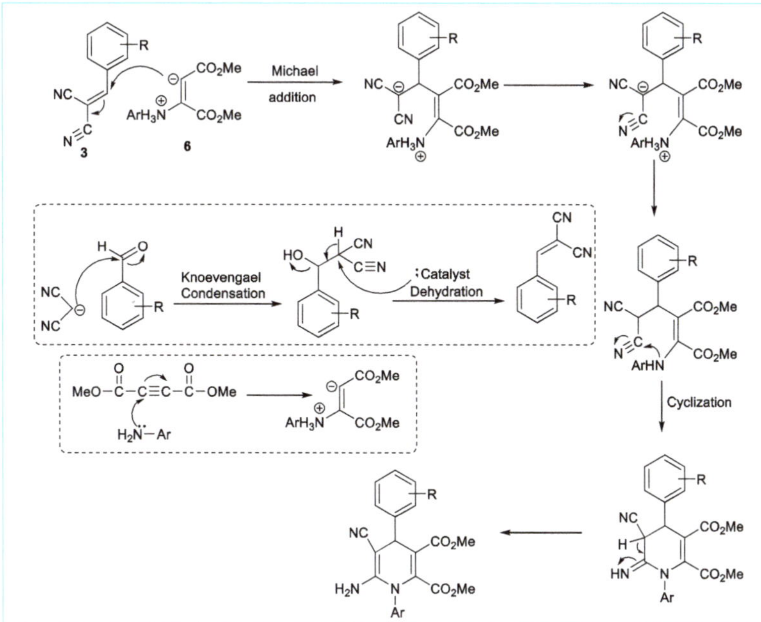

Scheme (26). Proposed mechanism.

Synthesis of Polysubstituted Pyridines

Gandhi and his team have outlined a one-pot, multicomponent synthesis method that is sustainable and eco-friendly. Using a Knoevenagel condensation followed by Michael addition, 5-amino-3-benzothiazol-2-yl-7-(phenyl)-2-phenyl-2,3-dihydro-thiazolo [4,5-*b*]pyridine-6-carbonitrile is synthesized in this method. The process is carried out at EtOH and includes malononitrile, substituted aromatic aldehydes, and NH_4OAc using magnesium oxide (MgO NPs) as a heterogeneous base catalyst (Scheme **27**) [104]. Synthesis of MgO nanoparticles begins with its synthesis using $Mg(NO_3)_2$ as a precursor. The target product is obtained with a low yield and requires synthesizing for more than 24 hours without a catalyst. However, they discovered that MgO demonstrated superior catalytic efficiency, resulting in product yields of 80-92% within approximately 6-8 hours in EtOH. This enhanced efficiency is attributed to MgO's basic properties and its ability to bind the substrate to active sites during the synthesis process.

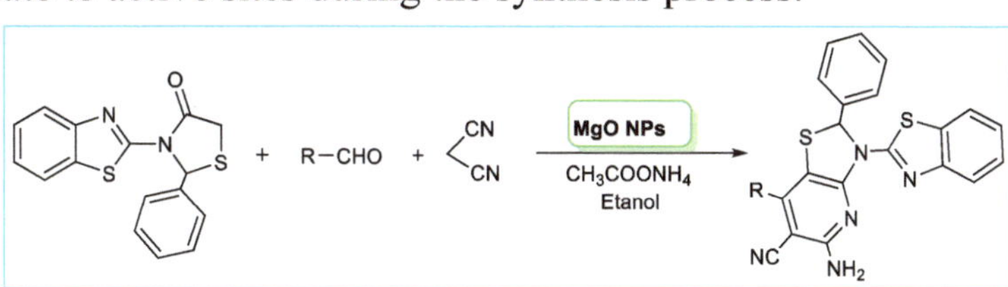

Scheme (27). Synthesis of dihydrothiazolo [4,5-*b*]pyridine-6-carbonitriles.

The substrate scope of the MgO NPs-catalyzed one-pot multicomponent reaction for synthesizing thiazolo [4,5-*b*]pyridine-6-carbonitrile derivatives was thoroughly examined using various substituted aromatic aldehydes. The reaction exhibited excellent versatility, tolerating a wide range of functional groups on the aromatic rings. For electron-donating groups, such as methoxy (-OCH_3) and ethyl (-CH_2CH_3), the reaction proceeded smoothly, yielding high outcomes typically in the range of 80–92%. Electron-withdrawing groups, like nitro (-NO_2) and bromo (-Br), also produced good to excellent yields, further demonstrating the robustness of the catalyst. Substituents at different positions on the aromatic ring, including ortho, meta, and para positions, did not negatively impact the reaction's efficiency.

Additionally, compounds with both bulky and smaller substituents were synthesized with high yields, indicating broad substrate compatibility. The yields for all reactions were consistently high, with most products obtained within 6–8 hours under mild conditions. Moreover, some sterically hindered substrates, such as 2,4-dimethyl- and 3-nitro-benzaldehydes, also yielded the desired thiazolo [4,5-*b*]pyridine-6-carbonitrile derivatives, with yields ranging from 85% to 92%. This suggests that the catalytic system is highly efficient and adaptable to a diverse range of substrates, which is an important characteristic for developing practical synthetic methodologies in organic chemistry.

At the first stage, the carbonyl carbon in 3-benzothiazol-2-yl-2-phenyl-thiazolidin-4-one function is activated by MgO-NPs, thus allowing for the formation of an imine, which reacts with ammonium acetate. Additionally, while the imine formation is developed, Knoevenagel condensation of substituted aromatic aldehyde and malononitrile happens simultaneously, and an arylidene is also created. Subsequently, the imine acting as a nucleophile undergoes Michael addition to the activated double bond in the arylidene intermediate. Aldol condensations of the produced adduct are then made during the intramolecular cyclization. The final reaction steps are oxidation and aromatization, which lead to the formation of the target compounds (Scheme **28**).

Synthesis of Novel Dihydropyridines

Hadebe *et al.* synthesized a sturdy catalyst consisting of MoO_3/ZrO_2 using a straightforward wet-impregnation technique. This catalyst proved highly active in producing new dihydropyridine derivatives in exceptional yields. The synthesis sequence was a one-pot, four-component interconnection reaction of dimethyl--amine pyridine, diethylbut-2-ynedioate, several benzaldehydes, and malononitrile in EtOH (Scheme **29**) [105]. Remarkably, these reactions proceeded swiftly at room temperature, completing in less than 18 minutes. Finally, the condensation reaction was performed with various base catalysis at 25°C, including Et_3N,

K_2CO_3, Cs_2CO_3, KOH, NaOH, pyridine, and piperidine, which all resulted in satisfactory yields of the desired derivatives. However, employing various acid catalysts like hydrochloric acid, perchloric acid, para-toluenesulfonic acid (*p*-TSA), and ethanoic acid yielded poor results. An enhanced yield was achieved by employing a 2.5 wt% MoO_3/ZrO_2 catalyst. The choice of solvent significantly influenced the reaction yield, with both polar and non-polar solvents being evaluated. At first, under conditions devoid of solvent, the reaction mixture exhibited no advancement even after 24 hours. The presence of a solvent proved crucial for the effective transformation of the target product. Polar-protic solvents such as EtOH, MeOH, and H_2O offered good yields in shorter reaction times. Based on its superior performance, EtOH was selected as the preferred solvent for further investigations into the reaction's generality.

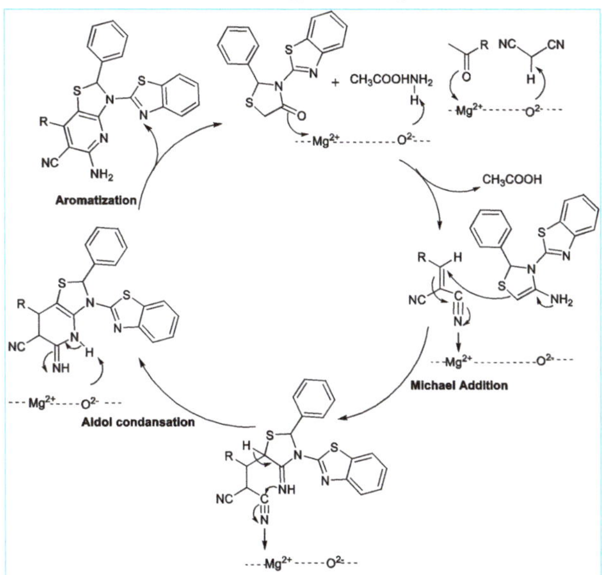

Scheme (28). Plausible mechanism.

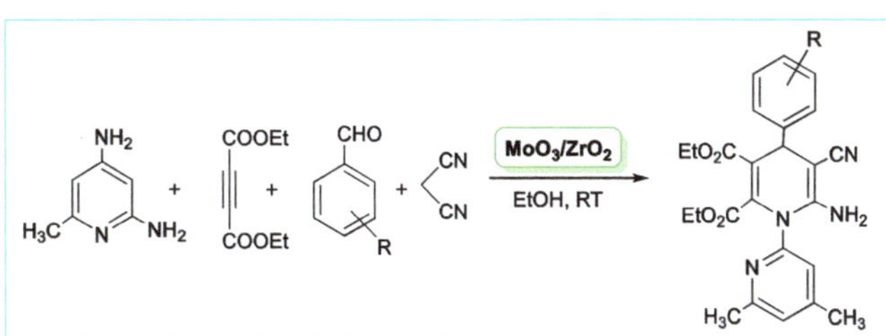

Scheme (29). Synthesis dihydropyridines.

The synthesized MoO_3/ZrO_2 catalyst was tested for its reactivity with various benzaldehyde derivatives under optimized reaction conditions, specifically at room temperature, with ethanol as the solvent. A wide range of substituted benzaldehydes, including both electron-donating and electron-withdrawing groups, was successfully employed, resulting in excellent yields (92–96%) of the desired dihydropyridine products within a short reaction time of less than 18 minutes. Notably, substrates with various functional groups such as methoxy, ethyl, bromo, nitro, and dimethylamino were well-tolerated, yielding high amounts of the target compounds. These results highlight the versatility and efficiency of the 2.5% MoO_3/ZrO_2 catalyst, which demonstrated superior catalytic activity for both electron-donating and electron-withdrawing groups, thus showcasing the broad substrate scope of the method. The yields remained consistent across different substrates, underscoring the robustness of the catalytic system.

Lewis acidic sites of the MoO_3/ZrO_2 catalyst are available for the active stage, especially for the interaction with the carbonyl group of the different aldehydes. As activated aldehydes can interact with the malononitrile compound **4**, the Knoevenagel condensed product **I** is formed. Lewis and Bronsted basic sites on the catalyst also participate in the condensation of 4,6-dimethylpyridin-2-amine, **1** and diethyl but-2-syndicate **2**, which leads to the formation of the enolate intermediate **II**. Further, the Michael addition process occurs with the interaction of **I** and **II**, so the Michael adduct **III** is formed. Finally, adduct **III** undergoes intramolecular cyclization **IV** and subsequent tautomerization, ultimately yielding the desired dihydropyridine derivatives (Scheme **30**).

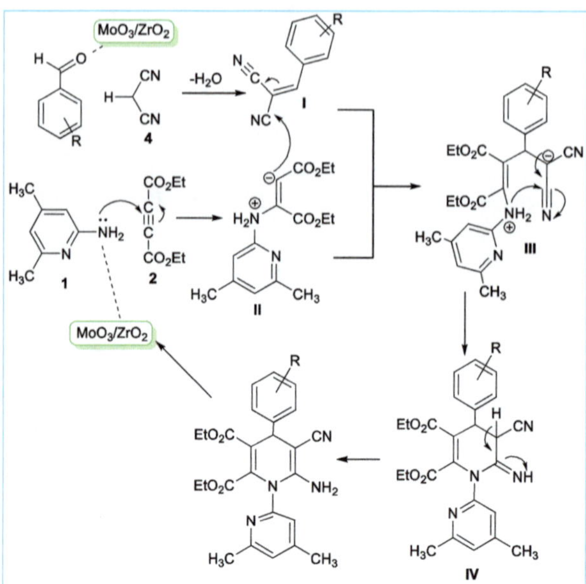

Scheme (30). Possible mechanism.

CONCLUDING REMARKS

In conclusion, the recent trends in Multicomponent Reactions (MCRs) for eco-compatible heterocyclic synthesis using solid base catalysis represent a significant step forward in sustainable organic chemistry. The advancements in solid base-catalyzed MCRs hold great promise for various applications in medicinal chemistry, materials science, and other interdisciplinary fields. Future research efforts should focus on further understanding the mechanistic facets of solid base catalysis, optimizing catalyst design, and expanding the scope of MCRs to access novel heterocyclic motifs. Overall, the recent trends in solid base-catalyzed MCRs underscore the potential of green catalytic approaches in addressing the challenges of modern organic synthesis while advancing towards more sustainable and environmentally friendly chemical processes.

ACKNOWLEDGEMENTS

The authors gratefully acknowledge the Veer Narmad South Gujarat University (VNSGU), Surat, India, for the financial assistance (UGC/17163/2023). DPP is thankful to the UGC for a NFOBC fellowship.

REFERENCES

[1] Jambhulkar, D.K.; Ugwekar, R.P.; Bhanvase, B.A.; Barai, D.P. A review on solid base heterogeneous catalysts: preparation, characterization and applications. *Chem. Eng. Commun.*, **2022**, *209*(4), 433-484. [http://dx.doi.org/10.1080/00986445.2020.1864623]

[2] Agarwal, B.; Kailasam, K.; Sangwan, R.S.; Elumalai, S. Traversing the history of solid catalysts for heterogeneous synthesis of 5-hydroxymethylfurfural from carbohydrate sugars: A review. *Renew. Sustain. Energy Rev.*, **2018**, *82*, 2408-2425. [http://dx.doi.org/10.1016/j.rser.2017.08.088]

[3] Olagunju, O.A.; Musonge, P.; Kiambi, S.L. Production and optimization of biodiesel in a membrane reactor, using a solid base catalyst. *Membranes (Basel)*, **2022**, *12*(7), 674. [http://dx.doi.org/10.3390/membranes12070674] [PMID: 35877877]

[4] Hattori, H. Solid base catalysts: generation of basic sites and application to organic synthesis. *Appl. Catal. A Gen.*, **2001**, *222*(1-2), 247-259. [http://dx.doi.org/10.1016/S0926-860X(01)00839-0]

[5] Kouzu, M.; Kasuno, T.; Tajika, M.; Sugimoto, Y.; Yamanaka, S.; Hidaka, J. Calcium oxide as a solid base catalyst for transesterification of soybean oil and its application to biodiesel production. *Fuel*, **2008**, *87*(12), 2798-2806. [http://dx.doi.org/10.1016/j.fuel.2007.10.019]

[6] Foroutan, R.; Mohammadi, R.; Esmaeili, H.; Mirzaee Bektashi, F.; Tamjidi, S. Transesterification of waste edible oils to biodiesel using calcium oxide@magnesium oxide nanocatalyst. *Waste Manag.*, **2020**, *105*, 373-383. [http://dx.doi.org/10.1016/j.wasman.2020.02.032] [PMID: 32120264]

[7] Bing, W.; Wei, M. Recent advances for solid basic catalysts: Structure design and catalytic performance. *J. Solid State Chem.*, **2019**, *269*, 184-194. [http://dx.doi.org/10.1016/j.jssc.2018.09.023]

[8] Li, Z. *Novel solid base catalysts for Michael additions.*, **2005**.

[9] Cundy, C.S.; Cox, P.A. The hydrothermal synthesis of zeolites: history and development from the earliest days to the present time. *Chem. Rev.*, **2003**, *103*(3), 663-702. [http://dx.doi.org/10.1021/cr020060i] [PMID: 12630849]

[10] Flanigen, E.M. Zeolites and molecular sieves an historical perspective. *Studies in surface science and catalysis*; Elsevier, **1991**, 58, pp. 13-34.

[11] Rabo, J.A.; Schoonover, M.W. Early discoveries in zeolite chemistry and catalysis at Union Carbide, and follow-up in industrial catalysis. *Appl. Catal. A Gen.*, **2001**, *222*(1-2), 261-275. [http://dx.doi.org/10.1016/S0926-860X(01)00840-7]

[12] Davis, R.J. New perspectives on basic zeolites as catalysts and catalyst supports. *J. Catal.*, **2003**, *216*(1-2), 396-405. [http://dx.doi.org/10.1016/S0021-9517(02)00034-9]

[13] Ono, Y. Solid base catalysts for the synthesis of fine chemicals. *J. Catal.*, **2003**, *216*(1-2), 406-415. [http://dx.doi.org/10.1016/S0021-9517(02)00120-3]

[14] Hattori, H. Solid base catalysts: generation, characterization, and catalytic behavior of basic sites. *J. Jpn. Petrol. Inst.*, **2004**, *47*(2), 67-81. [http://dx.doi.org/10.1627/jpi.47.67]

[15] Dusselier, M.; Davis, M.E. Small-pore zeolites: synthesis and catalysis. *Chem. Rev.*, **2018**, *118*(11), 5265-5329. [http://dx.doi.org/10.1021/acs.chemrev.7b00738] [PMID: 29746122]

[16] Yashima, T.; Ahmad, H.; Yamazaki, K.; Katsuta, M.; Hara, N. Alkylation on synthetic zeolites I. Alkylation of toluene with methanol. *J. Catal.*, **1970**, *16*(3), 273-280. [http://dx.doi.org/10.1016/0021-9517(70)90223-X]

[17] Niederberger, M.; Garnweitner, G. Organic reaction pathways in the nonaqueous synthesis of metal oxide nanoparticles. *Chemistry*, **2006**, *12*(28), 7282-7302. [http://dx.doi.org/10.1002/chem.200600313] [PMID: 16927442]

[18] Yokoi, T.; Kubota, Y.; Tatsumi, T. Amino-functionalized mesoporous silica as base catalyst and adsorbent. *Appl. Catal. A Gen.*, **2012**, *421-422*, 14-37. [http://dx.doi.org/10.1016/j.apcata.2012.02.004]

[19] Roelofs, J.C.A.A.; Lensveld, D.J.; van Dillen, A.J.; de Jong, K.P. On the structure of activated hydrotalcites as solid base catalysts for liquid-phase aldol condensation. *J. Catal.*, **2001**, *203*(1), 184-191. [http://dx.doi.org/10.1006/jcat.2001.3295]

[20] Yang, H.; Lu, Q.; Gao, F.; Shi, Q.; Yan, Y.; Zhang, F.; Xie, S.; Tu, B.; Zhao, D. One-Step Synthesis of Highly Ordered Mesoporous Silica Monoliths with Metal Oxide Nanocrystals in their Channels. *Adv. Funct. Mater.*, **2005**, *15*(8), 1377-1384. [http://dx.doi.org/10.1002/adfm.200500026]

[21] Wang, S.H.; Wang, Y.B.; Dai, Y.M.; Jehng, J.M. Preparation and characterization of hydrotalcite-like compounds containing transition metal as a solid base catalyst for the transesterification. *Appl. Catal. A Gen.*, **2012**, *439-440*, 135-141. [http://dx.doi.org/10.1016/j.apcata.2012.06.049]

[22] Vallet-Regí, M.; Balas, F.; Arcos, D. Mesoporous materials for drug delivery. *Angew. Chem. Int. Ed.*, **2007**, *46*(40), 7548-7558. [http://dx.doi.org/10.1002/anie.200604488] [PMID: 17854012]

[23] ALOthman, Z. Z. A. A review: fundamental aspects of silicate mesoporous materials. *Materials (Basel)*, **2012**, *5*(12), 2874-2902. [http://dx.doi.org/10.3390/ma5122874]

[24] Liu, W.; Zhu, L.; Jiang, Y.; Liu, X.Q.; Sun, L.B. Direct fabrication of strong basic sites on ordered nanoporous materials: exploring the possibility of metal–organic frameworks. *Chem. Mater.,* **2018**, *30*(5), 1686-1694.
[http://dx.doi.org/10.1021/acs.chemmater.7b05102]

[25] Xie, Y.; Sharma, K.K.; Anan, A.; Wang, G.; Biradar, A.V.; Asefa, T. Efficient solid-base catalysts for aldol reaction by optimizing the density and type of organoamine groups on nanoporous silica. *J. Catal.,* **2009**, *265*(2), 131-140.
[http://dx.doi.org/10.1016/j.jcat.2009.04.018]

[26] Albanese, A.; Tang, P.S.; Chan, W.C.W. The effect of nanoparticle size, shape, and surface chemistry on biological systems. *Annu. Rev. Biomed. Eng.,* **2012**, *14*(1), 1-16.
[http://dx.doi.org/10.1146/annurev-bioeng-071811-150124] [PMID: 22524388]

[27] Han, D.; Lee, D. An integral design and synthesis approach to heterogeneous catalysts: Tailoring the surface characteristics and mass and heat transport properties. *Chem. Eng. J.,* **2023**, *468*, 143657.
[http://dx.doi.org/10.1016/j.cej.2023.143657]

[28] Rinaldi, R.; Schüth, F. Design of solid catalysts for the conversion of biomass. *Energy Environ. Sci.,* **2009**, *2*(6), 610-626.
[http://dx.doi.org/10.1039/b902668a]

[29] Serrano, D.P.; Melero, J.A.; Morales, G.; Iglesias, J.; Pizarro, P. Progress in the design of zeolite catalysts for biomass conversion into biofuels and bio-based chemicals. *Catal. Rev., Sci. Eng.,* **2018**, *60*(1), 1-70.
[http://dx.doi.org/10.1080/01614940.2017.1389109]

[30] Chinthala, M.; Balakrishnan, A.; Venkataraman, P.; Manaswini Gowtham, V.; Polagani, R.K. Synthesis and applications of nano-MgO and composites for medicine, energy, and environmental remediation: a review. *Environ. Chem. Lett.,* **2021**, *19*(6), 4415-4454.
[http://dx.doi.org/10.1007/s10311-021-01299-4]

[31] Chen, S.S.; Yu, I.K.M.; Cho, D.W.; Song, H.; Tsang, D.C.W.; Tessonnier, J.P.; Ok, Y.S.; Poon, C.S. Selective glucose isomerization to fructose *via* a nitrogen-doped solid base catalyst derived from spent coffee grounds. *ACS Sustain. Chem.& Eng.,* **2018**, *6*(12), 16113-16120.
[http://dx.doi.org/10.1021/acssuschemeng.8b02752]

[32] Tanaka, K.; Toda, F. Solvent-free organic synthesis. *Chem. Rev.,* **2000**, *100*(3), 1025-1074.
[http://dx.doi.org/10.1021/cr940089p] [PMID: 11749257]

[33] Mazumder, N.A.; Rano, R. An efficient solid base catalyst from coal combustion fly ash for green synthesis of dibenzylideneacetone. *J. Ind. Eng. Chem.,* **2015**, *29*, 359-365.
[http://dx.doi.org/10.1016/j.jiec.2015.04.015]

[34] Li, C.; Hu, X.; Feng, W.; Wu, B.; Wu, K. A supported solid base catalyst synthesized from green biomass ash for biodiesel production. *Energy Sources A Recovery Util. Environ. Effects,* **2018**, *40*(2), 142-147.
[http://dx.doi.org/10.1080/15567036.2017.1405121]

[35] Liu, K.; Wang, R.; Yu, M. An efficient, recoverable solid base catalyst of magnetic bamboo charcoal: Preparation, characterization, and performance in biodiesel production. *Renew. Energy,* **2018**, *127*, 531-538.
[http://dx.doi.org/10.1016/j.renene.2018.04.092]

[36] Marwaha, A.; Dhir, A.; Mahla, S.K.; Mohapatra, S.K. An overview of solid base heterogeneous catalysts for biodiesel production. *Catal. Rev., Sci. Eng.,* **2018**, *60*(4), 594-628.
[http://dx.doi.org/10.1080/01614940.2018.1494782]

[37] Reddy, B.M.; Sreekanth, P.M.; Yamada, Y.; Kobayashi, T. Surface characterization and catalytic activity of sulfate-, molybdate- and tungstate-promoted Al2O3–ZrO2 solid acid catalysts. *J. Mol. Catal. Chem.,* **2005**, *227*(1-2), 81-89.

[http://dx.doi.org/10.1016/j.molcata.2004.10.011]

[38] K.; YAMAGUCHI, T.; TAKESHITA, T. Solid bases and their catalytic activity. *J.Res. Instit. Catal. Hokkaido Uni.,* **1968**, *16*(1), 425-447.

[39] Jothiramalingam, R.; Wang, M.K. Review of recent developments in solid acid, base, and enzyme catalysts (heterogeneous) for biodiesel production *via* transesterification. *Ind. Eng. Chem. Res.,* **2009**, *48*(13), 6162-6172.
[http://dx.doi.org/10.1021/ie801872t]

[40] Choudhury, H.A.; Chakma, S.; Moholkar, V.S. Mechanistic insight into sonochemical biodiesel synthesis using heterogeneous base catalyst. *Ultrason. Sonochem.,* **2014**, *21*(1), 169-181.
[http://dx.doi.org/10.1016/j.ultsonch.2013.04.010] [PMID: 23742888]

[41] Munyentwali, A.; Li, H.; Yang, Q. Review of advances in bifunctional solid acid/base catalysts for sustainable biodiesel production. *Appl. Catal. A Gen.,* **2022**, *633*, 118525.
[http://dx.doi.org/10.1016/j.apcata.2022.118525]

[42] Busca, G. Bases and basic materials in industrial and environmental chemistry: a review of commercial processes. *Ind. Eng. Chem. Res.,* **2009**, *48*(14), 6486-6511.
[http://dx.doi.org/10.1021/ie801878d]

[43] Saracco, G.; Neomagus, H.W.J.P.; Versteeg, G.F.; Swaaij, W.P.M. High-temperature membrane reactors: potential and problems. *Chem. Eng. Sci.,* **1999**, *54*(13-14), 1997-2017.
[http://dx.doi.org/10.1016/S0009-2509(99)00009-3]

[44] Pines, H. Base-catalyzed reactions of hydrocarbons and related compounds. *Elsevier,* **2012**.

[45] Hattori, H. Solid base catalysts: fundamentals and their applications in organic reactions. *Appl. Catal. A Gen.,* **2015**, *504*, 103-109.
[http://dx.doi.org/10.1016/j.apcata.2014.10.060]

[46] Fürstner, A. Gold catalysis for heterocyclic chemistry: a representative case study on pyrone natural products. *Angew. Chem. Int. Ed.,* **2018**, *57*(16), 4215-4233.
[http://dx.doi.org/10.1002/anie.201707260] [PMID: 28862364]

[47] Ciufolini, M.A. Synthetic studies on heterocyclic natural products. *Farmaco,* **2005**, *60*(8), 627-641.
[http://dx.doi.org/10.1016/j.farmac.2005.01.007] [PMID: 16095596]

[48] Taylor, A.P.; Robinson, R.P.; Fobian, Y.M.; Blakemore, D.C.; Jones, L.H.; Fadeyi, O. Modern advances in heterocyclic chemistry in drug discovery. *Org. Biomol. Chem.,* **2016**, *14*(28), 6611-6637.
[http://dx.doi.org/10.1039/C6OB00936K] [PMID: 27282396]

[49] Balaban, A.T.; Oniciu, D.C.; Katritzky, A.R. Aromaticity as a cornerstone of heterocyclic chemistry. *Chem. Rev.,* **2004**, *104*(5), 2777-2812.
[http://dx.doi.org/10.1021/cr0306790] [PMID: 15137807]

[50] Zhou, L.; Lokman Hossain, M.; Xiao, T. Synthesis of N-containing heterocyclic compounds using visible-light photoredox catalysis. *Chem. Rec.,* **2016**, *16*(1), 319-334.
[http://dx.doi.org/10.1002/tcr.201500228] [PMID: 26751828]

[51] Asif, M.; Imran, M. *A chapter on synthesis of various heterocyclic compounds by environmentally friendly green chemistry technologies*; , **2021**.
[http://dx.doi.org/10.1016/B978-0-12-822446-5.00004-6]

[52] Sharma, S.; Gangal, S.; Rauf, A. Green chemistry approach to the sustainable advancement to the synthesis of heterocyclic chemistry. *Rasayan J. Chem.,* **2008**, *1*(4), 693-717.

[53] Gupta, S.S.; Kumari, S.; Kumar, I.; Sharma, U. Eco-friendly and sustainable synthetic approaches to biologically significant fused N-heterocycles. *Chem. Heterocycl. Compd.,* **2020**, *56*(4), 433-444.
[http://dx.doi.org/10.1007/s10593-020-02678-5]

[54] Kandepi, V.V.K.M.; Narender, N. Synthesis of N-heterocyclic compounds over zeolite molecular sieve catalysts: an approach towards green chemistry. *Catal. Sci. Technol.,* **2012**, *2*(3), 471-487.

[http://dx.doi.org/10.1039/C2CY00162D]

[55] Daştan, A.; Kulkarni, A.; Török, B. Environmentally benign synthesis of heterocyclic compounds by combined microwave-assisted heterogeneous catalytic approaches †. *Green Chem.,* **2012**, *14*(1), 17-37. [http://dx.doi.org/10.1039/C1GC15837F]

[56] Ison, E.A.; Ison, A. Synthesis of well-defined copper N-heterocyclic carbene complexes and their use as catalysts for a "click reaction": A multistep experiment that emphasizes the role of catalysis in green chemistry. *J. Chem. Educ.,* **2012**, *89*(12), 1575-1577. [http://dx.doi.org/10.1021/ed300243s]

[57] Sunderhaus, J.D.; Martin, S.F. Applications of multicomponent reactions to the synthesis of diverse heterocyclic scaffolds. *Chemistry,* **2009**, *15*(6), 1300-1308. [http://dx.doi.org/10.1002/chem.200802140] [PMID: 19132705]

[58] Cores, Á.; Clerigué, J.; Orocio-Rodríguez, E.; Menéndez, J.C. Multicomponent reactions for the synthesis of active pharmaceutical ingredients. *Pharmaceuticals (Basel),* **2022**, *15*(8), 1009. [http://dx.doi.org/10.3390/ph15081009] [PMID: 36015157]

[59] Afshari, R.; Shaabani, A. Materials functionalization with multicomponent reactions: state of the art. *ACS Comb. Sci.,* **2018**, *20*(9), 499-528. [http://dx.doi.org/10.1021/acscombsci.8b00072] [PMID: 30106275]

[60] Aguirre-Díaz, L.M.; Gándara, F.; Iglesias, M.; Snejko, N.; Gutiérrez-Puebla, E.; Monge, M.Á. Tunable catalytic activity of solid solution metal-organic frameworks in one-pot multicomponent reactions. *J. Am. Chem. Soc.,* **2015**, *137*(19), 6132-6135. [http://dx.doi.org/10.1021/jacs.5b02313] [PMID: 25843316]

[61] Calvo-Flores, F.G.; Mingorance-Sánchez, C. Deep eutectic solvents and multicomponent reactions: Two convergent items to green chemistry strategies. *ChemistryOpen,* **2021**, *10*(8), 815-829. [http://dx.doi.org/10.1002/open.202100137] [PMID: 34402596]

[62] Buskes, M.J.; Coffin, A.; Troast, D.M.; Stein, R.; Blanco, M.J. Accelerating drug discovery: Synthesis of complex chemotypes *via* multicomponent reactions. *ACS Med. Chem. Lett.,* **2023**, *14*(4), 376-385. [http://dx.doi.org/10.1021/acsmedchemlett.3c00012] [PMID: 37077380]

[63] Hulme, C.; Ayaz, M.; Martinez-Ariza, G.; Medda, F.; Shaw, A. *Recent advances in multicomponent reaction chemistry: Applications in small molecule drug discovery.,* **2015**, 145-187. [http://dx.doi.org/10.1002/9781118771723.ch6]

[64] Martins, M.A.P.; Frizzo, C.P.; Moreira, D.N.; Zanatta, N.; Bonacorso, H.G. Ionic liquids in heterocyclic synthesis. *Chem. Rev.,* **2008**, *108*(6), 2015-2050. [http://dx.doi.org/10.1021/cr078399y] [PMID: 18543878]

[65] Wang, Z.; Domling, A. Multicomponent reactions in medicinal chemistry *Multicomponent Reactions towards Heterocycles: Concepts and Applications,* **2022**, 91-137. [http://dx.doi.org/10.1002/9783527832439.ch3]

[66] Müller, T.J.; Orru, R.V.; Chebanov, V.A.; Sakhno, Y.I.; Saraev, V.E.; Muravyova, E.A.; Andrushchenko, A.Y.; Desenko, S.M.; Akhmetova, V.; Khabibullina, G. Multi-component reactions in heterocyclic chemistry. In: *MCR 2009: Proceedings of the 4th International Conference on Multi-Component Reactions and Related Chemistry*; Springer: Ekaterinburg, Russia, **2010**; pp. 31-73.

[67] Ibarra, I.A.; Islas-Jácome, A.; González-Zamora, E. Synthesis of polyheterocycles *via* multicomponent reactions. *Org. Biomol. Chem.,* **2018**, *16*(9), 1402-1418. [http://dx.doi.org/10.1039/C7OB02305G] [PMID: 29238790]

[68] Zhu, J. Recent developments in the isonitrile-based multicomponent synthesis of heterocycles. *Eur. J. Org. Chem.,* **2003**, *2003*(7), 1133-1144. [http://dx.doi.org/10.1002/ejoc.200390167]

[69] Choudhury, L.H.; Parvin, T. Recent advances in the chemistry of imine-based Multicomponent Reactions (MCRs). *Tetrahedron,* **2011**, *67*(43), 8213-8228.

[http://dx.doi.org/10.1016/j.tet.2011.07.020] [PMID: 32287421]

[70] de Graaff, C.; Ruijter, E.; Orru, R.V.A. Recent developments in asymmetric multicomponent reactions. *Chem. Soc. Rev.,* **2012**, *41*(10), 3969-4009.
[http://dx.doi.org/10.1039/c2cs15361k] [PMID: 22546840]

[71] Rocha, R.O.; Rodrigues, M.O.; Neto, B.A.D. Review on the Ugi multicomponent reaction mechanism and the use of fluorescent derivatives as functional chromophores. *ACS Omega,* **2020**, *5*(2), 972-979.
[http://dx.doi.org/10.1021/acsomega.9b03684] [PMID: 31984252]

[72] Liu, Z.Q. Ugi and Passerini reactions as successful models for investigating multicomponent reactions. *Curr. Org. Chem.,* **2014**, *18*(6), 719-739.
[http://dx.doi.org/10.2174/13852728196661402010027l7]

[73] Ugi, I.; Heck, S. The multicomponent reactions and their libraries for natural and preparative chemistry. *Comb. Chem. High Throughput Screen.,* **1970**, *4*(1), 1-34.
[http://dx.doi.org/10.2174/1386207013331291] [PMID: 11281825]

[74] Sehlinger, A.; Meier, M. A. Passerini and Ugi multicomponent reactions in polymer science *Multi-component and sequential reactions in polymer synthesis,* **2015**, 61-86.

[75] Wan, J.P.; Liu, Y. Synthesis of dihydropyrimidinones and thiones by multicomponent reactions: strategies beyond the classical Biginelli reaction. *Synthesis,* **2010**, *2010*(23), 3943-3953.
[http://dx.doi.org/10.1055/s-0030-1258290]

[76] Funicello, M.; Cerminara, I.; Chiummiento, L.; Lupattelli, P.; Felluga, F.; Berti, F. Biginelli reaction and β-secretase inhibition: a multicomponent reaction as a friendly educational approach to bioactive compounds. *J. Chem. Educ.,* **2021**, *98*(5), 1756-1761.
[http://dx.doi.org/10.1021/acs.jchemed.0c01298]

[77] Wu, H.; Fu, C.; Zhao, Y.; Yang, B.; Wei, Y.; Wang, Z.; Tao, L. Multicomponent copolycondensates *via* the simultaneous hantzsch and biginelli reactions. *ACS Macro Lett.,* **2015**, *4*(11), 1189-1193.
[http://dx.doi.org/10.1021/acsmacrolett.5b00637] [PMID: 35614835]

[78] Ruijter, E.; Orru, R. Multicomponent reactions in drug discovery and medicinal chemistry. *Drug Discov. Today. Technol.,* **2018**, *29*, 1-2.
[http://dx.doi.org/10.1016/j.ddtec.2018.11.002] [PMID: 30471667]

[79] Saquib, M.; Khan, M.F.; Ansari, M.I.; Khan, I.; Hussain, M.K.; Singh, J. Multicomponent Reactions for Generation of Molecular Libraries in Anticancer Drug Discovery.*Multicomponent Reactions*; CRC Press, **2017**, pp. 117-138.
[http://dx.doi.org/10.1201/9781315369754-5]

[80] Javanbakht, S.; Shaabani, A. Multicomponent reactions-based modified/functionalized materials in the biomedical platforms. *ACS Appl. Bio Mater.,* **2020**, *3*(1), 156-174.
[http://dx.doi.org/10.1021/acsabm.9b00799] [PMID: 35019432]

[81] Singh, M.S.; Chowdhury, S. Recent developments in solvent-free multicomponent reactions: a perfect synergy for eco-compatible organic synthesis. *RSC Advances,* **2012**, *2*(11), 4547-4592.
[http://dx.doi.org/10.1039/c2ra01056a]

[82] Burange, A.S.; Gopinath, C.S. Catalytic applications of hydrotalcite and related materials in multi - component reactions: Concepts, challenges and future scope. *Sustain. Chem. Pharm.,* **2021**, *22*, 100458.
[http://dx.doi.org/10.1016/j.scp.2021.100458]

[83] Bhaskaruni, S.V.H.S.; Maddila, S.; Gangu, K.K.; Jonnalagadda, S.B. A review on multi-component green synthesis of N-containing heterocycles using mixed oxides as heterogeneous catalysts. *Arab. J. Chem.,* **2020**, *13*(1), 1142-1178.
[http://dx.doi.org/10.1016/j.arabjc.2017.09.016]

[84] Lima, C.G.S.; Moreira, N.M.; Paixão, M.W.; Corrêa, A.G. Heterogenous green catalysis: Application of zeolites on multicomponent reactions. *Curr. Opin. Green Sustain. Chem.,* **2019**, *15*, 7-12.

[http://dx.doi.org/10.1016/j.cogsc.2018.07.006]

[85] Pramanik, M.; Bhaumik, A. Phosphonic acid functionalized ordered mesoporous material: a new and ecofriendly catalyst for one-pot multicomponent Biginelli reaction under solvent-free conditions. *ACS Appl. Mater. Interfaces,* **2014**, *6*(2), 933-941.
[http://dx.doi.org/10.1021/am404298a] [PMID: 24372168]

[86] D'Oca, C.R.; Naciuk, F.F.; Silva, J.C.; Guedes, E.P.; Moro, C.C.; D'Oca, M.G.; Santos, L.S.; Natchigall, F.M.; Russowsky, D. New multicomponent reaction for the direct synthesis of β-aryl-γ-nitroesters promoted by hydrotalcite-derived mixed oxides as heterogeneous catalyst. *J. Braz. Chem. Soc.,* **2017**, *28*, 285-298.

[87] Mokhtar, M.; Saleh, T.S.; Ahmed, N.S.; Al-Thabaiti, S.A.; Al-Shareef, R.A. An eco-friendly N-sulfonylation of amines using stable and reusable Zn–Al–hydrotalcite solid base catalyst under ultrasound irradiation. *Ultrason. Sonochem.,* **2011**, *18*(1), 172-176.
[http://dx.doi.org/10.1016/j.ultsonch.2010.05.001] [PMID: 20627793]

[88] Chelladurai, K.; Rajamanickam, M. Environmentally benign neem biodiesel synthesis using Nano-Z--Mg-Al hydrotalcite as solid base catalysts *Journal of Catalysts,* **2014**.
[http://dx.doi.org/10.1155/2014/326575]

[89] Ansari, A.; Ali, A.; Asif, M.; Shamsuzzaman, S. Microwave-assisted MgO NP catalyzed one-pot multicomponent synthesis of polysubstituted steroidal pyridines. *New J. Chem.,* **2018**, *42*(1), 184-197.
[http://dx.doi.org/10.1039/C7NJ03742B]

[90] Antonyraj, C.A.; Kannan, S. Hantzsch pyridine synthesis using hydrotalcites or hydrotalcite-like materials as solid base catalysts. *Appl. Catal. A Gen.,* **2008**, *338*(1-2), 121-129.
[http://dx.doi.org/10.1016/j.apcata.2007.12.028]

[91] Chandralekha, E.; Thangamani, A.; Valliappan, R. Efficient atom-economical solvent-free one-pot multicomponent synthesis of 2-amino-4H-benzo[b]pyrans catalyzed by solid base SiO_2-OK as a reusable catalyst. *Res. Chem. Intermed.,* **2015**, *41*(4), 1951-1966.
[http://dx.doi.org/10.1007/s11164-013-1322-x]

[92] Eskandari, K.; Karami, B.; Pourshojaei, Y.; Asadipour, A. An eco-compatible, three-component synthesis of acyl-substituted bis(pyrazolyl)methanes by employing recyclable silica sodium carbonate as an efficient and environmentally benign catalyst in aqueous medium. *Monatsh. Chem.,* **2018**, *149*(6), 1075-1081.
[http://dx.doi.org/10.1007/s00706-017-2106-6]

[93] Gutierrez, L.F.; Nope, E.; Rojas, H.A.; Cubillos, J.A.; Sathicq, Á.G.; Romanelli, G.P.; Martínez, J.J. New application of decaniobate salt as basic solid in the synthesis of 4H-pyrans by microwave assisted multicomponent reactions. *Res. Chem. Intermed.,* **2018**, *44*(9), 5559-5568.
[http://dx.doi.org/10.1007/s11164-018-3440-y]

[94] Lal, J.; Sharma, M.; Gupta, S.; Parashar, P.; Sahu, P.; Agarwal, D.D. Hydrotalcite: A novel and reusable solid catalyst for one-pot synthesis of 3,4-dihydropyrimidinones and mechanistic study under solvent free conditions. *J. Mol. Catal. Chem.,* **2012**, *352*, 31-37.
[http://dx.doi.org/10.1016/j.molcata.2011.09.009]

[95] Sheik Mansoor, S.; Syed Shafi, S.; Zaheer Ahmed, S. An efficient one-pot multicomponent synthesis of 3,4-dihydropyrimidine-2-(1H)-ones/thiones/imines *via* a Lewis base catalyzed Biginelli-type reaction under solvent-free conditions. *Arab. J. Chem.,* **2016**, *9*, S846-S851.
[http://dx.doi.org/10.1016/j.arabjc.2011.09.018]

[96] Safaei-Ghomi, J.; Ghasemzadeh, M.; Mehrabi, M. Calcium oxide nanoparticles catalyzed one-step multicomponent synthesis of highly substituted pyridines in aqueous ethanol media. *Sci. Iran.,* **2013**, *20*(3), 549-554.

[97] Rajesh, U.C.; Manohar, S.; Rawat, D.S. Hydromagnesite as an Efficient Recyclable Heterogeneous Solid Base Catalyst for the Synthesis of Flavanones, Flavonols and 1,4-Dihydropyridines in Water. *Adv. Synth. Catal.,* **2013**, *355*(16), 3170-3178.

[http://dx.doi.org/10.1002/adsc.201300555]

[98] Safari, E.; Hasaninejad, A. One-pot, multi-component synthesis of novel bis-spiro pyranopyrazole derivatives in the presence of DABCO as an efficient and reusable solid base catalyst. *ChemistrySelect,* **2018**, *3*(12), 3529-3533. [http://dx.doi.org/10.1002/slct.201800410]

[99] Sheibani, H.; Esfandiarpoor, Z. Multicomponent synthesis of pyridazine and pyridazinoquinazoline derivatives in the presence of catalysts such as magnesium oxide (MgO) and 12-tungstophosphoric acid (PW). *J. Heterocycl. Chem.,* **2011**, *48*(5), 1122-1125. [http://dx.doi.org/10.1002/jhet.355]

[100] Toorbaf, M.; Moradi, L. Preparation of GO/SiO_2 /PEA as a new solid base catalyst for the green synthesis of some spirooxindole derivatives. *RSC Advances,* **2021**, *11*(35), 21840-21850. [http://dx.doi.org/10.1039/D1RA02850B] [PMID: 35478825]

[101] Kale, S.R.; Surve, S.K. One-pot multicomponent synthesis of highly substituted pyridines using hydrotalcite as a solid base and reusable catalyst. *Current Chemistry Letters,* **2021**, *10*(3), 169-174. [http://dx.doi.org/10.5267/j.ccl.2021.1.001]

[102] Tanuraghaj, H.M.; Farahi, M. A novel sodium carbonate-catalyzed regioselective synthesis of pyrano[2,3-h]coumarins using a three-component reaction. *Tetrahedron Lett.,* **2019**, *60*(7), 557-559. [http://dx.doi.org/10.1016/j.tetlet.2019.01.030]

[103] Shabalala, S.; Maddila, S.; van Zyl, W.E.; Jonnalagadda, S.B. A facile, efficacious and reusable Sm2O3/ZrO2 catalyst for the novel synthesis of functionalized 1,4-dihydropyridine derivatives. *Catal. Commun.,* **2016**, *79*, 21-25. [http://dx.doi.org/10.1016/j.catcom.2016.02.017]

[104] Gandhi, D.; Agarwal, S. MgO NPs Catalyzed Eco-friendly Reaction: A Highly Effective and Green Approach for the Multicomponent One-pot Synthesis of Polysubstituted Pyridines using 2-Aminobenzothiazole. *J. Heterocycl. Chem.,* **2018**, *55*(12), 2977-2984. [http://dx.doi.org/10.1002/jhet.3384]

[105] Hadebe, N.P.; Kerru, N.; Tukulula, M.; Jonnalagadda, S.B. A sustainable molybdenum oxide loaded on zirconia (MoO_3/ZrO_2) catalysed multicomponent reaction to synthesise novel dihydropyridines. *Sustain. Chem. Pharm.,* **2022**, *25*, 100578. [http://dx.doi.org/10.1016/j.scp.2021.100578]

CHAPTER 6

Historical and Recent Advances in Solid Base Catalysts for Drug Synthesis

Urvashi Garg[1], **Abubakar Suraja Ibrahim**[1], **Noopur Srivastava**[1,*] and **Nisha Saxena**[2]

[1] *Department of Chemistry & Biochemistry, Sharda School of Engineering & Science, Sharda University, Greater Noida, UP-201310, India*

[2] *PG Department of Chemistry, M.R.M. College (Lalit Narayan Mithila University), Darbhanga, Bihar-846004, India*

Abstract: Pines and Haag were the first to demonstrate the use of heterogeneous base catalysts in the isomerization of alkenes using sodium metal dispersed on alumina. Following this, a number of work has been performed on solid bases as catalyst materials. Metal oxides, zeolites, supported alkali metal complexes, clay minerals, waste solid base catalysts, mesoporous solid base catalysts, *etc*., are examples of solid base catalysts. Precipitation, co-precipitation, sol-gel, hydrothermal, impregnation, vapor phase deposition, and sonochemical techniques can all be used to create them. The majority of materials referred to as solid bases exhibit catalytic activity when water and carbon dioxide are removed from their surfaces. The degree of pre-treatment conditions influences the surface basic sites' characteristics. In addition to eliminating carbon dioxide and water, pre-treatment involves the rearrangement of bulk and surface atoms, which modifies the kind and quantity of basic sites as the pre-treatment temperature rises. Double bond isomerization, the addition of anion and proton to various double bonds and alcohol decomposition, hydrogenation, amination, dehydrocyclodimerization, aldol addition, nitroaldol reaction, michael addition, conjugate addition of alcohol, cyanoethylation, and Tischenko reaction are among the reactions that solid base catalysts can catalyze. Solid base catalysts play a crucial function in industrial processes due to their varied application.

Keywords: Alcohol decomposition, Catalytic activities, Conjugate addition of alcohol, Cyanoethylation, Double bond isomerization, Heterogeneous base catalyst, Isomerization of alkenes, Sodium metal dispersed on alumina, Solid base catalysts, Tischenko reaction.

* **Corresponding author Noopur Srivastava:** Department of Chemistry & Biochemistry, Sharda School of Engineering & Science, Sharda University, Greater Noida, UP-201310, India; E-mail: noopursriv@gmail.com

Ravi Tomar, Vivek Mishra, Kapil Arya & Garima Pandey (Eds.)

INTRODUCTION

The development of efficient and sustainable catalysts is a crucial aspect of modern organic synthesis for various industrial processes, including the pharmaceutical industry [1]. Traditional acids and bases catalysts pose significant environmental issues due to their toxicity, waste generation, and high energy consumption. As a result, there is a growing need for eco-friendly and sustainable chemical processes, driving researchers to seek out effective and environmentally friendly replacement of conventional acids and bases in reactions [2]. Reducing environmental impact and boosting sustainability in organic processes requires substituting solid catalysts for liquid acids and bases [3]. Solid base catalysts have emerged as a promising class of materials, offering advantages such as recyclability, reusability, and reduced environmental impact [2]. This highlights the significance of catalysts in enabling efficient and sustainable production of various essential products, including medicines, that are vital to our well-being and daily lives [4]. The most notable heterogeneous basic catalysts are typically derived from three main categories: alkaline earth metals, transition metals, and metal oxides. These classes of materials have shown significant potential in facilitating various chemical reactions as effective heterogeneous basic catalysts [5]. The ability of solid base catalysts to effectively catalyze the isomerization of unsaturated molecules containing heteroatoms, such as N and O, gives them an edge over solid acid catalysts. Solid base catalysts are a useful option for these particular reactions since they are less prone to deactivation (or poisoning) when exposed to heteroatom-containing reactants, compared to solid acid catalysts [6].

Catalysts

A catalyst is a material that, depending on the reaction, can speed up or slow down a reaction's pace, while staying in the same composition once the reaction is complete. It modifies the reaction's activation energy without changing the equilibrium. Catalysts are effective for reactions that are thermodynamically feasible, such as those with a negative change in Gibb's free energy under the given reaction conditions. When selecting a catalyst for a reaction, the focus often lies in its ability to influence, rather than enhance, the reaction rate. These principles apply universally across various types of catalytic reactions, whether they are homogeneous, heterogeneous, or enzymatic. Catalysts are commonly classified as either solid acid or solid base catalysts [7].

Solid Base Catalyst

A solid base catalyst is a type of catalyst that operates by donating electrons and displaying basic properties due to the presence of exposed oxygen atoms on its surface. These oxygen atoms interact attractively with Hydrogen ions (H^+), enabling the abstraction of H^+ from strongly acidic molecules [8].

Solid Base Catalysts Preparation Methods

Precipitation and Co-precipitation

This method involves separating the desired substance from a uniform mixture by inducing precipitation. The precipitation process is initiated by manipulating temperature conditions and introducing acidic or alkaline solutions, causing the substance of interest to precipitate out of the solution, allowing for its isolation and collection. Various factors that influence the formation of the precipitate include temperature, pH levels, the composition of raw materials, type of solvent used, *etc.* As a result, the transition from a liquid solution to a solid state unfolds in four distinct stages [9].

Sol-Gel Method

The first step in the sol-gel process is the formation of a stable colloidal solution (sol) that has very pure finely distributed particles. The sol is formed by hydrolysis and partial condensation of the precursor, and then it passes through polycondensation to produce a gel. Evaporation is used to remove extra liquid from the gel, producing a dry product called a xerogel. After that, this xerogel is heated to enhance its qualities and provide the finished product with the necessary features [10].

Hydrothermal Method

In a hydrothermal process, a homogeneous aqueous solution, including alkali and precursor chemicals, is heated to a high temperature in a sealed autoclave for a predetermined amount of time. This controlled environment facilitates chemical reactions and material formation, allowing for the synthesis of unique materials with specific properties. The materials dissolve and then reform into crystals under the influence of the elevated temperature and pressure. The resulting solid material is recovered through filtration, which is then rinsed and subsequently dried [10].

Impregnation Method

The impregnation technique involves applying aqueous precursors to a support material, typically an oxide support. The process entails:

1. Contacting the support with a solution containing precursors of the active catalyst elements.

2. Allowing the support to absorb the precursor solution for a specified time.
3. Drying to remove excess liquid from the support.
4. Calcining to activate the catalyst.
5. Reducing the catalyst to its active form [1].

Vapor Phase Deposition Method

Thin film coating involves depositing atoms or molecules from a vapor phase onto a substrate. The substrate is placed in a chamber where vaporized material is applied to its surface. Chemical Vapor Deposition (CVD) and Physical Vapor Deposition (PVD) are the two main techniques. The choice between CVD and PVD depends on whether the coating process is driven by chemical reactions or physical interactions [1].

Early Development of Solid Based Catalysts

Pines and Haag were the pioneers in the solid base catalysts research, they demonstrated that sodium metal deposited on alumina could effectively catalyze the migration of alkenes double bond. This groundbreaking work laid the foundation for further investigations, and the study of solid base catalysts gained significant momentum in the 1970s, leading to a surge in research and development in this field [11]. Researchers have discovered that certain single-component metal oxides can exhibit solid base catalytic properties without the need for alkali metals like Na and K. Previous studies have extensively explored the synthesis and characterization of solid basic catalysts, investigating factors such as composition, size of the particle, surface area, and morphology. The progress made in catalysis and material chemistry has led to a surge in research focused on designing and synthesizing solid base catalysts, resulting in the creation of a wide spectrum of new catalysts. These innovations span from simple metal hydroxides and oxides to complex multicomponent metal hydroxides and oxides, showcasing the rapid advancement in this field. The development of these solid base catalysts has the potential to revolutionize various industries and enable more efficient and sustainable chemical processes. These catalysts have diverse purposes in various processes of synthesis of fine chemicals, driving continued investigation into their potential, Fig. (**1**) [12].

Recent Development in Solid Base Catalysts

Solid heterogeneous catalysts have garnered significant interest in various industries due to their unique characteristics, including high specific surface area, easy separability, same pore size distribution, and very strong alkalinity. These properties enable solid base catalysts to facilitate a wide range of processes in industrial settings, like condensation, dehydration, amination, hydrogenation, alkylation, esterification, polymerization, and isomerization. The solid base catalysts' important performance metrics include:

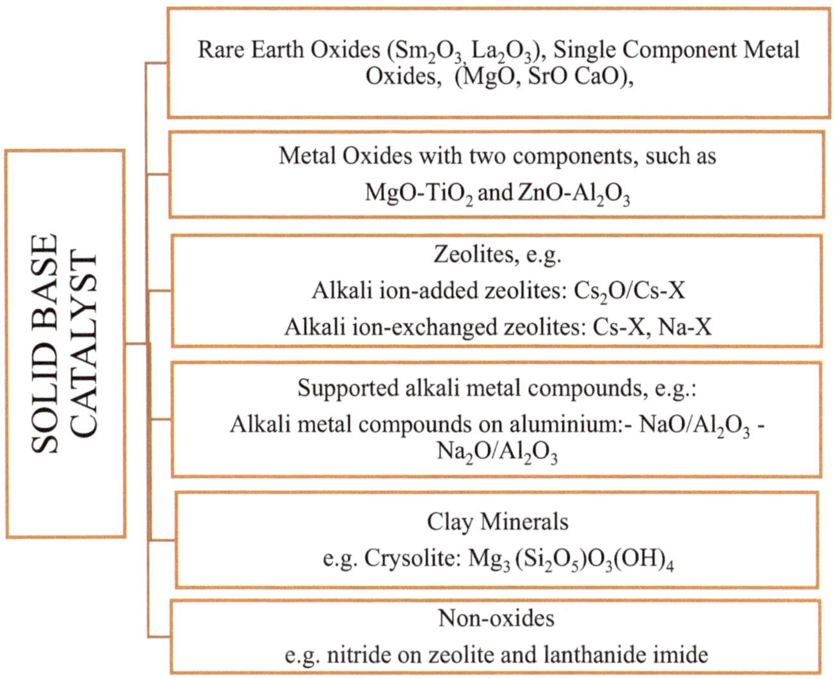

Fig. (1). Solid Base Catalysts [1].

- Activity: The capacity to transform reactants into products, while making effective use ofenergy and processing time.
- Selectivity: The capacity to minimize the production of byproducts, while selectively accelerating the conversion of reactants into particular desired products.
- Stability: The ability to maintain selectivity and activity over some time without deactivating [1].

Characterstics of Solid Base Catalysts

Notable solid base catalysts' characteristics include:

a. Easy product separation and catalyst reuse.
b. Environmental friendliness.
c. Possession, as opposed to liquid ones, of both basic and acidic sites on the surface.
d. High productivity and effectiveness in flow reactor systems.

e. Reaction mixture neutralization, leading to reduction in waste generation.
f. Ability to perform high-temperature reactions without solvent evaporation concerns.
g. Elimination of the need for base-tolerant materials in reactor systems.
h. Unique capabilities for novel reactions. These advantages make solid base catalysts an attractive option for various industrial applications, as shown in Fig. (**2**) [13].

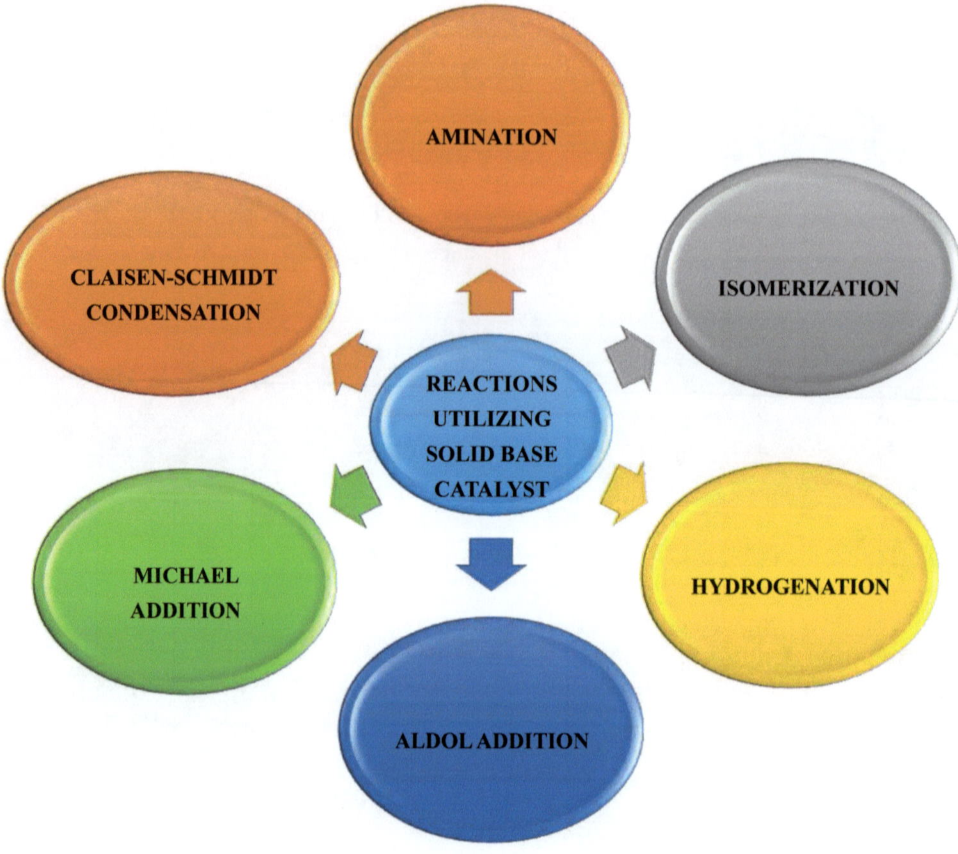

Fig. (2). Some reactions utilizing solid base catalysts [14].

Hydrogenation

Hydrogenation, a versatile chemical process, is extensively utilized across various industries, including food, petrochemical, and pharmaceutical sectors. In the food industry, hydrogenation is used to produce margarine, among various other products. In the petrochemical and pharmaceutical industries, it plays a crucial role in the manufacture of various chemicals and medicines. The most prevalent industrial application of hydrogenation is the reduction of carbon-carbon double bonds, a reaction that saturates unsaturated compounds with hydrogen, leading to the formation of valuable chemicals and materials [15].

Aldol Addition

The aldol addition is a valuable and efficient method for synthesizing β-hydroxy carbonyl and 1, 3-diol of carbonyl compounds, like aldehyde and ketone or two aldehydes. In the past five years, significant advancements have been made in the development of enantioselective aldol reactions. The progress made in this field encompasses the utilization of preformed and stereospecific stable enolates, such as metal enolates (for example, borane, silyl, titanium, and tin enolates), alongside the innovation of "*in situ* formed labile enolate synthons," which entail a reversible amalgamation of two entities, like keto-enamine or keto-enolate. These developments have facilitated the enhancement of both the efficacy and selectivity of aldol reactions [16].

Michael Addition

This basic organic chemistry reaction adds a nucleophile to α, β-unsaturated carbonyl molecules, resulting in the formation of a carbon-carbon single bond. This reaction is driven by base catalysts, and has several industrial applications, including the synthesis of monomers, coatings, corrosion inhibitors, and medicines. The Michael addition is a versatile method for synthesizing a variety of compounds, making it a crucial process in multiple industries [3].

SOLID BASE CATALYSTS IN DRUG SYNTHESIS

Metal Oxides

Solid base catalysts offer practical advantages, such as easy separation from reaction mixtures, straightforward regeneration, and non-corrosiveness to reactor systems. They consist of various metal oxides, including rare earth oxides, alkaline earth oxides, and other metal oxides. Surface reactions and geometric factors influence their catalytic behavior, and they are known to catalyze reactions typically associated with bases in homogeneous systems. The basic nature of these catalysts is indicated by the color changes of adsorbed acid-base indicators, poisoning effects by acidic molecules, similar activity to homogeneous base catalysts, and evidence of anionic intermediates in the reactions they facilitate [8].

Examples

Magnesium Oxide (MgO) shows basic characteristics on its surface. It is believed that reactions involving MgO start with the basic center of the catalyst, abstracting an allylic proton. This leads to the formation of carbanion species, which promotes the formation of an endo-cyclic double bond (Fig. **3**) [17].

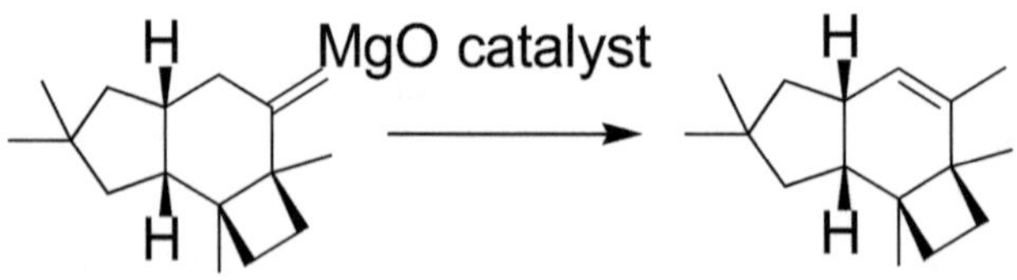

1
Protoilludene

1
Protoilludene

Fig. (3). Reaction catalyzed by solid basic catalyst [17].

Protoilludene (2), a significant sesquiterpene, functions as a starting material for numerous antimicrobial chemicals and medicinal compounds [18].

Illudadiene (Fig. **4**) and its derivatives are of significant interest in drug synthesis due to their unique chemical structures and potential biological activities, including anticancer properties, antimicrobial activity, potential for chemical modifications, use as synthetic intermediates, insights from mechanistic studies, and development as lead compounds, highlighting their potential in creating novel pharmaceuticals [18 - 21].

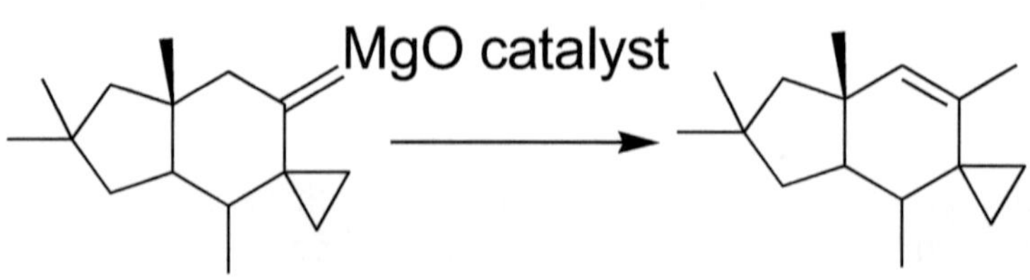

3
Illudadiene

4
Illudadiene

Fig. (4). Reaction Catalyzed by Solid Basic Catalyst [18 - 21]..

Ruiz *et al*. (2018) evaluated the pharmacological activities of 2-carene (6) (Fig. **5**), highlighting its anti-inflammatory and antimicrobial effects. The study examined its potential as a new therapeutic agent, particularly in pharmaceutical formulations aimed at treating inflammatory conditions, and combating microbial infections. The research underscores 2-carene's (6) promising role in drug development, suggesting its application in formulations targeting these health challenges [22].

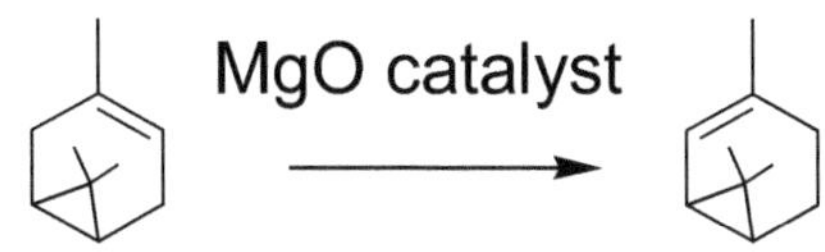

5
3-Carene

6
2-Carene

Fig. (5). Conversion of 3-Carene to 2-Carene Using MgO [22]..

Zeolites

Because of their well-defined structure and ion exchange capabilities, zeolites are used as solid base catalysts, facilitating effective catalysis in a variety of organic synthesis applications. Their basic strength, influenced by the exchangeable alkali metal cations (Li < Na < K < Rb < Cs), dictates their catalytic performance, with CsX zeolites exhibiting the highest basicity. This property makes them particularly effective in reactions, such as Knoevenagel condensation and selective methylation using Dimethyl Carbonate (DMC), where they demonstrate high selectivity towards specific functional group transformations. Zeolites offer environmental and economic advantages over liquid bases by reducing corrosion and environmental impact, and facilitating easier catalyst recovery and reuse. Their thermal stability and selectivity in complex reactions underscore their utility as versatile solid base catalysts in industrial processes [23].

Examples

A critical application of base catalysis using zeolites involves the synthesis of 4-methylthiazole (7), which is utilized as a systemic fungicide [11]. For instance, cesium-loaded ZSM-5 zeolite demonstrated outstanding performance characteristics, including activity, selectivity, and longevity. The reaction was conducted in the vapor phase at a temperature of 427°C in the presence of water vapor, utilizing zeolites known for their high hydrothermal stability (Fig. **6**) [23].

N,N'-Dimethylaniline (8) plays a crucial role in the manufacturing of penicillins and cephalosporins, by scavenging acidic impurities. However, it has been detected as a contaminant in commercial preparations of these antibiotics, with levels as high as 1500 parts per million (ppm), highlighting the need for improved purification processes to minimize its presence in the final products (Fig. **7**) [24 - 27].

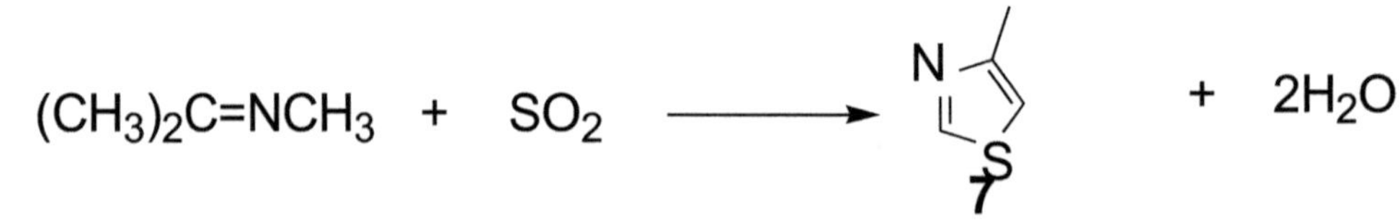

4-methyl-1,3-thiazole

Fig. (6). Example of reaction catalyzed by Zeolite [23]..

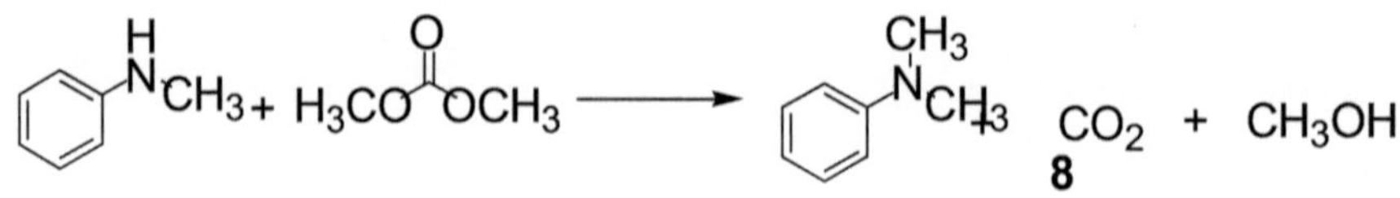

N-methyl aniline dimethyl carbonate ***N.N'*-dimethyl aniline**

Fig. (7). Synthesis of *N, N'*-Dimethylaniline [24 - 27]..

Supported alkali metal compounds

Potassium-based compounds like KNO_3, $KHCO_3$, and K_2CO_3 on alumina show exceptional catalytic activity for isomerizing 1-butene at 273 K and 2,3-dimethylbut-1-ene. Despite partially decomposing upon heating, the source of their basic sites is still unknown. On the other hand, alumina-containing alkali metal hydroxides, such as KOH/Al_2O_3, $RbOH/Al_2O_3$ and $CsOH/Al_2O_3$, exhibit remarkable catalytic activity for 2,3-dimethyl but-1-ene isomerization even at 201 K after high-temperature pretreatment. These catalysts are prized for their easy preparation and strong basic properties, making them desirable for various applications [28 - 31].

Examples

A solid base catalyst was developed by impregnating potassium onto mixed oxides (K_2O/Al_2O_3-CaO), enabling the efficient one-step synthesis of various derivatives, such as pyrano [2,3-d]pyrimidine, benzylidene malononitrile derivatives, and benzylidene barbituric acid derivatives. The K_2O/Al_2O_3-CaO catalyst notably facilitated the Knoevenagel condensation reaction, producing benzylidene malononitrile and benzylidene barbituric derivatives with high effectiveness. This catalyst has the potential to be used in a variety of organic synthesis processes due to its high performance in stimulating these crucial reactions [32].

Pyrano [2,3-d]pyrimidine derivatives (Fig. **8**) are known for their extensive biological and pharmacological properties, including:

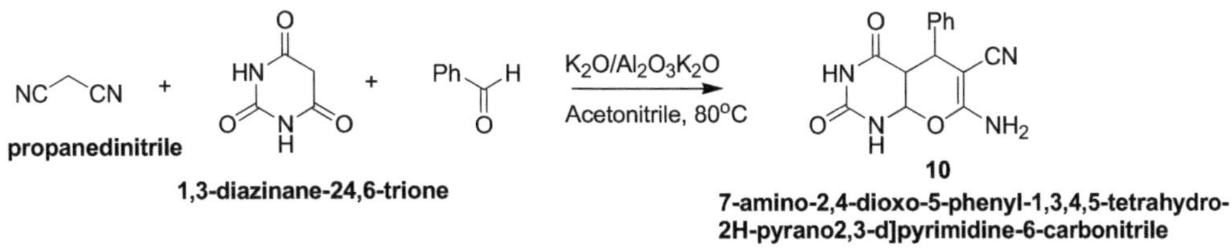

Fig. (8). Example of Reaction Catalyzed by supported alkali metal compounds [33]..

- Antitumor effects.
- Antibacterial and antimicrobial activities.
- Antioxidant properties.
- Antifungal effects.
- Vasodilatory actions.
- Anti-inflammatory properties.
- Antidiabetic effects.
- Cardiotonic activities.

Because of these unique characteristics, pyrano [2, 3-d]pyrimidine derivatives (Fig. **8**) are attractive prospects for the creation of novel medications and treatments for a range of illnesses [33].

Clay Minerals

Clay minerals, particularly montmorillonite and Hydrotalcite (HT) are versatile catalysts with various applications in chemical reactions. Montmorillonite, like K-30, serves as a base catalyst in transesterification reactions for biodiesel production from waste cooking oil, demonstrating high efficiency and reusability. On the other hand, HT, or hydrotalcite, exhibits anionic properties, and is utilized for a wide range of reactions, including dehydrogenation, polymerization, transesterification, and more complex reactions such as epoxidation and oxidation of aromatic compounds. Its structural flexibility allows for modifications by replacing Mg and Al ions with other metals, enhancing its catalytic performance and selectivity. Supported HT catalysts, prepared *via* methods like anion exchange or impregnation, offer improved stability and efficiency in various chemical processes. These characteristics make them valuable in sustainable and eco-friendly catalytic applications [34].

Example

Acid-activated montmorillonite, a type of clay, is used as a template to synthesize Gold Nanoparticles (AuNPs) with unique properties. The AuNPs, with diameters less than 8 nm, are hosted within the nanopores of the clay and exhibit a high surface area (327-579 m^2/g) and large pore volumes (~0.7 cm^3/g). Oriented in the (111) plane, these single-crystal AuNPs exhibit superior catalytic activity for A3 coupling reactions, yielding high-yielding (82-94%) and completely selective propargyl amines under moderate conditions (Fig. **9**). The nanocatalysts are a viable and sustainable solution for a range of applications, since they may be reused several times without experiencing a noticeable loss of activity [35].

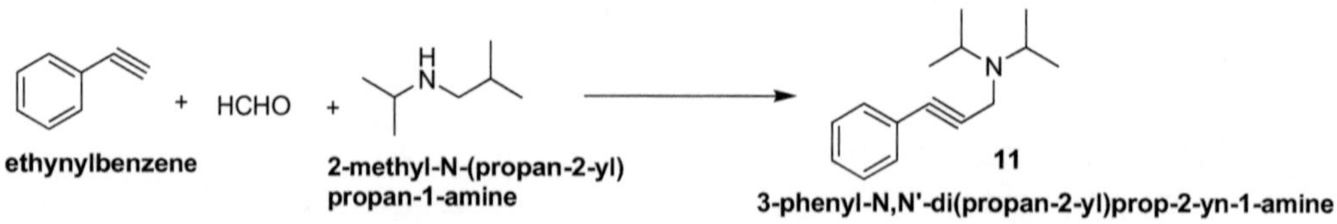

Fig. (9). Example of reaction catalyzed by clay minerals basic solid catalyst [31].

Propargyl amines serve as crucial intermediates in the synthesis of biologically active compounds, such as 1, 4-oxazepane and 1, 4-diazepane. 1, 4-diazepane, in particular, finds application as an antidepressant, antiplatelet agent, anticonvulsant, and various other therapeutic uses due to its pharmacological properties [36].

Mesoporous solid base catalysts

Mesoporous solid bases are highly valued in sustainable catalysis due to their efficient mass transfer, minimal corrosion, and ease of separation. Significant strides have been made over the past decade, expanding their applications in organic synthesis, fine chemicals, energy, and environmental catalysis. This review outlines recent advancements categorized by support type, featuring exemplary cases and cooperative catalysts incorporating acid and metal functionalities. It emphasizes the design and fabrication principles of mesostructured basic materials, and delves into the mechanisms behind basic site formation in diverse mesoporous systems [38].

Examples

A novel solid base catalyst, Guanine-Functionalized Mesoporous Silica (SBA-16-G), has been successfully synthesized. This catalyst has various benefits, such as the efficient promotion of the one-pot, three-component synthesis of physiologically active pyran-annulated heterocyclic compounds:

- Metal-free composition.
- Easy synthesis and isolation.
- Recyclability (four cycles without any activity loss) [39].

The class of compounds, known as 3, 4-dihydropyrano [3, 2-c]chromenes (Fig. **10**) (12) and their derivatives, are highly sought after due to their diverse and valuable biological properties, which include:

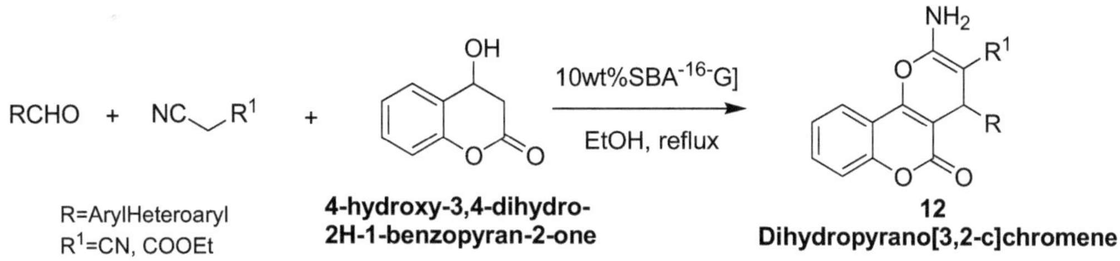

Fig. (10). Example of reaction catalyzed by mesoporous solid base catalysts [40].

- Diuretic effects.
- Emetic properties.
- Anti-HIV activity.
- Anti-tumor properties.
- Anti-Alzheimer's activity.
- Anti-bacterial effects.
- Anti-malarial properties [40].

Eco-Friendly Production of Novel Chemical

Isomerization reactions can convert eugenol (13) into valuable compounds like isoeugenol (14,15), anethole, and isosafrole (Fig. **11**). Isoeugenol, in particular, is used in pharmaceuticals and fragrance formulations for creating floral scents. Traditionally, this isomerization process employs KOH in alcoholic solutions as a homogeneous catalyst. However, research by Kishore and Kannan has demonstrated that eugenol can be isomerized to isoeugenol using hydrotalcite as a catalyst, offering a potentially more efficient and sustainable method [41].

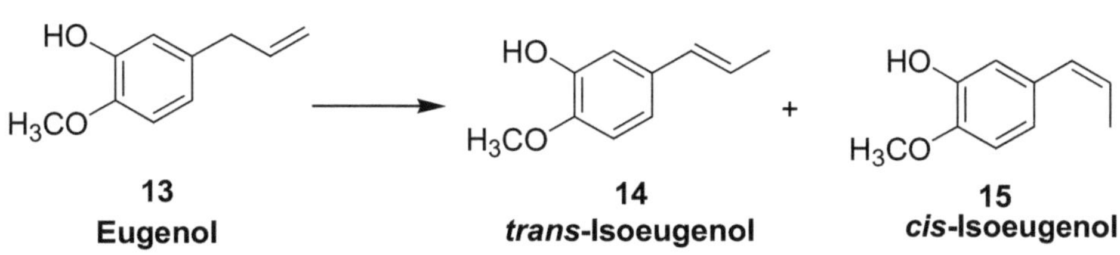

Fig. (11). Example of isomerization reaction using solid based catalyst [41].

Among various hydrotalcite-like materials, combinations of MgAl and NiAl have shown high activity. Notably, a ternary hydrotalcite made of Mg, Ni, and Al achieved an impressive 94% conversion at 473 K, with a *cis/trans* ratio of 16/84, showcasing exceptional catalytic performance. In contrast, calcined hydrotalcite exhibited lower activity, underscoring the importance of the specific composition and structure of the hydrotalcite material for effective catalysis [42, 43].

Synthesis Approach for Pseudoionones

Ionone and methylionones, which are essential intermediates in the synthesis of flavors and fragrances, are produced through the acid-catalyzed ring closure of pseudo ionone and methylpseudoionone. Pseudoionone (16) is produced *via* aldol condensation of citral and acetone, and it also acts as a precursor for the synthesis of vitamin A. On the other hand, methylpseudoionones are the product of the interaction between citral and methyl ethyl ketone. Specifically, citral's aldehyde group combines with the methyl group to make n-methylpseudoionone (17), while the methylene group reacts to form isopseudoionone (18).

Research has focused on the condensation of citral and acetone to synthesize pseudoionone, highlighting important optimization parameters, including the reaction temperature and acetone-to-citral molar ratio. Significant yields of pseudoionone have been achieved using basic alumina as a catalyst under reflux conditions for 4 hours [44]. Additionally, Sodium-doped Magnesium Oxide (Na^+-doped MgO) has been identified for its catalytic activity in pseudoionone synthesis, achieving optimal conditions with a 100% yield (Fig. **12**). These parameters include a temperature of 323 K (50°C), an acetone-to-citral molar ratio of 25, and a catalyst-to-citral weight ratio of 0.4 [45].

Various investigations have shown that optimized MgO-Al_2O_3 mixed oxides, which are produced by calcining hydrotalcite, have improved catalytic performance. For example, a 60% yield and 60-80% selectivity were obtained with an acetone/citral molar ratio of 1 and 5 weight percent catalyst loading, calcination at 723 K, aging at 333 K, and Mg/Al molar ratio of 4 [46].

Rehydrated hydrotalcite catalysts were investigated for synthesizing methyl pseudoionones, achieving 65% citral conversion and 90% selectivity at 273 K, with notable improvement to 76% yield at 293 K [47 - 49].

Climent *et al.* reported that MgO and calcined hydrotalcite are effective catalysts for pseudoionone synthesis, with rehydrated calcined hydrotalcite showing exceptional performance—98% yield and 99% selectivity—at a molar ratio of 2.4, 333 K, and 1-hour reaction time using 40 wt% catalysts [49]. The acetone/citral ratio of 2.8, 333 K, and 15 minutes were the ideal conditions for attaining 96% conversion and 99% selectivity after the sonication treatment of hydrotalcite [50]. Liquid-phase rehydration of hydrotalcite, as studied by Abelló *et al.*, led to a surface area increase to 400 m^2/g, achieving 99.9% citral conversion and 100% pseudoionone selectivity at a 4.4:1 citral/acetone ratio, 333 K, and 1-hour reaction time. However, regeneration through calcination at 723 K was necessary for reuse [51, 52].

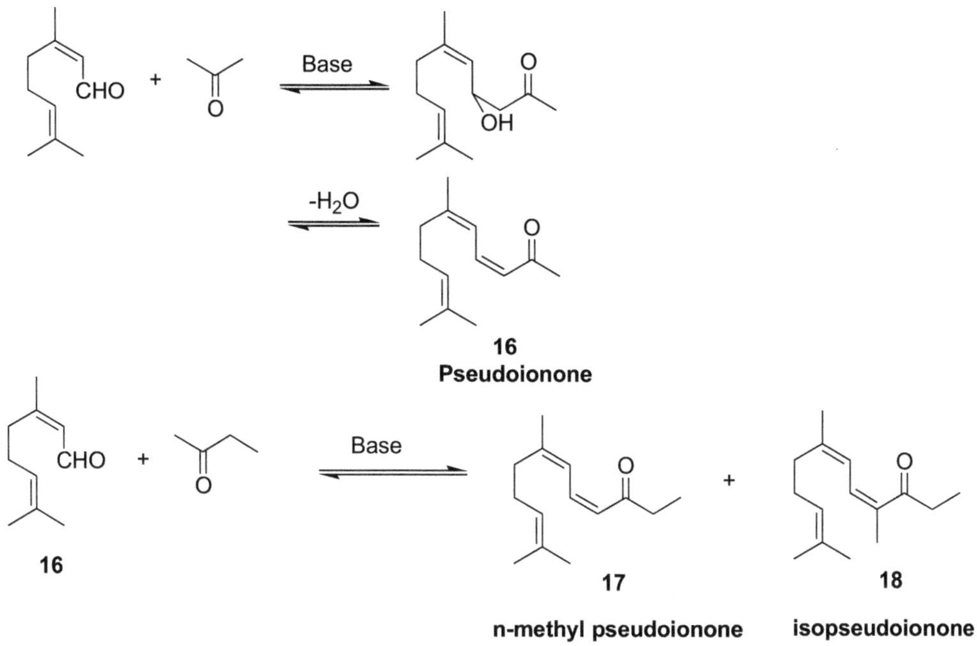

Fig. (12). Synthesis approach for pseudoionones [44].

Production of 2-Pentylidene-Cyclopentanone

The synthesis of 2-pentylidene-cyclopentanone (19), an important precursor in perfume ingredients, involves an aldol condensation reaction between cyclopentanone and valeraldehyde (Fig. **13**). Usually, this reaction is carried out in a batch reactor at 403 K using basic or acidic catalysts. ZrPON demonstrated the best catalytic efficiency of all the catalysts examined, turning 94% of valeraldehyde into the required product in under two hours [53].

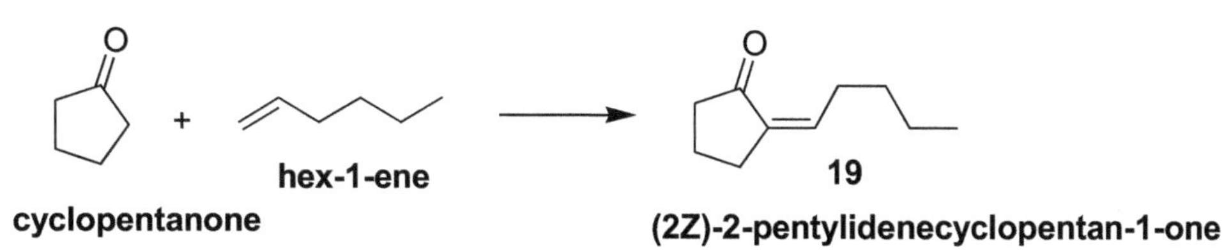

Fig. (13). Production of 2-pentylidene-cyclopentanone [53].

FLAVANONE SYNTHESIS

The synthesis of flavanone is a crucial step in the production of many good chemicals and medications. It is typically synthesized *via* the Claisen-Schmidt condensation reaction between benzaldehyde and 2-hydroxyacetophenone(20). This process involves the initial formation of 2'-hydroxychalcone (21), which then isomerize to flavanone (22).

The MgO-Al_2O_3 mixed oxides obtained from calcined hydrotalcite (Mg/Al = 3) were used as a catalyst in a solvent-free system by Climent *et al.* to study this process. After 1.5 hours at 443 K, they obtained yields of 24% for flavanone and 54% for chalcone (Fig. **14**) [54]. In this reaction, high surface-area MgO (590 m^2/g) has also demonstrated remarkable catalytic efficiency, yielding 90% flavanone in ethanol and 100% in DMSO after 12 hours, along with full benzaldehyde conversion. This efficiency is ascribed to the large number of active sites that MgO's large surface area provides [55].

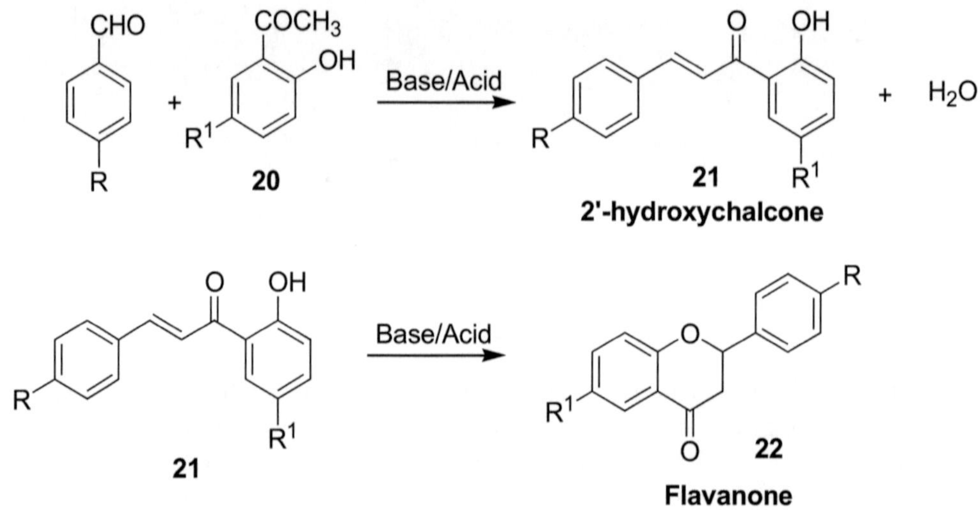

Fig. (14). Flavanone synthesis [54].

Mesoporous silica materials functionalized with amino groups have been explored as heterogeneous catalysts for flavanone synthesis under solvent-free conditions. These materials exhibited promising catalytic activity, highlighting their potential in organic synthesis [56 - 58].

Moreover, MgO-Al_2O_3 mixed oxides, derived from hydroxide, have demonstrated efficacy in the condensation reaction, leading to Vesidryl (2,4,4'-trimethoxychalcone), and achieving an 85% yield after 20 hours at 443 K in air

[59]. These studies underscore the diverse applications of different catalyst systems in the synthesis of valuable intermediates, like flavanone and related compounds.

More Examples of Application of Solid Base Catalysts in Organic Synthesis

Solid-base catalysts—which extend beyond the more widely used solid acids—are increasingly recognised for their potential in chemical synthesis. Promising candidates include Layered Double Hydroxides (LDHs) and hydrotalcite clays. The formula for these artificial materials $[Mg_{8-x}Al_x(OH)_{16}(CO_3)_{x/2}].H_2O$ features a layered structure where carbonate anions between layers balance the positive charge. Upon calcination, LDHs transform into strongly basic mixed magnesium-aluminum oxides. These oxides exhibit high efficacy as solid-base catalysts in various reactions, including aldol, Knoevenagel, and Claisen-Schmidt condensations, as well as the epoxidation of olefins using H_2O_2 [60].

An alternative strategy involves anchoring organic bases onto materials like mesoporous silicas. For instance, functionalizing silicas with 3-aminopropyl(trimethoxy)silane creates aminopropyl-silica, which catalyzes Knoevenagel condensations effectively. Grafting guanidine bases onto surfaces yields even more effective solid-base catalysts. This involves lycidylating triazabicyclo[4.4.0]dec-5-ene (TBD) onto mesoporous MCM-41 to immobilize it (Fig. **15**). This creates a flexible catalyst that may be used for Michael additions, Robinson annulations, and Knoevenagel condensations. Moreover, a "ship-in-a-bottle" guanidine catalyst is produced by encasing big guanidines like *N,N',N"*-tricyclohexyl guanidine inside the super cages of the hydrophobic zeolite Y. With the help of this catalyst, acetone, and benzaldehyde may combine *via* aldol reaction to produce 4-phenyl-4-hydroxybutan-2-one [60]. These methods demonstrate the potential and wide range of uses for solid-base catalysts in organic synthesis.

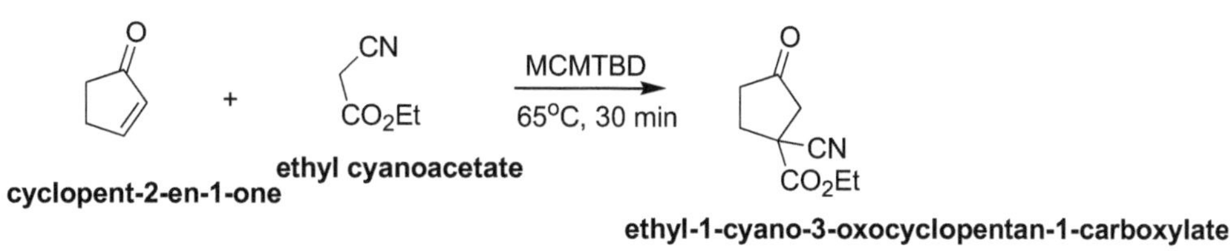

Fig. (15). Example of reaction catalyzed by MCM-TBD Solid Basic Catalysts [60].

The synthesis of methylpseudoionones, which are precursors to methyl ionones, involves the aldol condensation of citral (24) and Methyl Ethyl Ketone (MEK) using solid base catalysts. Amorphous Aluminophosphate (ALPO), KF-Alumina, and several types of activated Al-Mg hydrotalcites were among the catalysts that were assessed. Out of these, calcined or calcined-rehydrated activated hydrotalcites showed the highest activity and selectivity in the production of methylpseudoionones (Fig. **16**) (25,26).

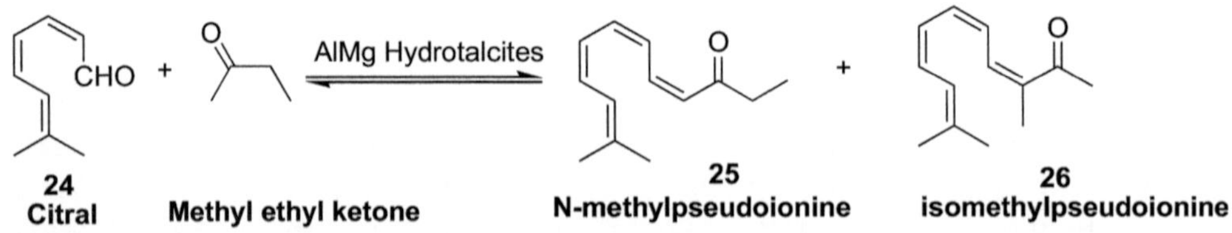

Fig. (16). Synthesis of methylpseudoionones using solid basic catalysts [61].

The study examined how the water content of rehydrated samples, and the chemical makeup of calcined hydrotalcites affected the response. The MEK: citral molar ratio and reaction temperature were two important variables that were systematically changed for each activated catalyst. Higher reaction temperatures typically resulted in faster rates of methyl pseudoionone production, albeit the particular activated hydrotalcite utilized affected selectivity.

Reaction rate and selectivity trends were different between calcined and rehydrated hydrotalcites, due to differences in basicity and adsorption characteristics. The type and distribution of basic sites on the catalyst surface vary, which is the reason for these changes [61].

The uncoordinated amino groups present in zinc carboxylate MOFs, IRMOF-3 in particular, have shown great efficacy as solid-base catalysts in the Knoevenagel condensation process between benzaldehyde and ethyl cyanoacetate.

This particular MOF catalyst has demonstrated superior performance compared to counterparts lacking amino groups, efficiently producing 4-aryl-*4H*-chromenes and substituting dihydropyrimidines. These substances are used as calcium channel blockers and inducers of apoptosis, respectively, demonstrating the catalytic flexibility and possible uses of zinc carboxylate MOFs with uncoordinated amino groups.

MOFs (Metal-Organic Frameworks) have also found successful applications as catalysts in the Henry, or nitro aldol condensation reactions, involving nitroalkanes and carbonyl compounds. These reactions are pivotal for forming carbon-carbon bonds, crucial in the synthesis of important pharmaceuticals, such as norepinephrine and Phenibut. Phenibut, specifically, acts as a GABA receptor agonist, and is utilized in the treatment of various neurological disorders. The ability of MOFs to catalyze these reactions underscores their importance in pharmaceutical and fine chemical synthesis, offering efficient pathways to valuable compounds for medical and industrial applications [62].

Benzimidazole (29) and its analogs represent a pivotal group of bioactive substances within the pharmaceutical sector, showcasing a wide array of biological and pharmacological characteristics. Thakore and collaborators devised an innovative technique for synthesizing 2-substituted benzimidazole utilizing polymer vesicles loaded with copper nanoparticles (CuNPs@vesicles) as nanoreactors. This eco-friendly technique uses a one-step cascade process that begins with 2-nitrophenol. CuNPs@vesicles facilitate the conversion of 2-nitroaniline to o-phenylenediamine, a precursor that is then utilized in the synthesis of benzimidazole. Additionally, the researchers looked at how various substituents affected aldehydes, and discovered that substituents with Electron-Withdrawing Ggroups (EWG) needed shorter reaction times than substituents with Electron-Releasing Groups (ERG). This approach enables a more efficient and sustainable synthesis of benzimidazoles (Fig. **17**). It is noteworthy that the entire sequence of reactions takes place in an aqueous environment under normal conditions, while the catalyst is capable of being rejuvenated and reused up to five times, underscoring the robustness and durability of the vesicular nanostructures [63].

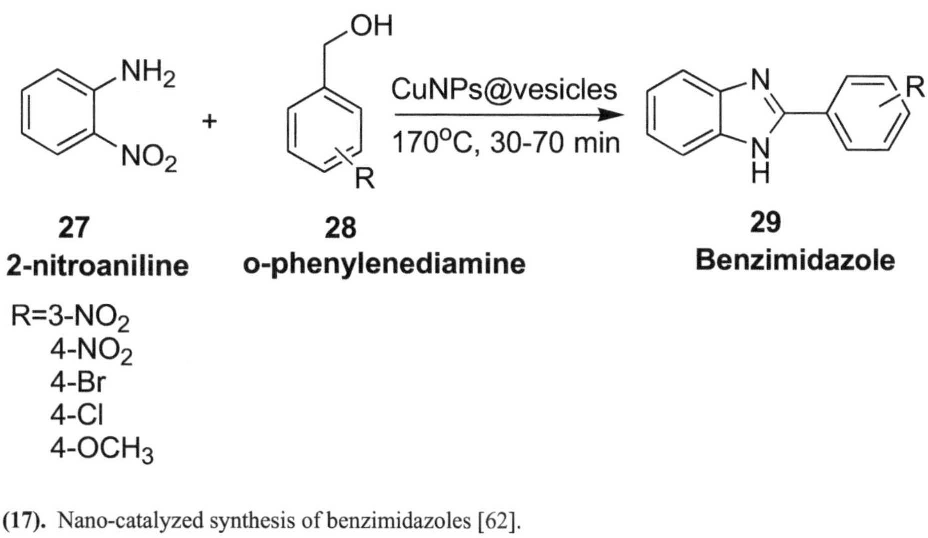

Fig. (17). Nano-catalyzed synthesis of benzimidazoles [62].

CONCLUSION

Solid base catalysts, including metal oxides, alkali metal compounds, and modified mesoporous materials, offer significant advantages in organic synthesis due to their high reactivity, selectivity, and environmental friendliness. These catalysts play a crucial role in various applications, such as hydrogenation, aldol condensation, and isomerization reactions.

In the synthesis of pseudo ionone synthesis from citral and acetone, solid base catalysts such as sodium-doped magnesium oxide and MgO-Al_2O_3 have demonstrated high efficiency and selectivity, making them suitable for industrial-scale production. These catalysts facilitate the reaction under mild conditions, reducing energy consumption and minimizing by-product formation.

Furthermore, the development of solid base catalysts for the synthesis of bioactive compounds from 2-carene and its derivatives highlights the potential of these materials in pharmaceutical applications. The tailored properties of solid bases, including surface area, pore size, and acid-base functionality, enable precise control over reaction pathways, leading to high yields of desired products.

Overall, the advancements in solid base catalyst technology underscore their importance in sustainable and efficient chemical processes, contributing to the development of greener and more economical methods in organic synthesis. Continued research and optimization of these catalysts will further enhance their applications and impact across various industries, including pharmaceuticals, fine chemicals, and materials science.

REFERENCES

[1] Jambhulkar, D.K.; Ugwekar, R.P.; Bhanvase, B.A.; Barai, D.P. A review on solid base heterogeneous catalysts: preparation, characterization and applications. *Chem. Eng. Commun.*, **2022**, *209*(4), 433-484. [http://dx.doi.org/10.1080/00986445.2020.1864623]

[2] You, X.; Chen, L.; He, S.; Zhang, G. *Solid acid – base catalysts based on layered double hydroxides applied for green catalytic transformations,* **2024**.

[3] Figueras, F.; Lakshmi Kantam, M.; Manoranjan Choudary, B. Solid base catalysts in organic synthesis. *Curr. Org. Chem.*, **2006**, *10*(13), 1627-1637. [http://dx.doi.org/10.2174/138527206778249658]

[4] Adeyeye Nafiu, S.; Ajeebi, A.M.; Alghamdi, H.S.; Aziz, M.A.; Nasiruzzaman Shaikh, M. Solid-supported catalysts for organic functional group transformations. *Asian J. Org. Chem.*, **2023**, *12*(4), e202300051. [http://dx.doi.org/10.1002/ajoc.202300051]

[5] Saleh, J.N.; Khalid, A. Synthesis, characterization, and biological activity evaluation of some new pyrimidine derivatives by solid base catalyst AL2O3-OBa. *Cent. Asian J. Med. Nat. Sci.*, **2023**, *4*(2), 231-239.

[6] Hattori, A.; Hattori, H.; Tanabe, K. Double-bond migration of allylamine to enamine over basic oxide catalysts. *J. Catal.*, **1980**, *65*(2), 245-252. [http://dx.doi.org/10.1016/0021-9517(80)90302-4]

[7] Hattori, H. Solid base catalysts: Generation, characterization, and catalytic behavior of basic sites. *J. Jpn. Petrol. Inst.*, **2004**, *47*(2), 67-81. [http://dx.doi.org/10.1627/jpi.47.67]

[8] Pines, H.; Vesely, J.A.; Ipatieff, V.N. Migration of double bonds in olefinic and diolefinic hydrocarbons catalyzed by sodium. dehydrogenation of d-limonene to p-cymene [2]. *J. Am. Chem. Soc.*, **1955**, *77*(2), 347-348. [http://dx.doi.org/10.1021/ja01607a031]

[9] Ngamcharussrivichai, C.; Totarat, P.; Bunyakiat, K. Ca and Zn mixed oxide as a heterogeneous base catalyst for transesterification of palm kernel oil. *Appl. Catal. A Gen.*, **2008**, *341*(1-2), 77-85. [http://dx.doi.org/10.1016/j.apcata.2008.02.020]

[10] Lee, D.W.; Yoo, B.R. Advanced metal oxide (supported) catalysts: Synthesis and applications. *J. Ind. Eng. Chem.*, **2014**, *20*(6), 3947-3959. [http://dx.doi.org/10.1016/j.jiec.2014.08.004]

[11] Hattori, H. Heterogeneous Basic Catalysis. *Chem. Rev.*, **1995**, *95*(3), 537-558. [http://dx.doi.org/10.1021/cr00035a005]

[12] Bing, W.; Wei, M. Recent advances for solid basic catalysts: Structure design and catalytic performance. *J. Solid State Chem.*, **2019**, *269*, 184-194. [http://dx.doi.org/10.1016/j.jssc.2018.09.023]

[13] Hattori, H. Solid base catalysts: Fundamentals and their applications in organic reactions. *Appl. Catal. A Gen.*, **2015**, *504*, 103-109. [http://dx.doi.org/10.1016/j.apcata.2014.10.060]

[14] Hattori, H. Solid base catalysts: Generation of basic sites and application to organic synthesis. *Appl. Catal. A Gen.*, **2001**, *222*(1-2), 247-259. [http://dx.doi.org/10.1016/S0926-860X(01)00839-0]

[15] Busacca CA, Fandrick DR, Song JJ, Senanayake CH. The growing impact of catalysis in the pharmaceutical industry. *Advanced Synthesis & Catalysis*. **2011** Aug; *353*(11□12):1825-64.

[16] Saito, S.; Yamamoto, H. Design of acid-base catalysis for the asymmetric direct Aldol reaction. *Acc. Chem. Res.*, **2004**, *37*(8), 570-579. [http://dx.doi.org/10.1021/ar030064p] [PMID: 15311956]

[17] Hattori, H. Catalysis by basic metal oxides. *Mater. Chem. Phys.*, **1988**, *18*(5-6), 533-552. [http://dx.doi.org/10.1016/0254-0584(88)90021-1]

[18] Yang, L.; Wang, C.; Zhou, J.; Kim, S.W. Combinatorial engineering of hybrid mevalonate pathways in Escherichia coli for protoilludene production. *Microb. Cell Fact.*, **2016**, *15*(1), 14-24. [http://dx.doi.org/10.1186/s12934-016-0409-7] [PMID: 26785630]

[19] McMorris, T.C.; Anchel, M.; Newman, D.J. Illudins and their derivatives: Chemistry, biology, and synthetic studies. *Prog. Chem. Org. Nat. Prod.*, **1996**, *67*, 135-222.

[20] Arnone, A.; Assante, G.; Camarda, L.; Nasini, G.; Vurro, M. Illudins: Structure, biosynthesis, and biological activity. *Nat. Prod. Rep.*, **2004**, *21*, 218-229.

[21] Mehta, G.; Pan, S.C. Synthetic and mechanistic aspects of illudins and illudalane sesquiterpenes: applications to natural product synthesis. *Tetrahedron,* **2005**, *61*, 4707-4734.

[22] Albaqami JJ, Hamdi H, Narayanankutty A, Visakh NU, Sasidharan A, Kuttithodi AM, Famurewa AC, Pathrose B. Chemical composition and biological activities of the leaf essential oils of Curcuma longa, Curcuma aromatica and Curcuma angustifolia. *Antibiotics*. **2022** Nov 3; *11*(11):1547.

[23] Ono, Y.; Baba, T. Selective reactions over solid base catalysts. *Catal. Today,* **1997**, *38*(3), 321-337. [http://dx.doi.org/10.1016/S0920-5861(97)81502-5]

[24] Gortsema, F.P.; Beshty, B.; Friedman, J.J.; Matsumoto, D.; Sharkey, J.J.; Wildman, G.; Pan, S.H. Preparation of pharmaceutical intermediates using zeolite catalysis: Preparation of 4-methylthiazole with ZSM-5.*Catalysis of Organic Reactions*; CRC Press, **2020**, pp. 445-460. [http://dx.doi.org/10.1201/9781003067184-30]

[25] IARC working group on the evaluation of carcinogenic risks to humans. N, N-Dimethylaniline.*Occupational Exposures of Hairdressers and Barbers and Personal Use of Hair Colourants; Some Hair Dyes, Cosmetic Colourants, Industrial Dyestuffs and Aromatic Amines*; International Agency for Research on Cancer, **1993**.

[26] Margosis, M. GLC determination of N,N-dimethylaniline in penicillins. *J. Pharm. Sci.*, **1977**, *66*(11), 1634-1636.
[http://dx.doi.org/10.1002/jps.2600661136] [PMID: 915748]

[27] Akutu K, Kabashima H, Seki T, Hattori H. Nitroaldol reaction over solid base catalysts. *Applied Catalysis A: General*. **2003** Jul 10; *247*(1):65-74.

[28] Cram, D.J. *Fundamentals of Carbanion Chemistry*; Elsevier: Amsterdam, **2012**, 4, .

[29] Zhu J, Chun Y, Wang Y, Xu Q. Attempts to create new shape-selective solid strong base catalysts. *Catalysis today*. **1999** Jun 1; *51*(1):103-11.

[30] Yamaguchi, T.; Wang, Y.; Komatsu, M.; Ookawa, M. Preparation of new solid bases derived from supported metal nitrates and carbonates. *Catal. Surv. Jpn.*, **2002**, *5*(2), 81-89.
[http://dx.doi.org/10.1023/A:1015108902323]

[31] Ono, Y. Solid base catalysts for the synthesis of fine chemicals. *J. Catal.*, **2003**, *216*(1-2), 406-415.
[http://dx.doi.org/10.1016/S0021-9517(02)00120-3]

[32] Vaid, R.; Sharma, N.; Gupta, M.; Chowhan, B. Nanosized calcined mixed oxides-supported alkali metal (K_2O/Al_2O_3–CaO) as solid base catalyst: preparation and investigation of its catalytic efficiency in the synthesis of benzylidene malononitriles/barbiturates and in pyrano[2,3-d]pyrimidines. *J. Indian Chem. Soc.*, **2022**, *19*(7), 2855-2870.
[http://dx.doi.org/10.1007/s13738-022-02501-2]

[33] Parmar, M.P.; Vala, R.M.; Patel, H.M. Importance of hybrid catalysts toward the synthesis of 5*H*-pyrano[2,3-*d*]pyrimidine-2-ones/2,4-diones (Thiones). *ACS Omega*, **2023**, *8*(2), 1759-1816.
[http://dx.doi.org/10.1021/acsomega.2c05349] [PMID: 36687108]

[34] Védrine JC. Acid–base characterization of heterogeneous catalysts: An up-to-date overview. *Research on Chemical Intermediates*. **2015** Dec; *41*:9387-423.

[35] Borah, B.J.; Borah, S.J.; Saikia, K.; Dutta, D.K. Efficient one-pot synthesis of propargylamines catalysed by gold nanocrystals stabilized on montmorillonite. *Catal. Sci. Technol.*, **2014**, *4*(11), 4001-4009.
[http://dx.doi.org/10.1039/C4CY00666F]

[36] Manujyothi, R.; Aneeja, T.; Anilkumar, G. Solvent-free synthesis of propargylamines: An overview. *RSC Advances*, **2021**, *11*(32), 19433-19449.
[http://dx.doi.org/10.1039/D1RA03324G] [PMID: 35479216]

[37] Jeganathan, M.; Dhakshinamoorthy, A.; Pitchumani, K. One-Pot synthesis of propargylamines using ag(i)-exchanged k10 montmorillonite clay as reusable catalyst in water. *ACS Sustain. Chem.& Eng.*, **2014**, *2*(4), 781-787.
[http://dx.doi.org/10.1021/sc400450t]

[38] Sun, L.B.; Liu, X.Q.; Zhou, H.C. Design and fabrication of mesoporous heterogeneous basic catalysts. *Chem. Soc. Rev.*, **2015**, *44*(15), 5092-5147.
[http://dx.doi.org/10.1039/C5CS00090D] [PMID: 26051787]

[39] Gupta, R.; Layek, S.; Pathak, D.D. Synthesis and characterization of guanine-functionalized mesoporous silica [SBA-16-G]: a metal-free and recyclable heterogeneous solid base catalyst for synthesis of pyran-annulated heterocyclic compounds. *Res. Chem. Intermed.*, **2019**, *45*(3), 1619-1637.
[http://dx.doi.org/10.1007/s11164-018-3693-5]

[40] Pendalwar, S.S.; Chakrawar, A.V.; Chavan, A.S.; Bhusare, S.R. An efficient synthesis of dihydropyrano[3,2-c]chromene derivatives *via* green approach. *Pharma Chem.*, **2016**, *8*(3), 143-145.

[41] Jinesh, C.M.; Antonyraj, C.A.; Kannan, S. Isomerization of eugenol and alkenyl aromatics of perfumery interest over Ni-containing layered double hydroxides as solid base catalysts. *Catal. Today,* **2009**, *141*(1-2), 176-181. [http://dx.doi.org/10.1016/j.cattod.2008.03.023]

[42] Kishore, D.; Kannan, S. Isomerization of eugenol and safrole over MgAl hydrotalcite, a solid base catalystElectronic supplementary information (ESI) available: Figs. S1 and S2: PXRD patterns. See http://www.rsc.org/suppdata/gc/b2/b207865a/. *Green Chem.*, **2002**, *4*(6), 607-610. [http://dx.doi.org/10.1039/b207865a]

[43] Kishore, D.; Kannan, S. Double bond migration of eugenol to isoeugenol over as-synthesized hydrotalcites and their modified forms. *Appl. Catal. A Gen.,* **2004**, *270*(1-2), 227-235. [http://dx.doi.org/10.1016/j.apcata.2004.05.008]

[44] Vatakencherry, P.A.; Pushpakumarian, K.N. *Chem. Ind.,* **1987**, 163.

[45] Jablonskí, E.L.; Vilella, I.M.J.; Maina, S.C.; de Miguel, S.R.; Scelza, O.A. *Catal. Commun.,* **2005**, *6*(7), 18-29.

[46] Noda Pérez, C.; Pérez, C.A.; Henriques, C.A.; Monteiro, J.L.F. Hydrotalcites as precursors for Mg,Al-mixed oxides used as catalysts on the aldol condensation of citral with acetone. *Appl. Catal. A Gen.,* **2004**, *272*(1-2), 229-240. [http://dx.doi.org/10.1016/j.apcata.2004.05.045]

[47] Roelofs, J.C.A.A.; van Dillen, A.J.; de Jong, K.P. Base-catalyzed condensation of citral and acetone at low temperature using modified hydrotalcite catalysts. *Catal. Today,* **2000**, *60*(3-4), 297-303. [http://dx.doi.org/10.1016/S0920-5861(00)00346-1]

[48] Roelofs, J.; van Dillen, A; de Jong, K. Synthesis of pseudoionones by acid and base solid catalysts. *Catal. Lett.,* **2002**, *79*, 157-163. [http://dx.doi.org/10.1023/A:1015364526587]

[49] Climent, M.; Corma, A.; Iborra, S. Synthesis of pseudoionones by acid and base solid catalysts *Catal. Lett.,* **2002**, *79*(1/4), 157-163. [http://dx.doi.org/10.1023/A:1015364526587]

[50] Climent, M.; Corma, A.; Iborra, S.; Epping, K.; Velly, A. Increasing the basicity and catalytic activity of hydrotalcites by different synthesis procedures. *J. Catal.,* **2004**, *225*(2), 316-326. [http://dx.doi.org/10.1016/j.jcat.2004.04.027]

[51] Abelló, S.; Medina, F.; Tichit, D.; Pérez-Ramírez, J.; Groen, J.C.; Sueiras, J.E.; Salagre, P.; Cesteros, Y. Aldol condensations over reconstructed Mg-Al hydrotalcites: structure-activity relationships related to the rehydration method. *Chemistry,* **2005**, *11*(2), 728-739. [http://dx.doi.org/10.1002/chem.200400409] [PMID: 15584078]

[52] Abelló, S.; Vijaya-Shankar, D. Pérez-Ramírez. *J. Appl. Catal. A,* **2007**, *342*(1-2), 119.

[53] Hasni, M.; Prado, G.; Rouchaud, J.; Grange, P.; Devillers, M.; Delsarte, S. Liquid phase aldol condensation of cyclopentanone with valeraldehyde catalysed by oxynitrides possessing tuneable acid–base properties. *J. Mol. Catal. Chem.,* **2006**, *247*(1-2), 116-123. [http://dx.doi.org/10.1016/j.molcata.2005.11.023]

[54] Climent, M.J.; Corma, A.; Iborra, S.; Primo, J. Base catalysis for fine chemicals production: claisen-schmidt condensation on zeolites and hydrotalcites for the production of chalcones and flavanones of pharmaceutical interest. *J. Catal.,* **1995**, *151*(1), 60-66. [http://dx.doi.org/10.1006/jcat.1995.1008]

[55] Drexler, M.T.; Amiridis, M.D. Kinetic investigation of the heterogeneous synthesis of flavanone over MgO. *Catal. Lett.,* **2002**, *79*(1/4), 175-181. [http://dx.doi.org/10.1023/A:1015320711566]

[56] Choudary, B.M.; Ranganath, K.V.S.; Yadav, J.; Lakshmi Kantam, M. Synthesis of flavanones using

nanocrystalline MgO. *Tetrahedron Lett.,* **2005**, *46*(8), 1369-1371. [http://dx.doi.org/10.1016/j.tetlet.2004.12.078]

[57] Wang, X.; Cheng, S. Solvent-free synthesis of flavanones over aminopropyl-functionalized SBA-15. *Catal. Commun.,* **2006**, *7*(9), 689-695. [http://dx.doi.org/10.1016/j.catcom.2006.02.021]

[58] Wang, X.; Tseng, Y.; Chan, J.; Cheng, S. Catalytic applications of aminopropylated mesoporous silica prepared by a template-free route in flavanones synthesis. *J. Catal.,* **2005**, *233*(2), 266-275. [http://dx.doi.org/10.1016/j.jcat.2005.04.007]

[59] Climent, M.J.; Corma, A.; Iborra, S.; Velty, A. Activated hydrotalcites as catalysts for the synthesis of chalcones of pharmaceutical interest. *J. Catal.,* **2004**, *221*(2), 474-482. [http://dx.doi.org/10.1016/j.jcat.2003.09.012]

[60] Sheldon, R.A. Atom efficiency and catalysis in organic synthesis. *Pure Appl. Chem.,* **2000**, *72*(7), 1233-1246. [http://dx.doi.org/10.1351/pac200072071233]

[61] Climent, M.J.; Corma, A.; Iborra, S.; Velty, A. Synthesis of methylpseudoionones by activated hydrotalcites as solid base catalysts. *Green Chem.,* **2002**, *4*(5), 474-480. [http://dx.doi.org/10.1039/b205532p]

[62] Cirujano, F.G. Engineered MOFs and enzymes for the synthesis of active pharmaceutical ingredients. *ChemCatChem,* **2019**, *11*(23), 5671-5685. [http://dx.doi.org/10.1002/cctc.201900131]

[63] Keri, R.S.; Adimule, V.; Kendrekar, P.; Sasidhar, B.S. The nano-based catalyst for the synthesis of benzimidazoles. *Top. Catal.,* **2022**, *65*(1-4), 123-130. [http://dx.doi.org/10.1007/s11244-022-01562-0]

CHAPTER 7

Historical and Recent Advances in Solid Base Catalysts for C-C, C-N, C-S Bond Formation

Vishwa K. Patel[1] and **Satish Kumar Singh**[1,*]

[1] *Department of Chemistry, Government College Daman (Affiliated to Veer Narmad South Gujarat University, Surat), Daman, U. T. of Dadra & Nagar Haveli and Daman & Diu, India*

Abstract: The formation of C-C and carbon-heteroatom (N and S) represents a fundamental process in organic chemistry, playing a pivotal role in synthesizing a wide array of organic compounds, with applications spanning from pharmaceuticals to materials science. Conventional catalytic approaches for C-C bond formation frequently depend on costly and toxic transition metal catalysts, which raise environmental and economic apprehensions. However, with the advent of metal oxides, ion-exchange resins and graphene-based carbocatalysts present a promising alternative owing to their distinctive attributes, chemical stability, conductivity, and high surface area. In environmentally sustainable and cost-effective processes, this catalytic method efficiently enables the formation of a wide range of C–N and C–S containing compounds across diverse organic synthesis pathways. This book chapter emphasizes the advancement of carbon-carbon and carbon-heteroatom (N & S) formation through various types of solid base catalysts.

Keywords: C-C bond formation, C-N bond formation, C-S bond formation, Coupling reactions, Solid base catalysis.

INTRODUCTION

Solid base catalysts have emerged as essential components in contemporary catalysis, providing environmentally friendly alternatives to conventional acidic catalysts [1]. In contrast to their liquid counterparts, solid base catalysts offer numerous benefits, including ease of separation, recyclability, and reduced environmental footprint [2]. Solid base catalysts are characterized by the existence of basic sites on a solid support [3]. The active sites on solid base catalysts generally comprise basic functional groups, such as hydroxides, oxides, or amine groups, which can accept protons or donate electrons during catalytic reactions [4]. Common materials utilized as solid bases include metal oxides such as MgO,

* **Corresponding author Satish Kumar Singh:** Department of Chemistry, Government College Daman (Affiliated to Veer Narmad South Gujarat University, Surat), Daman, U. T. of Dadra & Nagar Haveli and Daman & Diu, India; E-mail: drsatish.singh@ddd.gov.in

Ravi Tomar, Vivek Mishra, Kapil Arya & Garima Pandey (Eds.)

CaO, and Al_2O_3, along with various porous materials, such as zeolites and mesoporous silica [5 - 7]. Solid base catalysts facilitate chemical reactions by abstracting protons or donating electron pairs to substrates, thereby initiating catalytic cycles [8]. The basicity of a solid base catalyst is defined by its capability to adsorb and activate reactant molecules, resulting in the formation of intermediates with modified reactivity [9]. Commonly used solid base materials include metal oxides, zeolites, layered double hydroxides, and supported organic bases, each offering unique advantages depending on the specific reaction requirements [10 - 12]. The preparation of solid base catalysts involves chemical composition and customizing the surface properties to enhance catalytic performance. Techniques like impregnation, precipitation, sol-gel, and ion exchange are utilized to deposit basic species onto support materials, or to modify the surface chemistry of the catalyst [13].

Additionally, these catalysts participate in a crucial role in biomass conversion processes, including the production of biodiesel and the upgrading of biomass-derived intermediates into value-added products [14]. Moreover, solid base catalysts find applications in environmental endeavors, such as pollutant degradation, acid gas removal, and water purification, playing a role in reducing environmental pollution, and promoting sustainability [15]. Solid base catalysts are integral to green and sustainable chemistry, providing versatile tools for facilitating efficient, selective, and environmentally benign chemical transformations [16 - 18].

Solid base catalysts facilitate the synthesis of C-C/C-heteroatoms (N and S), which is an essential procedure in the synthesis of many organic molecules with uses in materials science and medicine (Fig. **1**).

Solid base catalysts play a crucial role in the formation of C-C bond reactions, which are primary in organic synthesis for constructing complex molecules. These catalysts facilitate various types of C-C bond formations, including Knoevenagel condensation, aldol condensation, Michael addition, Claisen condensation, and more. Solid base catalysts are involved in some of these reactions and transesterification, enabling the synthesis of pharmaceuticals, biofuels, and fine chemicals [19, 20]. Aldol condensation is a process that produces a β-hydroxy carbonyl molecule by condensing two carbonyl compounds with the help of a base catalyst. Solid base catalysts, such as metal oxides (like CaO, MgO) or amine-functionalized materials, which are supported amines on silica, are commonly employed to catalyze this reaction. They deprotonate the α-carbon of one carbonyl compound, enhancing its reactivity towards nucleophilic attack [21]. In the Claisen condensation, two ester molecules undergo condensation to produce a β-keto ester with the help of a base catalyst. Solid base catalysts, such as alkali earth metal oxides or alkaline, facilitate this reaction by extracting a proton from one of the ester molecules, thereby enabling nucleophilic attack by the carbonyl group of the other ester [22]. In the Michael addition, a new carbon-carbon bond is formed when a nucleophile is added to an *α,β*-unsaturated carbonyl molecule. Solid base catalysts, such as metal oxides and supported amines, can catalyze Michael addition reactions via activating the nucleophile, and aiding its addition to the electrophilic double bond [23, 24].

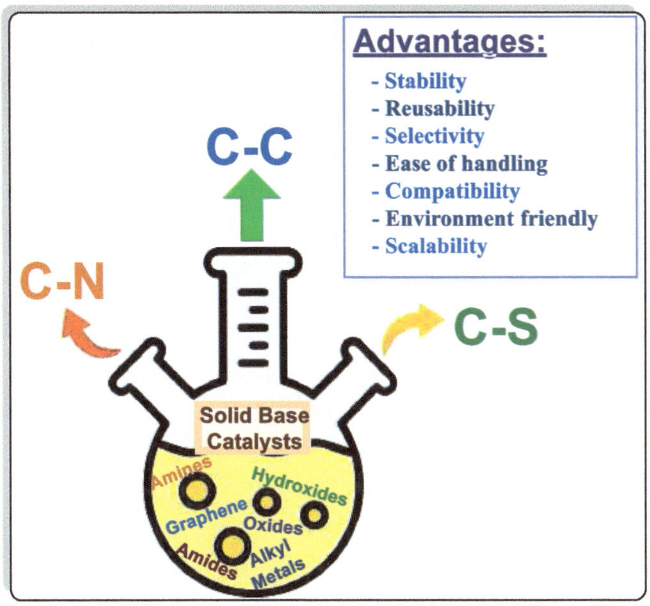

Fig. (1). Solid base catalyzed formation of C-C and C-heteroatoms bond.

In chemistry, a "C–N coupling" refers to a chemical reaction in which a carbon atom forms a bond with a nitrogen atom" [25, 26] In C-N bond formation processes, solid base catalysts have shown great promise as a viable substitute for conventional homogenous catalysts [27, 28]. There are various benefits to solid base catalysts over their liquid equivalents, because they are sustainable, stable, and allow for facile separation [29, 30]. The solid catalyst basic sites aid in the deprotonation process, and the follow-on nucleophile is able to interact with different electrophiles to generate the appropriate C-N bond [31]. Current developments in the sphere of solid base catalysis for the production of carbon-nitrogen bonds have centered on creating new types of catalysts, increasing catalytic efficiency, and broadening the variety of substrates that can be engaged [32 - 34].

A chemical process known as "carbon-sulfur coupling" creates a carbon-sulfur bond between these two molecules [35, 36]. Solid base catalysts have demonstrated significant potential as an effective replacement for traditional homogeneous catalysts in C-S bond production processes [37, 38]. Solid base catalysts generally produce less waste and exhibit more environmental sustainability, compared to liquid bases, hence encouraging the application of green chemistry principles [39]. Molecular Organic Frameworks (MOFs) have demonstrated potential as solid base catalysts for C-S bond production, since they have adaptable pore frameworks and huge surface areas, which enhance catalytic activity and selectivity [40]. Cascade reactions and nanostructured materials supported by solid base catalysts provide improved efficiency and recyclability [41, 42].

The different solid base catalysts employed in C-C, C-N and C-S bond formation reactions are shown in this chart, which is arranged according to the types, and includes examples of particular catalysts from each category (Fig. **2**).

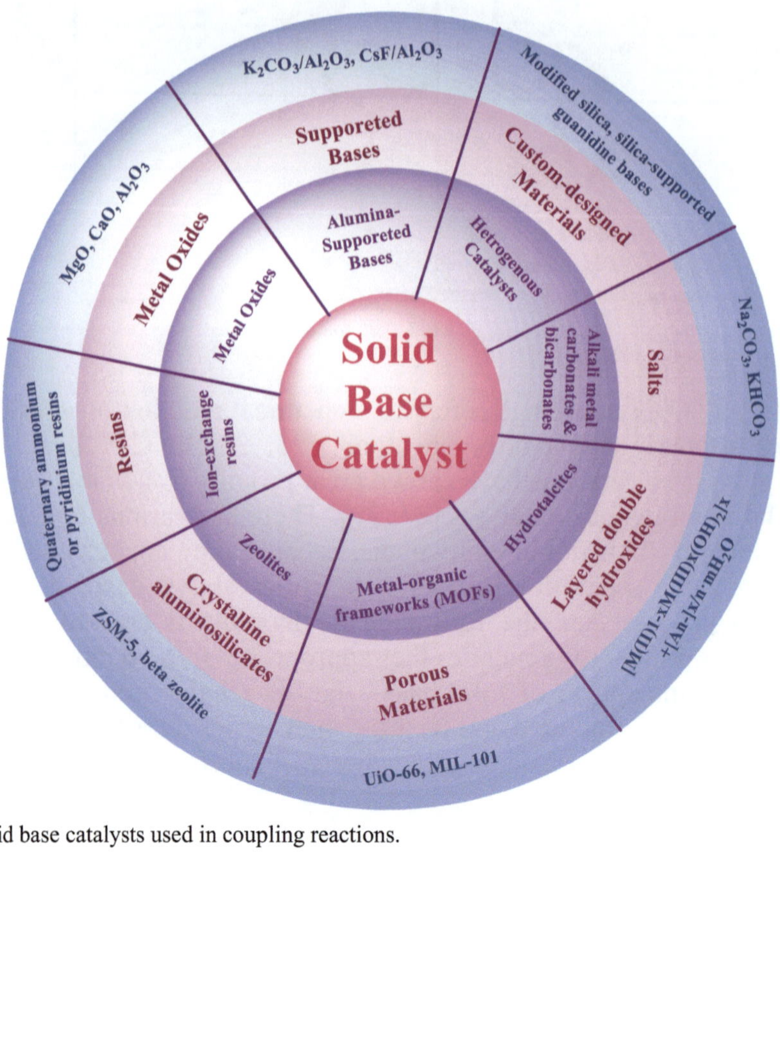

Fig. (2). Solid base catalysts used in coupling reactions.

Solid base catalysts can be classified into several categories based on their material composition and functional characteristics. The primary categories include:

- **Metal Oxides**: These include alkaline earth oxides such as MgO, CaO, and Al_2O_3. Their basic sites, often in the form of surface hydroxides or oxygen anions, provide strong basicity, making them ideal for catalytic applications, such as Knoevenagel condensation, aldol condensation, and Claisen condensation [43].
- **Porous Materials**: Materials like zeolites, mesoporous silica, and layered double hydroxides offer a high surface area and tunable pore size, enabling efficient adsorption and activation of reactants [44].
- **Supported Organic Bases**: These catalysts incorporate basic functional groups, such as amines, onto inert supports like silica or alumina. Supported organic bases are widely employed in processes requiring selective activation of nucleophiles [45].
- **Metal-Organic Frameworks (MOFs)**: MOFs are gaining attention for their versatility. Their large surface areas and tunable pore structures make them promising for reactions, such as C-S bond formation [46].
- **Nanostructured Materials**: These materials, often engineered at the nanoscale, provide enhanced reactivity and recyclability due to their high surface- area- to-volume ratio [47].

The catalytic processes involving solid base catalysts can be classified into the following classifications:

[1.1] C-C bond formation through solid base catalyzed reactions.

[1.2] C-N bond formation through solid base catalyzed reactions.

[1.3] C-S bond formation through solid base catalyzed reactions.

This can be sub-categorized into the following three types of reactions:

- **Condensation Reactions**: Reactions like Knoevenagel condensation and aldol condensation involve the activation of carbonyl compounds through proton abstraction, facilitating nucleophilic attack. Solid base catalysts, such as MgO and CaO are commonly used [48].
- **Addition Reactions**: Michael addition and Claisen condensation are critical for forming β-dicarbonyl or β-keto ester structures. These reactions utilize catalysts like supported amines and metal oxides to enhance nucleophilicity [49].

- **Alkylation Reactions**: This includes Friedel-Crafts alkylation and methylation processes, where aromatic substrates or carbonyl compounds are activated through solid base catalysts for nucleophilic attack using alkylating agents [50].

It is anticipated that ongoing research endeavors, focused on creating novel catalyst materials, comprehending reaction mechanisms, and broadening the range of suitable substrates, will propel the area forward and facilitate the production of intricate complex compounds for diverse uses.

C-C Bond Formation through Solid Base Catalyzed Reactions

The process of alkylation reactions includes the addition of an alkyl group to a substrate, which is usually an organic molecule that contains heteroatoms, such as nitrogen, oxygen, or functional groups like aromatics. Solid base catalysts can significantly enhance alkylation processes by promoting proton abstraction from the substrate, thereby activating it for nucleophilic attack by the alkylating agent [51, 52]. One of the challenges with Friedel-Crafts alkylation is the occurrence of polyalkylation and catalyst deactivation. To mitigate these issues, solid base catalysts—particularly supported amine compounds or metal oxides such as MgO—can be used as co-catalysts or alternative catalysts [53 - 57]. By removing a proton, they can activate the aromatic substrate, and increase its reactivity to the alkylating agent's electrophilic attack. In chemical synthesis, methylation reactions, particularly, the methylation of amines and phenols, are significant changes. The phenolic or amino group can be deprotonated through solid base catalysts, which increase the group's nucleophilicity for methylating agents, such as methyl iodide or dimethyl sulfate [58 - 60]. Solid base catalysts can facilitate alkylation processes involving carbonyl molecules (e.g., ketones, aldehydes, and esters) in addition to aromatic substrates. Through, in the removal of a proton from the α-carbon, the catalyst initiates the carbonyl molecule, producing an enolate intermediate that can be added nucleophilically to an alkylating agent [61, 62].

Efficient α-alkylation Using an Ir-oxalamidato Catalyst

Using an Ir–oxalamidato complex in combination with a solid base, like magnesium aluminometasilicate/$Ca(OH)_2$, significantly improved the catalytic performance in the α-alkylation of methyl ketones and primary alcohols. An Ir-oxalamidato catalyst was optimized by systematically varying substituents, in order to enhance catalyst productivity in the direct replacement of hydroxy groups (Scheme **1**) [63].

When combined through a solid base like $Ca(OH)_2$/magnesium aluminometa-silicate, Ir–oxalamidato complex facilitated the reaction between methylketones and primary alcohols with a turnover number of 10,000-40,000 utilizing [$Cp^*{}_2Ir_2$(μ-*N*,*N*'-bis(5-*tert*-butyl-4-hydroxy-2-methylphenyl)oxalamidato)Cl_2] as solid base catalyst. This process is capable of being a valuable method for synthesizing ketones with longer alkyl chains on an industrial scale.

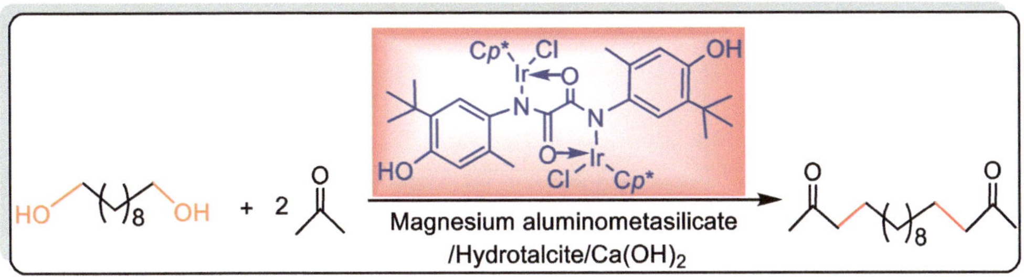

Scheme (1). *α*-Alkylation of ketones.

Efficient Generation of C-C Bond with Silica-supported Base Organocatalysts

Novel hybrid catalysts have been developed by grafting silyl-derivatives of compounds containing diamino, imidazolium, pyrazolium, pyrrolidine, and amino functionalities onto an M41S-type substrate. The organic bases were preserved inside the mesoporous silica structure, according to the analysis of the hybrids, resulting in hybrid materials with robust base characteristics. The compounds were shown to possess recyclability and catalytic activity as base catalysts for the generation of C-C bonds, specifically Michael additions and Knoevenagel condensations. The observed discrepancies in performance were explained based on the reaction mechanisms. In addition, the inclusion of silanol groups inside the support framework would facilitate a collaborative electrophilic activation. In contrast, for the Michael addition reaction, the ionic base catalysts performed better than the neutral base catalysts.

Synthetic procedures were utilized in the production of silica-supported base catalysts (**a-f**) with conditions (1) MCM-41, dry Toluene, 1h, RT; (2) silylated base precursor, 24h, reflux; (3) MCM-41, silylated base precursor, dry MeCN, 1h, RT; (4) TEA, reflux, 24h; (5) TMAOH, MeOH, 30 min. The base precursors were first silylated and, subsequently, attached to the MCM-41 support, resulting in the formation of hybrid base catalysts (Scheme **2a-f**).

Hybrid base catalysts were used to carry out the Knoevenagel condensation reaction between benzaldehyde, ethyl cyanoacetate, and diethyl malonate. These catalysts had an active site loading of 10 mol%. The neutral base catalysts **a-c** and ionic base catalysts **d-f** exhibited a high level of activity (Scheme **3**) [64].

The reaction of benzaldehyde with ethyl cyanoacetate yields 93%, 90%, 95%, 91%, 89%, and 91%, correspondingly when catalysts **a-f** are used. Similarly, when ethyl acetoacetate is used with catalyst **a-c**, it produces a yield of 65%, 63%, and 46%. However, when catalystd **d-f** are used, there is no yield percentage. Diethyl malonate yields 45%, 22%, and 18% when reacting with **a-c** catalysts, but no yield is obtained when using **d-f** catalysts.

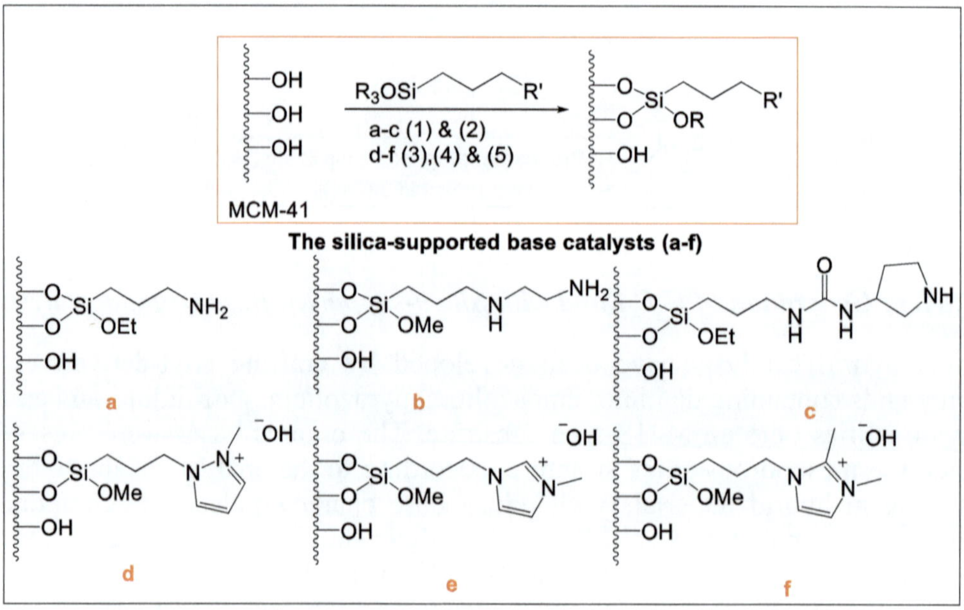

Scheme (2). Production of the silica-supported base.

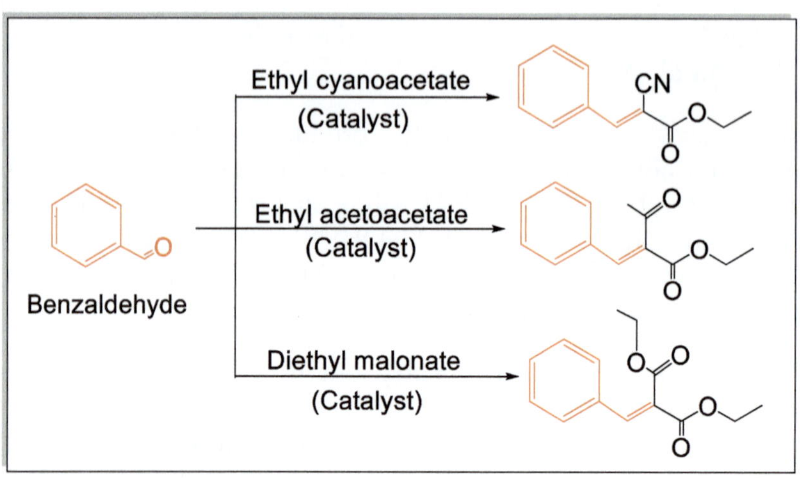

Scheme (3). Solid base catalyzed Knoevenagel reaction.

Base-promoted Synthesis of Medium-Sized Ring-Fused 4-quinolones

Quinones bearing an ortho-amino group reacted with cyclic β-ketoesters in the presence of cesium carbonate to afford 4-quinolones fused with medium-sized rings in excellent yields via a tandem sequence involving conjugate addition, C–C σ-bond cleavage, and nucleophilic aromatic substitution. A straightforward and effective procedure has been developed for the production of 4-quinolone derivatives by the base-promoted cyclic *β*-ketoesters, C-C σ-bond cleavage reaction under mild conditions (Scheme **4**) [65].

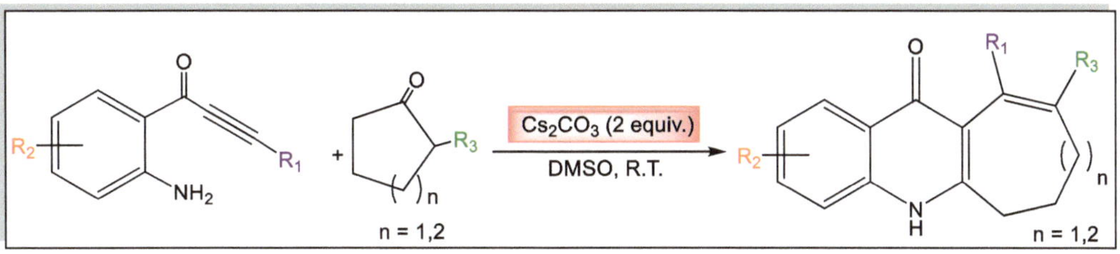

Scheme (4). Synthesis of 4-quinolone derivatives.

This example demonstrates, for the first time, the synthesis of 4-quinolones by base-promoted insertion procedures of carbon-carbon triple bonds into C-C σ-bonds. Mild and transition-metal-free reaction conditions are noteworthy aspects of this program.

Sulfones as Carbene Equivalents for the Efficient Synthesis of Orphaned Cyclopropanes

Although 1,1-dimethylcyclopropanes are found in many small compounds in nature, they are challenging to synthesize using modern metal-catalyzed carbene transfer techniques devoid of inducing opposing 1,2-hydride shifts. A novel mechanistically-based platform for the carbometalation reactions of dialkyl sulfonyl anions, which facilitate the transfer of 1,1-dialkylcarbene units to olefins, has been shown in Scheme **5** [66]. Between 23 °C to 70 °C, dialkyl sulfones react with arylbutadienes and styrenes with the help of $NaNH_2$ or *n*-BuLi in ethereal solvents to form the corresponding 1,1-dialkylcyclopropanes.

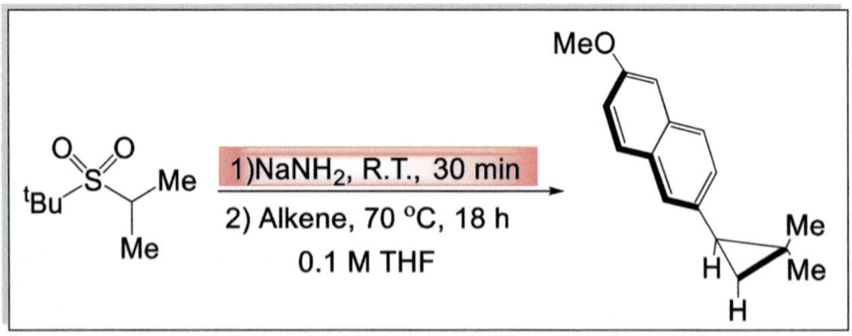

Scheme (5). Synthesis of orphaned cyclopropanes.

There have been 40 recorded instances of this reactivity, including 9 arylbutadienes (51–88% yields), 13 distinct sulfones (46–80% yields), and 16 distinct styrenes (89% yields). Furthermore, this approach has been claimed to be used in a sequential cyclopropanation reaction example. Preliminary mechanistic research suggests a stepwise anionic process initiated by the direct addition of sulfonyl anions to a C-C double bond.

Alkylation of Morita–Baylis–Hillman Carbonates Catalyzed by Cu and Lewis Base Cooperation

A wide range of metal and organocatalyzed annulation processes have made considerable use of (Morita-Baylis-Hillman) MBH adducts, which are easily obtainable chemicals that promote the synthesis of several carbo and heterocyclic compounds. Furthermore, allyl-alkylation agents for allylic alkylation processes have been extensively employed with MBH adduct. A method for asymmetric allylic alkylation involves synthesizing readily accessible azomethine ylides using MBH carbonates. This method employs a cooperative catalysis strategy including copper (I) and a Lewis base (Scheme **6**) [67].

Using co-catalyzed asymmetric allylic alkylation, the relevant amino acid derivatives were generated in up to 90% yields with up to 99% ee, and good to remarkable regioselectivity. The best reaction conditions were 5 mol% $Cu(MeCN)_4PF_6$, 6 mol% Lewis base, and 20 mol% DABCO in dichloromethane at a temperature of about -20 °C.

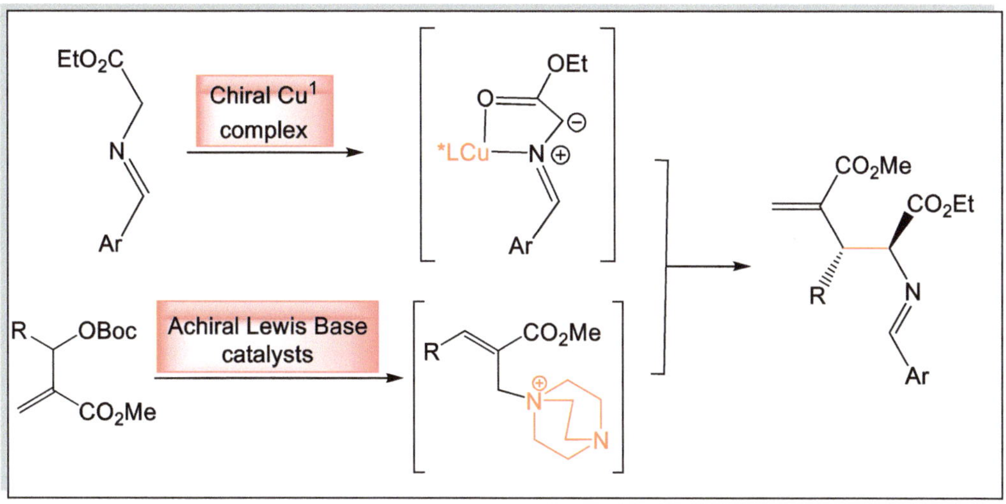

Scheme (6). Asymmetric allylic alkylation *via* cooperative catalysis.

C-N Bond Formation through Solid Base Catalyzed Reactions

The ability of solid base catalysts to effectively and selectively catalyze the production of necessary carbon-nitrogen bonds in organic compounds is both a distinctive property and a significant aspect of C-N bond formation [68 - 70]. Solid base catalysts provide various benefits over typical methods, which could call for hazardous chemicals or unfavorable surroundings [71, 72]. Their ability to catalyze a variety of C-N bond formation events, such as amidation, amination, and imination, among others, makes them special. Because of its adaptability, a variety of nitrogen-containing molecules, including agrochemicals, medicines, and materials with specific qualities, can be synthesized [73 - 78]. Furthermore, solid base catalysts frequently have great selectivity, which allows for exact control over reaction paths, and reduces the production of undesirable byproducts [79, 80]. Their significance for organic chemistry highlights how they influence innovation and development in a diversity of areas, influencing the development of contemporary synthesis techniques.

Catalysts for N-alkylation for Anilines Utilizing Benzylic Alcohols

The identification of a new class of air-stable manganese or rhenium complexes featuring bidentate bis(1,2,3-triazol-5-ylidene) ligands. These complexes have the generic formula [M(bis-triazolylidene)$(CO)_3$X] (M = manganese or rhenium; X = chlorine or bromine), and can have a direct link between the two triazolylidene ligands or a methylene linker. The complexes have been utilized as effective catalysts to promote the *N*-alkylation for anilines utilizing benzylic alcohols (Scheme 7) [81].

It has been shown that the Mn complex, which consists of a bis-triazolylidene ligand with a methylene bridge joining the two triazolylidene fragments, is an extraordinarily efficient catalyst. It produces the desired amine compound with a yield of 92% when a small amount of base (*t*-BuOK, 0.1 equiv) is present.

When Mn complexes with imidazolium- and triazolium-derived *N*-heterocyclic carbene ligands were compared, it was found that Mn triazolylidenes had better catalytic activity when there was less base loading. Furthermore, mechanistic investigations are being conducted with the goal of rationalizing the association between structure and activity.

Silica-supported Approach for the Preparation of Dimeric Pyridinium Bromide

It is possible to synthesize a dimeric substitution ester derivative of pyridinium salt by employing flexible linker units in a conventional or silica-supported technique. The solid-phase method is greatly advantageous to the usual process, since it is non-hazardous, does not require solvents, has a simple work-up method, and has a shorter reaction time. When oxizinone derivatives are produced using conventional techniques, a dimeric ester derivative of pyridinium salts that has been developed shows remarkable catalytic activity. Because they are more flexible and have a lower Lewis character than the extensible pentane moiety, *m*-xylene bridging dimeric pyridinium salts exhibit greater catalytic activity (Scheme **8**) [82].

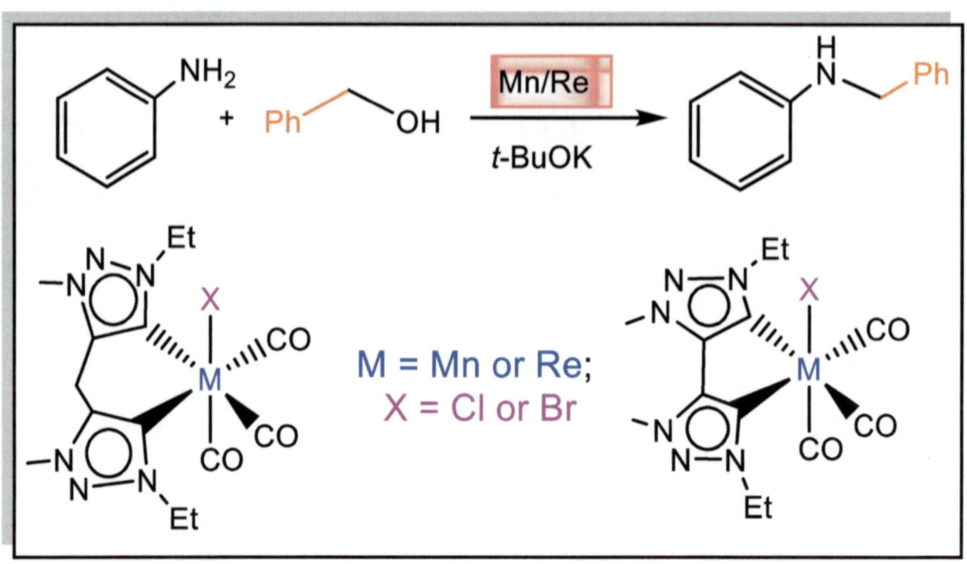

Scheme (7). *N*-Alkylation of amines with alcohols.

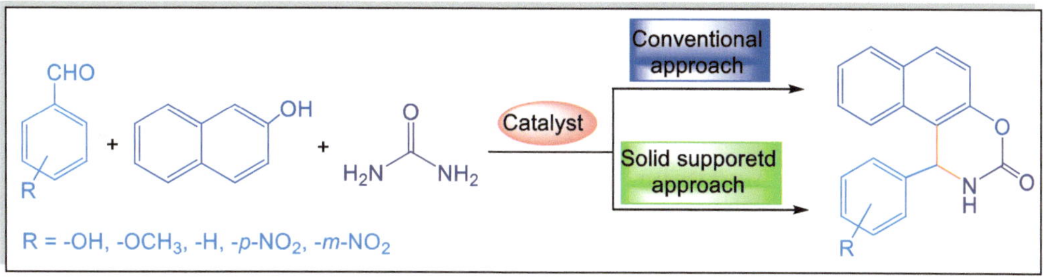

Scheme (8). Synthesis of substituted oxazinone derivatives.

The conventional method involves the use of a bridging dimeric pyridinium bromide in a mixture of MeCN and *m*-xylene. This method typically yields between 80% and 96% in a reaction time ranging from 3 to 120 minutes. A part of the Solid Supported Method (SSA) is using a muffled furnace set to 100°C for 20–45 minutes, resulting in a yield of 84–98%. The muffle furnace method is approximately 15 times more efficient compared to the usual process. The catalytic response of the dimeric ester derivative of pyridinium salts, which contains a bromide counter anion, is superior to other derivatives. This is because the larger size of the anion allows for effective binding, as it is further away from the cation. Therefore, a dimeric ester derivative of pyridinium cation with a bromide counter anion could exhibit superior Lewis properties.

N-Methylation of Nitroarenes Copper/Aluminum Oxide Catalyst

A double hydrolysis precipitation approach was used to create a copper/aluminum oxide catalyst. Using methanol and the Cu/Al_2O_3 catalyst, nitroarenes were successfully *N*-methylated in a continuous mode without the need for an external base or H_2. It is feasible to regulate the methylation process of nitroarenes and produce *N*-monomethylanilines or N, *N*-dimethylanilines as required by modifying the reaction conditions (Scheme **9**) [83].

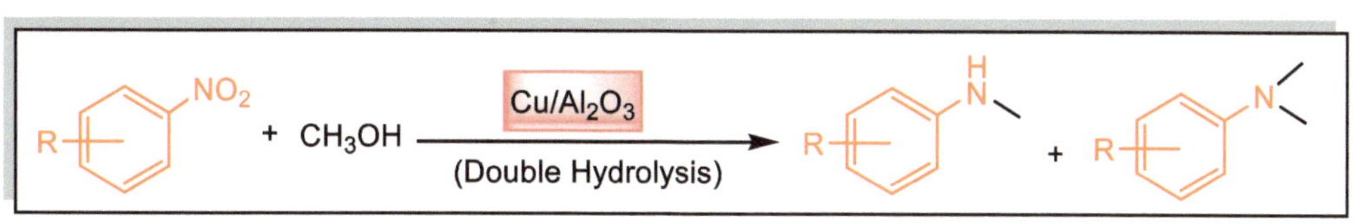

Scheme (9). Methylation of nitroarenes utilizing methanol.

The characterization revealed that the two-fold hydrolysis precipitation approach resulted in increased enrichment and dispersion of the active metal Cu on the surface, as well as a better-developed pore system in Cu/Al_2O_3. These factors contributed to the higher efficiency of catalysts compared to Cu/Al_2O_3-CP. The procedure exhibited the advantages of being cost-effective, environmentally friendly, highly controllable, and widely applicable.

Synthesis for N-alkylation of Amines with Alcohols

A newly developed bimetallic catalyst, $Fe_{10}Pd_1$/NC500, has been formed by integrating the bimetallic Metal-Organic Framework (MOF) NH_2-MIL-101 ($Fe_{10}Pd_1$). This catalyst demonstrates exceptional catalytic efficiency in employing a Hydrogen Auto-Transfer methodology (HAT) in the *N*-alkylation reaction of amines with alcohols (Scheme **10**) [84]. Optimal results were obtained at 120 °C using a molar ratio of 2:1 between alcohol and amine, without the requirement of any additional ingredient or solvent. Comparing bimetallic alloys to monometallic catalysts, the synergistic effects within them result in a considerable gain in catalytic performance. Furthermore, the $Fe_{10}Pd_1$/NC500 catalyst has demonstrated a wide variety of substrates and exceptional reusability. This study emphasizes the potential of bimetallic catalysts in sustainable chemistry.

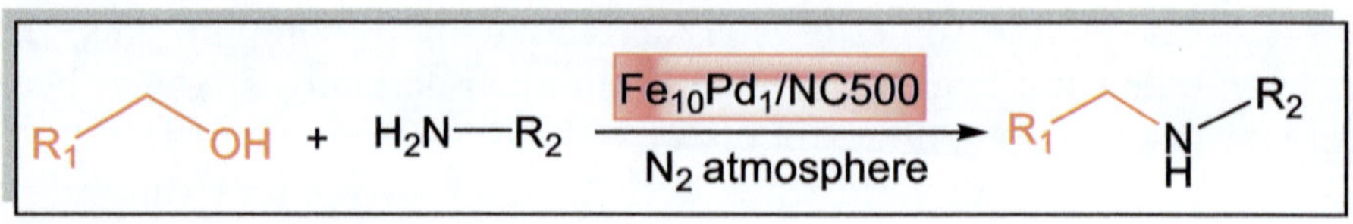

Scheme (10). *N*-Alkylation of amines using alcohols.

BH (Borrowing Hydrogen) or HAT process, which involves intermediate reaction, dehydrogenation, and hydrogenation, is widely acknowledged as a highly effective approach for one-pot synthesis, considering its environmental and economic advantages. Despite significant efforts in catalyst research for HAT reactions, the majority of reported catalysis systems still require huge quantities of noble metals, severe conditions, or external bases. Therefore, it is still important to explore an extremely effective and reusable heterogeneous catalyst.

N-Alkylation of Heterocycles with 1,3-Dibromopropane

Under mild circumstances, using Schiff base Ni(II), Co(II), Zn(II), and Cu(II) complexes of transition metals as a catalyst in 1:1 and 2:1 coupling ratios, the *N*-alkylation of heterocycles with 1,3-dibromopropane was studied. All complexes were found to function as effective catalysts, producing 78–92% of the product at a coupling ratio of 1:1, and 63–78% at a coupling ratio of 2:1. Compared to *N*-alkylation in a 2:1 coupling ratio, *N*-alkylation of heterocycles with 1,3-dibromopropane at a 1:1 coupling ratio is easier and yields better results (Scheme **11**) [85].

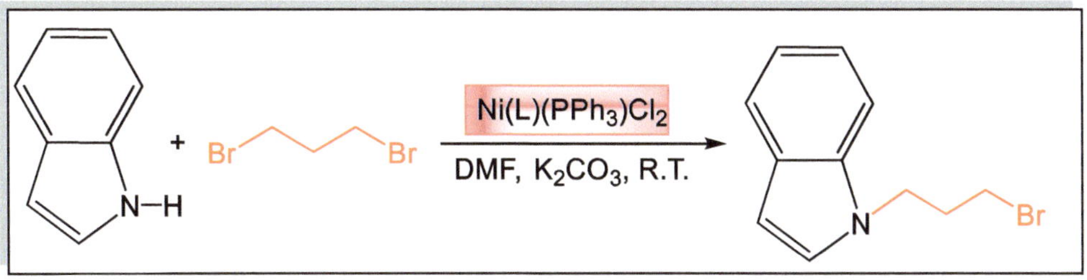

Scheme (11). *N*-Alkylation of indole with 1,3-dibromopropane.

For these reactions, the $[Ni(L)(PPh_3)_2Cl_2]$ complex is more appropriate. This catalytic method's features include broad substrate scope, facile synthesis of Schiff base complexes, mild catalytic reaction conditions, and extreme experiment simplicity.

C-S bond Formation through Solid Base Catalyzed Reactions

C-S coupling is one of the most significant reactions in chemical synthesis, which involves the production of Carbon-Sulfur (C-S) bonds. Solid base catalysts play a crucial role in facilitating and optimizing this transition [86 - 88]. Because of their well-known surface basicity, metal oxides such as MgO and CaO are effective catalysts for C-S coupling processes. They make it easier for thiols to deprotonate, and alkyl halides to activate, effectively forming C-S bonds [89, 90]. Ion-exchange resins and alkali metal carbonates/bicarbonates, such as Na_2CO_3 and $KHCO_3$, offer substituted solid base catalysts with unique catalytic environments and recyclable nature [91, 92]. Alkali metal carbonates/bicarbonates have found applications in the formation of sulfides, sulfur-containing heterocycles, and sulfones, contributing to the development of pharmaceuticals, agrochemicals, and functional materials [93 - 97]. When compared to their unsupported counterparts, bases supported on alumina, including potassium carbonate or cesium fluoride, offer improved catalytic activity and stability. Under benign reaction circumstances, these catalysts facilitate the production of C-S bonds by offering readily available basic sites for substrate activation [98 - 102]. Ion-exchange resins, modified with quaternary ammonium salt or pyridinium groups, are effective catalysts for C-S coupling processes, because they have a huge surface area, adjustable characteristics, and can be reused [103 - 105].

Preparation of Chiral Sulfoximines through Sulfur-selective Alkylation

In pharmaceutical chemistry, chiral sulfoximines are currently regarded as interesting bioisosteres. However, there is a lack of developed procedures for stereoselectively preparing chiral sulfoximines. An accessible chiral sulfinamide can be stereospecifically *S*-alkylated to produce chiral sulfoximines asymmetrically, under practical conditions (Scheme **12**) [106].

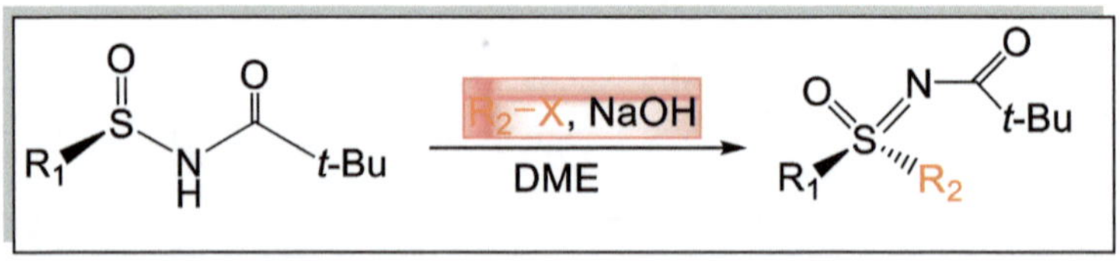

Scheme (12). Sulfur-selective alkylation.

Using the sodium salt of sulfinamide to create the *in situ*-generated crown ether complex allowed for the analysis of its X-ray crystal structure, which provided insights into the reaction's significant regioselectivity and the role of the pivaloyl group on the nitrogen atom. *S*-alkylation of sulfinamides developed under realistic circumstances based on the crystal structure analysis, avoiding the need for costly reagents like NaH and 15-crown-5.

S-Alkylation of Aryl thioureas and 2-mercaptobenzothiazole

A practical and extremely effective technique was offered for *S*-alkylation of arylthioureas. Using tetraalkylammonium salts as alkylation reagents, a range of 68 S-substituted aryl-isothioureas were synthesized in good to outstanding yields under transition-metal-free conditions (Scheme **13**) [107]. The approach exhibits straightforward and efficient performance, wide tolerance for different functional groups, high yields ranging from good to excellent, and readily accessible preliminary materials. This demonstrates its promise for synthesizing a wide range of pharmaceutically or biologically active molecules.

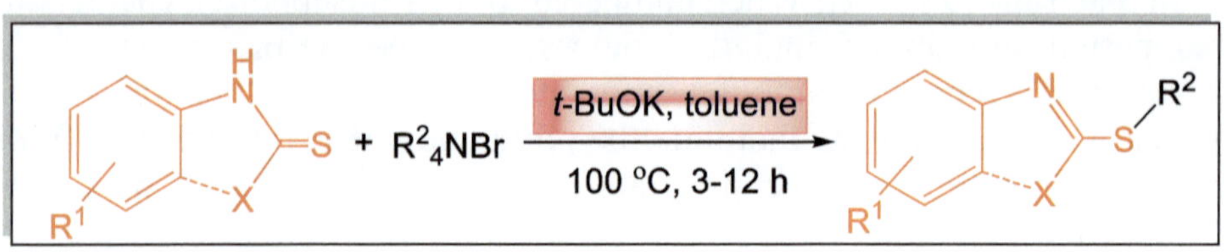

Scheme (13). *S*-Alkylation of arylthioureas and 2-mercaptobenzothiazoles.

Tetraalkylammonium salts were used as alkylation reagents to synthesize a series of 68 *S*-substituted aryl-isothioureas in good to outstanding yields under transition-metal-free conditions. The approach has easy-to-find starting ingredients, wide functional group tolerance, decent to outstanding yields, and straightforward performance.

Sulfilimine Formation via N-acyl Sulfenamides

The reaction of alkyl halides with *N*-acyl sulfenamides by sulfur alkylation produces sulfilimines with yields ranging from 47% to 98% (Scheme **14**) [108]. A wide range was created, encompassing several alkyl and aryl sulfenamides with differing *N*-acyl groups. Alkyl halides with varying steric and electronic characteristics, such as propargyl, benzyl, secondary, primary, and methyl halides, were shown to be effective inputs. Additionally, an example of asymmetric phase-transfer alkylation was showcased. An easily transformable sulfilimine compound was transformed into an *N*-acyl compound and one free sulfoximine compound, both of which are significant structures in the field of medicinal chemistry.

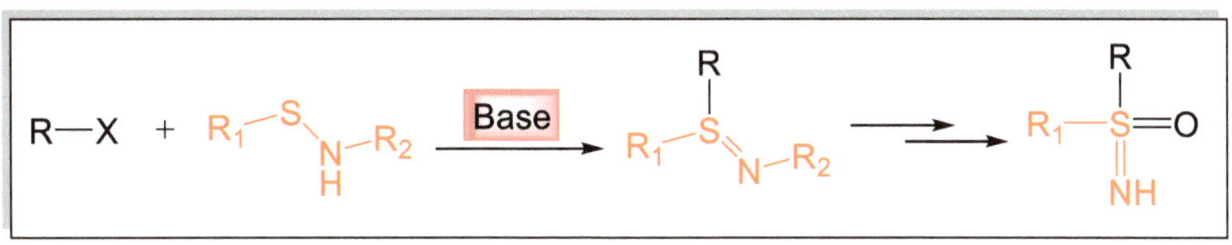

Scheme (14). Sulfilimine formation *via* *N*-acyl sulfonamides.

Polar solvents and counter ions exhibiting mostly ionic rather than covalent bonding interactions enhance the sulfur alkylation of sulfenamides. Based on empirical data, it may be concluded that a zwitterionic product yields by *S*-alkylation, but not by *N*-alkylation. Under simple reaction conditions, significant amounts of the sulfilimine compounds were frequently produced. It is easy to diversify the sulfilimine compounds into sulfoximines that are relevant to pharmaceuticals. Additionally, a unique catalytic asymmetric phase-transfer *S*-benzylation proof-of-concept for the asymmetric synthesis of sulfilimines is presented.

Synthesis of Sulfilimines using S-arylation of Sulfenamides

Sulfilimines play a crucial role as intermediary compounds in the synthesis of important structures found in pharmaceuticals and agrochemicals. Usually, these types of compounds are synthesized through the process of oxidation of thioethers, sulfur alkylation catalyzed by transition metals or enhanced by bases, and sulfur arylation catalyzed by transition metals. An effective and efficient method is using a base to introduce sulfur aryl groups in order to synthesize sulfilimines. Various *N*-aryl and *N*-acyl sulfenamides can react with different diaryliodonium salts effectively, resulting in the production of sulfilimines with exceptional chemoselectivities and large quantities (Scheme **15**) [109].

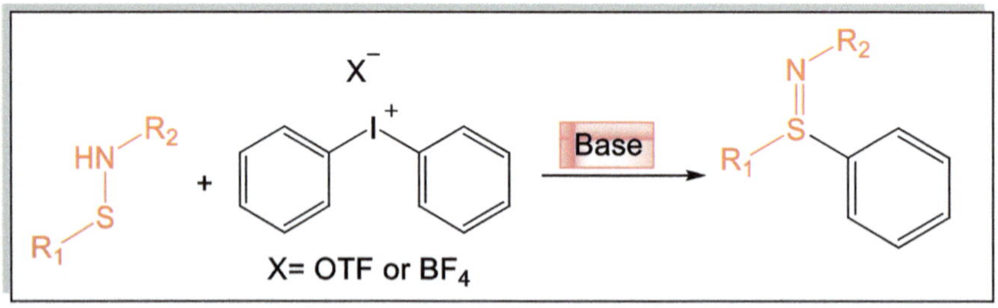

Scheme (15). Synthesis of sulfilimines.

The use of finely powdered solid inorganic bases has been suggested to avoid affecting reaction efficiency and yields. The resultant sulfilimines are adaptable synthons that are easily transformed into further significant sulfur-containing molecules, such as sulfoximines. Numerous noteworthy attributes of this reaction include gentle reaction conditions, straightforward procedures, a wide range of substrates, good yields, and strong functional group tolerance.

Sulfur-arylation of Sulfonamides

The sulfur-(hetero)arylation of sulfenamides, employing commercially accessible (hetero)aryl iodides *via* Ullmann-type coupling and inexpensive CuI(I) as the catalyst, demonstrated a wide range of reaction inputs, including aryl and alkyl sulfenamides, as well as extremely sterically mired aryl and heteroaryl iodides with five and six membered rings. The (hetero)arylation of S-methyl sulfenamides, including those involving complex aryl iodides, has been reported. This approach is applicable to a range of bioactive sulfur compounds in high oxidation states. The Smiles rearrangement of *S*-heteroaryl sulfilimines that lack electrons is also revealed (Scheme **16**) [110].

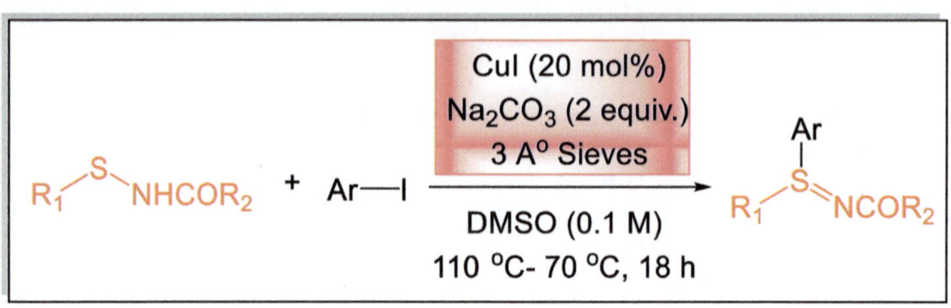

Scheme (16). Sulfur-arylation of sulfonamides.

The primary application of *S*-methyl sulfenamides in the production of C-S bonds was shown, and this could offer quick and easy access to sulfur compounds with

S-methyl substituents at high oxidation states. One-step synthesis of S-methyl sulfenamides has been developed using commercially available reagents, along with the first examples of their use in forming carbon–sulfur bonds for advanced aryl iodide-based drug intermediates and pharmaceuticals.

CONCLUDING REMARKS

In conclusion, solid base catalysts have played a crucial role in organic synthesis, as demonstrated by the past and present developments in these materials for the creation of C-C, C-N, and C-S bonds. The development of solid base catalysis, from early metal oxides to contemporary graphene-based carbocatalysts, has transformed the sector and provided more cost-effective, environmentally friendly, and efficient substitutes for conventional transition metal catalysts. The versatility of solid base catalysts enables the construction of diverse carbon-carbon and carbon-heteroatom bonds, crucial for the synthesis of pharmaceuticals, agrochemicals, and advanced materials. Solid base catalysts are expected to play a major role in promoting innovation and sustainability in organic synthesis, opening doors to the creation of new chemical processes with a wide range of industrial applications, as research into new catalytic systems and the optimization of current ones continue.

REFERENCES

[1] Hattori, H. Solid base catalysts: generation of basic sites and application to organic synthesis. *Appl. Catal. A Gen.,* **2001**, *222*(1-2), 247-259.
[http://dx.doi.org/10.1016/S0926-860X(01)00839-0]

[2] Ono, Y. Solid base catalysts for the synthesis of fine chemicals. *J. Catal.,* **2003**, *216*(1-2), 406-415.
[http://dx.doi.org/10.1016/S0021-9517(02)00120-3]

[3] Figueras, F. Base catalysis in the synthesis of fine chemicals. *Top. Catal.,* **2004**, *29*(3/4), 189-196.
[http://dx.doi.org/10.1023/B:TOCA.0000029802.75155.5a]

[4] Tanabe, K. *Solid acids and bases: their catalytic properties*; Elsevier, **2012**.

[5] Sun, L.B.; Liu, X.Q.; Zhou, H.C. Design and fabrication of mesoporous heterogeneous basic catalysts. *Chem. Soc. Rev.,* **2015**, *44*(15), 5092-5147.
[http://dx.doi.org/10.1039/C5CS00090D] [PMID: 26051787]

[6] Rath, D.; Rana, S.; Parida, K.M. Organic amine-functionalized silica-based mesoporous materials: an update of syntheses and catalytic applications. *RSC Advances,* **2014**, *4*(100), 57111-57124.
[http://dx.doi.org/10.1039/C4RA08005J]

[7] Velty, A.; Corma, A. Advanced zeolite and ordered mesoporous silica-based catalysts for the conversion of CO_2 to chemicals and fuels. *Chem. Soc. Rev.,* **2023**, *52*(5), 1773-1946.
[http://dx.doi.org/10.1039/D2CS00456A] [PMID: 36786224]

[8] Jambhulkar, D.K.; Ugwekar, R.P.; Bhanvase, B.A.; Barai, D.P. A review on solid base heterogeneous catalysts: preparation, characterization and applications. *Chem. Eng. Commun.,* **2022**, *209*(4), 433-484.
[http://dx.doi.org/10.1080/00986445.2020.1864623]

[9] Hattori, H. Solid base catalysts: generation, characterization, and catalytic behavior of basic sites. *J. Jpn. Petrol. Inst.,* **2004**, *47*(2), 67-81.
[http://dx.doi.org/10.1627/jpi.47.67]

[10] Bing, W.; Wei, M. Recent advances for solid basic catalysts: Structure design and catalytic performance. *J. Solid State Chem.,* **2019**, *269*, 184-194. [http://dx.doi.org/10.1016/j.jssc.2018.09.023]

[11] Jayakumar, M.; Karmegam, N.; Gundupalli, M.P.; Bizuneh Gebeyehu, K.; Tessema Asfaw, B.; Chang, S.W.; Ravindran, B.; Kumar Awasthi, M. Heterogeneous base catalysts: Synthesis and application for biodiesel production – A review. *Bioresour. Technol.,* **2021**, *331*125054 [http://dx.doi.org/10.1016/j.biortech.2021.125054] [PMID: 33832828]

[12] Reif, P.; Rosenthal, H.; Rose, M. Biomass-Derived Aromatics by Solid Acid-Catalyzed Aldol Condensation of Alkyl Methyl Ketones. *Adv. Sustain. Syst.,* **2020**, *4*(10)1900150 [http://dx.doi.org/10.1002/adsu.201900150]

[13] Munyentwali, A.; Li, H.; Yang, Q. Review of advances in bifunctional solid acid/base catalysts for sustainable biodiesel production. *Appl. Catal. A Gen.,* **2022**, *633*118525 [http://dx.doi.org/10.1016/j.apcata.2022.118525]

[14] Mukhtar, A.; Saqib, S.; Lin, H.; Hassan Shah, M.U.; Ullah, S.; Younas, M.; Rezakazemi, M.; Ibrahim, M.; Mahmood, A.; Asif, S.; Bokhari, A. Current status and challenges in the heterogeneous catalysis for biodiesel production. *Renew. Sustain. Energy Rev.,* **2022**, *157*112012 [http://dx.doi.org/10.1016/j.rser.2021.112012]

[15] Abrams, R.; Clayden, J. Photocatalytic difunctionalization of vinyl ureas by radical addition polar Truce–Smiles rearrangement cascades. *Angew. Chem. Int. Ed.,* **2020**, *59*(28), 11600-11606. [http://dx.doi.org/10.1002/anie.202003632] [PMID: 32227575]

[16] Sheldon, R.A.; Arends, I.; Hanefeld, U. *Green chemistry and catalysis*; John Wiley & Sons, **2007**. [http://dx.doi.org/10.1002/9783527611003]

[17] Formica, M.; Rozsar, D.; Su, G.; Farley, A.J.M.; Dixon, D.J. Bifunctional iminophosphorane superbase catalysis: applications in organic synthesis. *Acc. Chem. Res.,* **2020**, *53*(10), 2235-2247. [http://dx.doi.org/10.1021/acs.accounts.0c00369] [PMID: 32886474]

[18] Anastasiou DE. Teaching green chemistry and engineering through the epoxidation of poly- β-myrcene. *ChemPlusChem.* **2024**, *89*(8), e202400167

[19] Huang, R.; Chang, J.; Choi, H.; Vohs, J.M.; Gorte, R.J. Furfural upgrading by aldol condensation with ketones over solid-base catalysts. *Catal. Lett.,* **2022**, *152*(12), 3833-3842. [http://dx.doi.org/10.1007/s10562-022-03960-1]

[20] Anbu, N.; Hariharan, S.; Dhakshinamoorthy, A. Knoevenagel-Doebner condensation promoted by chitosan as a reusable solid base catalyst. *Molecular Catalysis,* **2020**, *484*110744 [http://dx.doi.org/10.1016/j.mcat.2019.110744]

[21] Wang, J.; Yan, T.; Tang, Y.; Miao, Y. High efficient Aldol condensation reaction utilizing modified calcium oxide as stable solid base catalyst. *Kinet. Catal.,* **2016**, *57*(4), 439-445. [http://dx.doi.org/10.1134/S0023158416040145]

[22] Jose, T.; Ftouni, J.; Bruijnincx, P.C.A. Structured hydroxyapatite composites as efficient solid base catalysts for condensation reactions. *Catal. Sci. Technol.,* **2021**, *11*(10), 3428-3436. [http://dx.doi.org/10.1039/D1CY00102G]

[23] Kabashima, H.; Tsuji, H.; Shibuya, T.; Hattori, H. Michael addition of nitromethane to α,β-unsaturated carbonyl compounds over solid base catalysts. *J. Mol. Catal. Chem.,* **2000**, *155*(1-2), 23-29. [http://dx.doi.org/10.1016/S1381-1169(99)00316-7]

[24] Xu, C.; Bartley, J.K.; Enache, D.I.; Knight, D.W.; Hutchings, G.J. High surface area MgO as a highly effective heterogeneous base catalyst for michael addition and knoevenagel condensation reactions. *Synthesis,* **2005**, (19), 3468-3476.

[25] Senra, D. J., CS Aguiar, L., & BC Simas, A. Recent progress in transition-metal-catalyzed cn cross-couplings: emerging approaches towards sustainability. *Curr. Org. Synth.,* **2011**, *8*(1), 53-78.

[http://dx.doi.org/10.2174/1570179117944076683]

[26] Yang, C.; Li, Z.; Xu, J.; Jiang, Y.; Zhu, W. Electrocatalytic C–N coupling for urea synthesis: a critical review. *Green Chem.*, **2024**, *26*(9), 4908-4933. [http://dx.doi.org/10.1039/D3GC04920E]

[27] Zhu, S.; Li, H.; Li, Y.; Huang, Z.; Chu, L. Exploring visible light for carbon-nitrogen and carbon–oxygen bond formation *via* nickel catalysis. *Org. Chem. Front.*, **2023**, *10*(2), 548-569. [http://dx.doi.org/10.1039/D2QO01700H]

[28] Singh, S.; Roy, V.J.; Dagar, N.; Sen, P.P.; Roy, S.R. Photocatalysis in Dual Catalysis Systems for Carbon-Nitrogen Bond Formation. *Adv. Synth. Catal.*, **2021**, *363*(4), 937-979. [http://dx.doi.org/10.1002/adsc.202001176]

[29] Wang, B.; Li, S.; Tian, S.; Feng, R.; Meng, Y. A new solid base catalyst for the transesterification of rapeseed oil to biodiesel with methanol. *Fuel*, **2013**, *104*, 698-703. [http://dx.doi.org/10.1016/j.fuel.2012.08.034]

[30] Zhang Z, Meng P, Luo H, Pei Z, Liu X. Lewis acid-base site-assisted *in situ* transesterification catalysis to produce biodiesel. *Catalysts*. **2024**, *14*(10), 731

[31] Li, X.; Yuan, X.; Hu, J.; Li, Y.; Bao, H. Radical decarboxylative carbon–nitrogen bond formation. *Molecules*, **2023**, *28*(10), 4249. [http://dx.doi.org/10.3390/molecules28104249] [PMID: 37241989]

[32] Ma, Y.; Li, H.; Yang, H.; Zhu, Y.; Zhao, L.; Li, M. Effects of solid acid and base catalysts on pyrolysis of rice straw and wheat straw biomass for hydrocarbon production. *Journal of the Energy Institute*, **2022**, *101*, 140-148. [http://dx.doi.org/10.1016/j.joei.2021.08.010]

[33] Lin, J.; Tian, W.; Guan, Z.; Zhang, H.; Duan, X.; Wang, H.; Sun, H.; Fang, Y.; Huang, Y.; Wang, S. Functional carbon nitride materials in photo-Fenton-like catalysis for environmental remediation. *Adv. Funct. Mater.*, **2022**, *32*(24)2201743 [http://dx.doi.org/10.1002/adfm.202201743]

[34] Zuo, Z.; Sha, Y.; Wang, P.; Da, Z. From bench to industry, the application of all-inorganic solid base materials in traditional heterogeneous catalysis: a mini review. *RSC Advances*, **2024**, *14*(11), 7468-7489. [http://dx.doi.org/10.1039/D4RA00335G] [PMID: 38440274]

[35] Lee, C.F.; Liu, Y.C.; Badsara, S.S. Transition-metal-catalyzed C-S bond coupling reaction. *Chem. Asian J.*, **2014**, *9*(3), 706-722. [http://dx.doi.org/10.1002/asia.201301500] [PMID: 24443103]

[36] Hosseinian, A.; Nezhad, P.D.K.; Ahmadi, S.; Rahmani, Z.; Monfared, A. A walk around the decarboxylative C-S cross-coupling reactions. *J. Sulfur Chem.*, **2019**, *40*(1), 88-112. [http://dx.doi.org/10.1080/17415993.2018.1515314]

[37] Schlögl, R. Carbons. In: *Handbook of heterogeneous catalysis*; Wiley-VCH, **2008**; pp. 357-427. [http://dx.doi.org/10.1002/9783527610044.hetcat0020]

[38] Zhang, Y.; Huang, C.; Mi, L. Metal–organic frameworks as acid- and/or base-functionalized catalysts for tandem reactions. *Dalton Trans.*, **2020**, *49*(42), 14723-14730. [http://dx.doi.org/10.1039/D0DT03025B] [PMID: 33047777]

[39] Yadav, G.D.; Wagh, D.P. Claisen-Schmidt Condensation using Green Catalytic Processes: A Critical Review. *ChemistrySelect*, **2020**, *5*(29), 9059-9085. [http://dx.doi.org/10.1002/slct.202001737]

[40] Zhang, Y.; Guan, W.; Song, H.; Wei, Y.; Jin, P.; Li, B.; Yan, C.; Pan, J.; Yan, Y. Coupled acid and base UiO-66-type MOFs supported on g-C3N4 as a bi-functional catalyst for one-pot production of 5-HMF from glucose. *Microporous Mesoporous Mater.*, **2020**, *305*110328 [http://dx.doi.org/10.1016/j.micromeso.2020.110328]

[41] Hernández-Giménez, A.M.; Hernando, H.; Danisi, R.M.; Vogt, E.T.C.; Houben, K.; Baldus, M.; Serrano, D.P.; Bruijnincx, P.C.A.; Weckhuysen, B.M. Deactivation and regeneration of solid acid and base catalyst bodies used in cascade for bio-oil synthesis and upgrading. *J. Catal.,* **2022**, *405*, 641-651. [http://dx.doi.org/10.1016/j.jcat.2021.09.029]

[42] Khalilzadeh, M.A.; Kim, S.Y.; Jang, H.W.; Luque, R.; Varma, R.S.; Venditti, R.A.; Shokouhimehr, M. Carbohydrate-based nanostructured catalysts: applications in organic transformations. *Mater. Today Chem.,* **2022**, *24*100869
[http://dx.doi.org/10.1016/j.mtchem.2022.100869]

[43] Li, J.P.H.; Adesina, A.A.; Kennedy, E.M.; Stockenhuber, M. A mechanistic study of the Knoevenagel condensation reaction: new insights into the influence of acid and base properties of mixed metal oxide catalysts on the catalytic activity. *Phys. Chem. Chem. Phys.,* **2017**, *19*(39), 26630-26644. [http://dx.doi.org/10.1039/C7CP04743F] [PMID: 28956036]

[44] Wang A, Ma Y, Zhao D. Pore engineering of porous materials: Effects and applications. *ACS nano.* **2024**, *18*(34), 22829-54

[45] Maurya, A.; Singh, S.; Pathak, N.P. The Importance of mesoporous materials (silica, alumina, and zeolite) as solid supports for metal complex catalysts in organic transformations. *J. Inorg. Organomet. Polym. Mater.,* **2024**, *2024*, 1-21.

[46] Shah, S.S.A.; Sohail, M.; Murtza, G.; Waseem, A.; Rehman, A.; Hussain, I.; Bashir, M.S.; Alarfaji, S.S.; Hassan, A.M.; Nazir, M.A.; Javed, M.S.; Najam, T. Recent trends in wastewater treatment by using Metal-Organic Frameworks (MOFs) and their composites: A critical view-point. *Chemosphere,* **2024**, *349*(349)140729
[http://dx.doi.org/10.1016/j.chemosphere.2023.140729] [PMID: 37989439]

[47] Niu, Y.Q.; Liu, J.H.; Aymonier, C.; Fermani, S.; Kralj, D.; Falini, G.; Zhou, C.H. Calcium carbonate: controlled synthesis, surface functionalization, and nanostructured materials. *Chem. Soc. Rev.,* **2022**, *51*(18), 7883-7943.
[http://dx.doi.org/10.1039/D1CS00519G] [PMID: 35993776]

[48] Ceaser R, Montané D, Medina F, Constantí M. Integration of deep eutectic solvents and hydrotalcites for biomass conversion and aldol condensation: Toward platform chemicals and jet fuel synthesis—a review. *ChemBioEng Reviews.* **2024**, *11*(4), e202400052

[49] Ispán, D.; Varga, B.; Balogh, S.; Zsirka, B.; Gömöry, Á.; Skoda-Földes, R. Claisen□schmidt condensation and domino claisen-schmidt condensation - michael addition of 16-formyl steroids in the presence of switchable polarity solvents. *ChemistrySelect,* **2021**, *6*(23), 5705-5710.
[http://dx.doi.org/10.1002/slct.202100886]

[50] Hazra, C.K.; Singh, S. Recent advancements in typical friedel–crafts alkylation reactions focused on targeting arene nucleophiles. *Synthesis,* **2024**, *56*(3), 368-388.
[http://dx.doi.org/10.1055/s-0042-1751492]

[51] Wieland, W.S.; Davis, R.J.; Garces, J.M. Solid base catalysts for side-chain alkylation of toluene with methanol. *Catal. Today,* **1996**, *28*(4), 443-450.
[http://dx.doi.org/10.1016/S0920-5861(96)00247-7]

[52] Wang, B.; Huang, W.; Wen, Y.; Zuo, Z.; Gao, Z.; Yin, L. Styrene from toluene by side chain alkylation over a novel solid acid-base catalyst. *Catal. Today,* **2011**, *173*(1), 38-43.
[http://dx.doi.org/10.1016/j.cattod.2011.05.038]

[53] Nishimura, Y.; Dasireddy, V.D.; Joseph, S.; Sugi, Y.; Vinu, A. Heterogeneous friedel–crafts alkylation. *Industrial Arene Chemistry: Markets, Technologies. Sustainable Processes and Cases Studies of Aromatic Commodities,* **2023**, *2*, 595-643.

[54] Wen, Z.; Cao, F.; Liu, J.B.; Xue, B. Shape-selective alkylation of toluene with dimethyl carbonate into p-Xylene over MgO/MCM-22 prepared by a novel pre-impregnation method. *Catal. Lett.,* **2024**, *154*(1), 170-180.

[http://dx.doi.org/10.1007/s10562-023-04289-z]

[55] Navarro de Miguel, J.C.; Chung, S.H.; Dikhtiarenko, A.; Li, T.; Patarroyo, J.; Ruiz-Martínez, J. Brønsted acid-site density controls the mechanistic cycle and product selectivity in the methanol-t--hydrocarbons reaction in bea zeolite. *ACS Catal.*, **2024**, *14*(8), 5989-6000. [http://dx.doi.org/10.1021/acscatal.3c06077]

[56] Chen, V.Y.; Kwon, O. Unified Approach to Furan Natural Products *via* Phosphine-Palladium catalysis. *Angew. Chem. Int. Ed.*, **2021**, *60*(16), 8874-8881. [http://dx.doi.org/10.1002/anie.202015232] [PMID: 33533120]

[57] van Vreeswijk, S.H.; Monai, M.; Oord, R.; Schmidt, J.E.; Parvulescu, A.N.; Yarulina, I.; Karwacki, L.; Poplawsky, J.D.; Weckhuysen, B.M. Detecting cage crossing and filling clusters of magnesium and carbon atoms in zeolite SSZ-13 with atom probe tomography. *JACS Au*, **2022**, *2*(11), 2501-2513. [http://dx.doi.org/10.1021/jacsau.2c00296] [PMID: 36465530]

[58] Bhongale, P.; Joshi, S.; Mali, N. A comprehensive review on catalytic *O*- alkylation of phenol and hydroquinone. *Catal. Rev., Sci. Eng.*, **2023**, *65*(2), 455-500. [http://dx.doi.org/10.1080/01614940.2021.1930490]

[59] Hong, Z.; Xiong, C.; Zhao, G.; Zhu, Z. Side-chain alkylation of toluene with methanol to produce styrene: an overview. *Catal. Sci. Technol.*, **2019**, *9*(24), 6828-6840. [http://dx.doi.org/10.1039/C9CY01581G]

[60] Afreen, G.; Upadhyayula, S. Alkylation of phenol and substituted phenols with C1–C4 alcohols/olefins as an upgrading route for bio-oil oxygenates: A review. *Renew. Sustain. Energy Rev.*, **2021**, *147*111189 [http://dx.doi.org/10.1016/j.rser.2021.111189]

[61] Martín-Aranda, R.M.; Cervantes, M.R.; López-Peinado, A.J.; López-González, J.D.D. Alkaline carbons as base catalysts: Alkylation of imidazole with alkyl halides. *J. Mol. Catal.*, **1993**, *85*(2), 253-264. [http://dx.doi.org/10.1016/0304-5102(93)80106-5]

[62] Bandini, M.; Melloni, A.; Umani-Ronchi, A. New catalytic approaches in the stereoselective friedel-crafts alkylation reaction. *Angew. Chem. Int. Ed.*, **2004**, *43*(5), 550-556. [http://dx.doi.org/10.1002/anie.200301679] [PMID: 14743405]

[63] Shimizu, H.; Maeda, H.; Nara, H. Highly productive α-alkylation of ketones with alcohols mediated by an ir–oxalamidato/solid base catalyst system. *Org. Process Res. Dev.*, **2020**, *24*(11), 2772-2779. [http://dx.doi.org/10.1021/acs.oprd.0c00362]

[64] Erigoni, A.; Hernández-Soto, M.C.; Rey, F.; Segarra, C.; Díaz, U. Highly active hybrid mesoporous silica-supported base organocatalysts for CC bond formation. *Catal. Today*, **2020**, *345*, 227-236. [http://dx.doi.org/10.1016/j.cattod.2019.09.041]

[65] Wang, M.; Ma, J.; Wang, J.; Lu, W.; Pan, B. Base promoted synthesis of 4-quinolones fused with medium□sized rings. *Eur. J. Org. Chem.*, **2023**, *26*(46)e202300519 [http://dx.doi.org/10.1002/ejoc.202300519]

[66] Johnson, J.D.; Teeples, C.R.; Akkawi, N.R.; Wilkerson-Hill, S.M. Efficient synthesis of orphaned cyclopropanes using sulfones as carbene equivalents. *J. Am. Chem. Soc.*, **2022**, *144*(32), 14471-14476. [http://dx.doi.org/10.1021/jacs.2c07063] [PMID: 35939335]

[67] Hu, Y.; Yan, Z.; Shi, W.; Liao, J.; Liu, M.; Pan, T.; Wang, W.; Wu, Y.; Hao, X.; Guo, H. Copper/lewis base cooperatively catalyzed asymmetric allylic alkylation of Morita–Baylis–Hillman carbonates with azomethine ylides. *Chem. Commun. (Camb.)*, **2021**, *57*(65), 8059-8062. [http://dx.doi.org/10.1039/D1CC02861H] [PMID: 34296236]

[68] Xu, J.; Shen, K.; Xue, B.; Li, Y.X. Microporous carbon nitride as an effective solid base catalyst for Knoevenagel condensation reactions. *J. Mol. Catal. Chem.*, **2013**, *372*, 105-113. [http://dx.doi.org/10.1016/j.molcata.2013.02.019]

[69] Kan-nari, N.; Okamura, S.; Fujita, S.; Ozaki, J.; Arai, M. Nitrogen-doped carbon materials prepared by ammoxidation as solid base catalysts for knoevenagel condensation and transesterification reactions. *Adv. Synth. Catal.,* **2010**, *352*(9), 1476-1484. [http://dx.doi.org/10.1002/adsc.201000029]

[70] Wang, Y.; Zhang, F.L.; Liu, Z.J.; Yao, Z.J. Half-sandwich iridium complexes with hydrazone ligands: Synthesis and catalytic activity in *N* -alkylation of anilines or nitroarenes with alcohols *via* hydrogen autotransfer. *Inorg. Chem.,* **2022**, *61*(27), 10310-10320. [http://dx.doi.org/10.1021/acs.inorgchem.2c00703] [PMID: 35767836]

[71] Liang, M.; Xu, Z.; Zhou, T.; Chen, L.; Chen, J. Construction of biomass-based amines *via* Ir-mediated *N* -alkylation: kinetic analysis. *Green Chem.,* **2024**, *26*(7), 3832-3852. [http://dx.doi.org/10.1039/D3GC04747D]

[72] Rajendran, A.; Rajendiran, M.; Yang, Z.F.; Fan, H.X.; Cui, T.Y.; Zhang, Y.G.; Li, W.Y. Functionalized silicas for metal-free and metal-based catalytic applications: A review in perspective of green chemistry. *Chem. Rec.,* **2020**, *20*(6), 513-540. [http://dx.doi.org/10.1002/tcr.201900056] [PMID: 31631504]

[73] Santos, A.S.; Silva, A.M.S.; Marques, M.M.B. Sustainable amidation reactions–recent advances. *Eur. J. Org. Chem.,* **2020**, *2020*(17), 2501-2516. [http://dx.doi.org/10.1002/ejoc.202000106]

[74] Xu, L.; Zhang, J.; Zong, Q.J.; Wang, L.; Xu, T.; Gong, J.; Liu, Z.H.; Li, B.Z.; Yuan, Y.J. High-solid ethylenediamine pretreatment to fractionate new lignin streams from lignocellulosic biomass. *Chem. Eng. J.,* **2022**, *427*130962 [http://dx.doi.org/10.1016/j.cej.2021.130962]

[75] Singh, H.; Sen, C.; Suresh, E.; Panda, A.B.; Ghosh, S.C. C–H amidation and amination of arenes and heteroarenes with amide and amine using Cu-MnO as a reusable catalyst under mild conditions. *J. Org. Chem.,* **2021**, *86*(4), 3261-3275. [http://dx.doi.org/10.1021/acs.joc.0c02603] [PMID: 33522804]

[76] Albkuri, Y.M. RanguMagar, A.B., Brandt, A., Wayland, H.A., Chhetri, B.P., Parnell, C.M., Szwedo, P., Parameswaran-Thankam, A. and Ghosh, A. C–N Cross-coupling reactions of amines with aryl halides using amide-based pincer nickel (II) catalyst. *Catal. Lett.,* **2020**, *150*, 1669-1678. [http://dx.doi.org/10.1007/s10562-019-03062-5]

[77] Zhang, Z.B.; Yang, Y.; Yu, Z.X.; Xia, J.B. Lewis base-catalyzed amino-acylation of arylallenes *via* C–N bond cleavage: reaction development and mechanistic studies. *ACS Catal.,* **2020**, *10*(10), 5419-5429. [http://dx.doi.org/10.1021/acscatal.0c01000]

[78] Van Emelen, L.; Henrion, M.; Lemmens, R.; De Vos, D. C–N coupling reactions with arenes through C–H activation: the state-of-the-art *versus* the principles of green chemistry. *Catal. Sci. Technol.,* **2022**, *12*(2), 360-389. [http://dx.doi.org/10.1039/D1CY01827B]

[79] Sarmah, B.K.; Konwar, M.; Das, A. Site-selective deoxygenative amination of azine N-oxides with carbodiimides under catalyst-, activator-, base-, and solvent-free conditions. *J. Org. Chem.,* **2021**, *86*(15), 10762-10772. [http://dx.doi.org/10.1021/acs.joc.1c00741] [PMID: 34260234]

[80] Jeong, J.; Jung, H.; Kim, D.; Chang, S. Multidimensional screening accelerates the discovery of rhodium catalyst systems for selective intra- and intermolecular C–H amidations. *ACS Catal.,* **2022**, *12*(13), 8127-8138. [http://dx.doi.org/10.1021/acscatal.2c02612]

[81] Friães, S.; Gomes, C.S.B.; Royo, B. Bis-Triazolylidenes of manganese and rhenium and their catalytic application in N-alkylation of amines with alcohols. *Organometallics,* **2023**, *42*(14), 1803-1809. [http://dx.doi.org/10.1021/acs.organomet.3c00046]

[82] Tamilarasan, R.; Govindaraj, S.; Ganesan, K. Synthesis of dimeric pyridinium bromide under silica supported approach. *Synth. Commun.*, **2020**, *50*(8), 1190-1198. [http://dx.doi.org/10.1080/00397911.2020.1733611]

[83] Zhang, H.; Wang, J.; Liu, M.; Liu, D. Cu/Al2O3 catalyst prepared by a double hydrolysis method for a green, continuous and controlled N-methylation reaction of nitroarenes with methanol. *Appl. Surf. Sci.*, **2020**, *526*146708 [http://dx.doi.org/10.1016/j.apsusc.2020.146708]

[84] Wu, P.; Lu, G.; Cai, C. Synthesis of an Fe–Pd bimetallic catalyst for *N* -alkylation of amines with alcohols *via* a hydrogen auto-transfer methodology. *Green Chem.*, **2021**, *23*(1), 396-404. [http://dx.doi.org/10.1039/D0GC03725G]

[85] Hegade, S.; Jadhav, Y.; Chavan, S.; Mulik, G.; Gaikwad, G. Catalytic assay of Schiff base Co(II), Ni(II), Cu(II) and Zn(II) complexes for N-alkylation of heterocycles with 1,3-dibromopropane. *J. Chem. Sci.*, **2020**, *132*(1), 92. [http://dx.doi.org/10.1007/s12039-020-01791-4]

[86] Chen, H.Y.; Peng, W.T.; Lee, Y.H.; Chang, Y.L.; Chen, Y.J.; Lai, Y.C.; Jheng, N.Y.; Chen, H.Y. Use of base control to provide high selectivity between diaryl thioether and diaryl disulfide for C-S coupling reactions of aryl halides and sulfur and a mechanistic study. *Organometallics,* **2013**, *32*(19), 5514-5522. [http://dx.doi.org/10.1021/om400784w]

[87] Bhunia, S.; Pawar, G.G.; Kumar, S.V.; Jiang, Y.; Ma, D. Selected copper-based reactions for C−N, C−O, C−S, and C−C bond formation. *Angew. Chem. Int. Ed.,* **2017**, *56*(51), 16136-16179. [http://dx.doi.org/10.1002/anie.201701690] [PMID: 28544319]

[88] Hosseinian, A.; Ahmadi, S.; Nasab, F.A.H.; Mohammadi, R.; Vessally, E. Cross-dehydrogenative C–H/S–H coupling reactions. *Top. Curr. Chem. (Cham),* **2018**, *376*(6), 39. [http://dx.doi.org/10.1007/s41061-018-0217-0] [PMID: 30306362]

[89] Bandalla, S.; Dosarapu, V.; Bathula, G.B.; Ravula, M.; Yadagiri, J.; Gogoi, P.; Baithy, M.; Jonnalagadda, S.B.; Vasam, C.S. Highly efficient solvent-free oxidation of cyclohexanol to cyclohexanone over nanocrystalline CaO–MgO binary metal-oxide catalysts. *Molecular Catalysis,* **2022**, *533*112759 [http://dx.doi.org/10.1016/j.mcat.2022.112759]

[90] Wang, S.; Wang, I.; Li, Z. Metal oxide and nitrate-modified calcium gluconate-based sorbent with high stability for integrated carbon capture and utilization process. *Separ. Purif. Tech.,* **2024**, *330*125502 [http://dx.doi.org/10.1016/j.seppur.2023.125502]

[91] Jian, Y.C.; Löwemark, L.; Chien-Yi Liao, A. Application of ion-exchange resin sachets and XRF-CS in heavy metal pollution sources monitoring In: *EGU General Assembly Conference Abstracts*; , **2023**; p. 4067. [http://dx.doi.org/10.5194/egusphere-egu23-4067]

[92] Saleh, E.A.M.; Kassem, A.F.; Altalbawy, F.; Shoja, S.J.; Bokov, D.O.; Elawady, A.; Al-Rubaye, A.H.; Saud, A.; Al-Mashhadani, Z.I.; Nejad, M.S.G. Effective role of two layers of silica in the performance of Fe_3o_4@ $xSio_2$@ $ySio_2$@ bispyp-ni core-shell catalyst for cc and cs coupling reactions. In: *Silicon*; , **2024**; pp. 1-15.

[93] Zhang, R.; Ding, H.; Pu, X.; Qian, Z.; Xiao, Y. Recent advances in the synthesis of sulfides, sulfoxides and sulfones *via* CS bond construction from non-halide substrates. *Catalysts,* **2020**, *10*(11), 1339. [http://dx.doi.org/10.3390/catal10111339]

[94] Sundaravelu, N.; Sangeetha, S.; Sekar, G. Metal-catalyzed C–S bond formation using sulfur surrogates. *Org. Biomol. Chem.,* **2021**, *19*(7), 1459-1482. [http://dx.doi.org/10.1039/D0OB02320E] [PMID: 33528480]

[95] Guo, Y.; Zhuang, Z.; Feng, X.; Ma, Q.; Li, N.; Jin, C.; Yoshida, H.; Tan, J. Selective s-arylation of sulfenamides with arynes: Access to sulfilimines. *Org. Lett.,* **2023**, *25*(39), 7192-7197. [http://dx.doi.org/10.1021/acs.orglett.3c02785] [PMID: 37733632]

[96] Yang, Q.; Zhao, Y.; Ma, D. Cu-mediated Ullmann-type cross-coupling and industrial applications in route design, process development, and scale-up of pharmaceutical and agrochemical processes. *Org. Process Res. Dev.,* **2022**, *26*(6), 1690-1750. [http://dx.doi.org/10.1021/acs.oprd.2c00050]

[97] Xiang, H.; Qi, J.; He, Q.; Jiang, M.; Yang, C.; Deng, L. Synthesis of 2-C-substituted benzothiazoles via a copper-promoted domino condensation/S-arylation/heterocyclization process. *Org. Biomol. Chem.,* **2014**, *12*(26), 4633-4636. [http://dx.doi.org/10.1039/c4ob00564c] [PMID: 24850390]

[98] Zhang, Z.; Wang, T.; Xiang, P.; Du, Q.; Han, S. Syntheses, characterization, and application of tridentate phenoxyimino-phenoxy aluminum complexes for the coupling of terminal epoxide with CO2: from binary system to single component catalyst. *Catalysts,* **2021**, *11*(2), 145. [http://dx.doi.org/10.3390/catal11020145]

[99] Vessally, E.; Mohammadi, R.; Hosseinian, A.; Didehban, K.; Edjlali, L. S-arylation of 2-mercaptobenzazoles: a comprehensive review. *J. Sulfur Chem.,* **2018**, *39*(4), 443-463. [http://dx.doi.org/10.1080/17415993.2018.1436712]

[100] Li, J.; Yang, S.; Wu, W.; Jiang, H. Recent developments in palladium-catalyzed C–S bond formation. *Org. Chem. Front.,* **2020**, *7*(11), 1395-1417. [http://dx.doi.org/10.1039/D0QO00377H]

[101] Acharya, A.; Vijay Kumar, S.; Saraiah, B.; Ila, H. One-pot synthesis of functionalized benzo[b]thiophenes and their hetero-fused analogues *via* intramolecular copper-catalyzed S-arylation of in situ generated enethiolates. *J. Org. Chem.,* **2015**, *80*(5), 2884-2892. [http://dx.doi.org/10.1021/acs.joc.5b00032] [PMID: 25658978]

[102] Biswas, S.; Hughes, W.B.; De Angelis, L.; Haug, G.C.; Trevino, R.; Fremin, S.O.; Arman, H.D.; Larionov, O.V.; Doyle, M.P. The "cesium effect" magnified: exceptional chemoselectivity in cesium ion mediated nucleophilic reactions. *Chem. Sci. (Camb.),* **2024**, *15*(14), 5277-5283. [http://dx.doi.org/10.1039/D4SC00316K] [PMID: 38577370]

[103] Mazur, N.; Huinink, H.; Fischer, H.; Donkers, P.; Adan, O. Accelerating the reaction kinetics of K2CO3 through the addition of CsF in the view of thermochemical heat storage. *Sol. Energy,* **2022**, *242*, 256-266. [http://dx.doi.org/10.1016/j.solener.2022.07.023]

[104] Bhandari, V.M.; Sorokhaibam, L.G.; Ranade, V.V. Ion exchange resin catalyzed reactions—an overview. In: *Industrial Catalytic Processes for Fine and Specialty Chemicals*; , **2016**; pp. 393-426.

[105] Gelbard, G. Organic synthesis by catalysis with ion-exchange resins. *Ind. Eng. Chem. Res.,* **2005**, *44*(23), 8468-8498. [http://dx.doi.org/10.1021/ie0580405]

[106] Maeda, Y.; Hamada, S.; Aota, Y.; Otsubo, K.; Kano, T.; Maruoka, K. Practical asymmetric synthesis of chiral sulfoximines *via* sulfur-selective alkylation. *J. Org. Chem.,* **2022**, *87*(5), 3652-3660. [http://dx.doi.org/10.1021/acs.joc.1c02424] [PMID: 35075904]

[107] Xu, X.H.; Hao, E.J.; Shi, Z.; Dong, Z.B. Easy S-Alkylation of arylthioureas and 2-mercaptobenzothiazoles using tetraalkylammonium salts under transition-metal-free conditions. *J. Org. Chem.,* **2022**, *87*(15), 9675-9687. [http://dx.doi.org/10.1021/acs.joc.2c00728] [PMID: 35896442]

[108] Champlin, A.T.; Ellman, J.A. Preparation of sulfilimines by sulfur-alkylation of *N* -acyl sulfenamides with alkyl halides. *J. Org. Chem.,* **2023**, *88*(11), 7607-7614. [http://dx.doi.org/10.1021/acs.joc.3c00750] [PMID: 37221855]

[109] Zhou, Q.; Li, J.; Wang, T.; Yang, X. Base-promoted *S* -arylation of sulfenamides for the synthesis of sulfilimines. *Org. Lett.,* **2023**, *25*(23), 4335-4339. [http://dx.doi.org/10.1021/acs.orglett.3c01436] [PMID: 37267093]

[110] Greenwood, N.S.; Ellman, J.A. Sulfur-arylation of sulfenamides via ullmann-type coupling with (hetero)aryl Idides. *Org. Lett.,* **2023**, *25*(25), 4759-4764. [http://dx.doi.org/10.1021/acs.orglett.3c01874] [PMID: 37338140]

CHAPTER 8

Advanced Solid Base Catalysts for Depolymerization of Lignin and Transesterification of Glycerol

Mohan Varkolu[1,*], **T. Vinodkumar**[2], **Prasad Thota**[3] and **Anil Kumar Teotia**[3]

[1] *Department of Chemistry, Koneru Lakshmaiah Education Foundation, RVS Nagar, Azziz Nagar (PO), Moinabad Road, Hyderabad-500075, Telanagana, India*

[2] *Central Council for Research in Ayurvedic Sciences (CCRAS), Department of Chemistry, CSMCARI, Arumbakkam, Chennai-600106, India*

[3] *Microbiology Division, Indian Pharmacopoeia Commission, Ghaziabad, Uttar Pradesh, India*

Abstract: Due to environmental concerns and the depletion of fossil resources, the sustainable utilization of biomass for producing valuable chemicals and fuels has gained substantial attention in recent years. Advanced solid base catalysts have emerged as essential tools in this attempt, with high catalytic efficiency, improved selectivity, and low environmental impact. The present chapter provides an in-depth exploration of various advanced solid base catalysts like metal oxides, alkali metal-supported metal oxides, alkaline earth metal oxides, waste solid base catalysts, clay materials, and mesoporous solid base catalysts, *etc*., for transforming biomass and its derivatives into value-added products. Specific processes covered include transesterification of oils, aldol condensation of biomass-derived compounds, depolymerization of lignin, glycerol transesterification for carbonate production, and so on.

Keywords: Aldol condensation, Glycerol carbonate, Glycerol, Lignin depolymerization, Oils, Solid base catalysts, Transesterification.

INTRODUCTION

Feedstocks derived from biomass can be used to produce renewable chemicals and biofuels. They are commonly categorized into three primary groups: lignocellulose, starch, and triglycerides. Lignocellulosic biomass is particularly favored due to its non-competitive nature with food supplies. The conversion process involves various catalytic or chemical treatments, including pyrolysis,

* **Corresponding author Mohan Varkolua:** Department of Chemistry, Koneru Lakshmaiah Education Foundation, RVS Nagar, Azziz Nagar (PO), Moinabad Road, Hyderabad-500075, Telanagana, India; E-mails: mohan.iict@gmail.com; mohan@klh.edu.in

Ravi Tomar, Vivek Mishra, Kapil Arya & Garima Pandey (Eds.)

hydrothermal treatment, liquefaction, hydrolysis, and fermentation. These treatments yield liquid fractions rich in oxygenated molecules like alcohols, aldehydes, acids, and furans, serving as crucial molecules for synthesizing a variety of chemicals and fuels, such as additives, solvents, and monomers.

Currently, there is a predominant focus on converting biomass into liquid fuels, with less emphasis on chemical and fine chemical production. However, the evolving bio-economy emphasizes the importance of co-producing biofuels alongside valuable chemicals. To address this shift, innovative processing strategies for second-generation biomass at unified biorefineries are being proposed. These strategies involve primary treatment, fractionation, and separation, followed by downstream processing into fuels, fuel additives, and chemicals.

In response to the demands of the evolving bio-economy, the development of novel catalytic methods is crucial for competitive and efficient production from biomass. While homogeneous catalyst systems have been traditionally utilised for biomass conversion due to cost-effectiveness, there is a notable shift towards exploring heterogeneous catalysts [1]. Solid catalysts offer advantages, such as reusability and the absence of a neutralization step, minimizing waste product formation, representing a significant avenue for environmentally conscious biomass utilization.

Heterogeneous catalysts, particularly solid base catalysts, have the ability to tailor active properties for achieving selective products, making them promising candidates for executing multistep processes. These catalysts enable the creation of robust multifunctional catalysts, facilitating collaboration among active sites across different stages of a given cascade process [2, 3].

In contemporary applications, solid base catalysts, including alkali-modified zeolites and various metal oxides, are widely employed for base-catalyzed reactions in biomass transformation. Aldol condensation, a crucial C—C bond-forming reaction, plays a significant role in biomass utilization for chemical and fuel production. Despite the high hydrogen consumption in deoxygenation steps, aldol condensation allows for the selective synthesis of linear alkanes suitable for liquid fuels.

While approximately 90% of biodiesel is obtained through the transesterification of vegetable oils *via* homogeneous bases as catalysts, the drawbacks in separation and purification steps have led to a growing interest in heterogeneous catalysts. Solid base catalysts, in particular, play a pivotal role in enhancing biodiesel production. This chapter extensively discusses and examines the transesterification of oils and glycerol, depolymerization of lignin, and aldol condensation using solid base catalysts.

CLASSIFICATION, CHARACTERIZATION, AND PROPERTIES OF BIOMASS FEEDSTOCKS

Classification of Biomass

Biomass feedstocks can be broadly classified into three main categories, which are lignocellulosic, animal, and aquatic biomass. Lignocellulosic Biomass (LCB) is the most common and widely researched category, consisting of plant-derived materials, such as wood, crop residues, and dedicated energy crops. Animal biomass includes organic waste from animals, such as manure. In contrast, aquatic biomass refers to materials obtained from aquatic sources like algae and aquatic plants. Lignocellulosic biomass is primarily composed of cellulose, hemicellulose, and lignin, and its composition is shown in Table **1**.

Table 1. Composition of common LCB [4, 5].

Lignocellulosic Biomass (LCB) Material	Cellulose %	Hemicellulose %	Lignin %
Hardwood stems	40–55	24–40	18–25
Softwood stems	45–50	25–35	25–35
Almond shells	29.0–31.1	28.0–38.0	27.7–35
Corncobs	45	35	15
Grasses	25–40	35–50	10–30
Wheat straw	30	50	15
Flax straw	36.70	34.40	28.90
Leaves	15–32	80–85	0
Sugar cane bagasse	32–44	27–32	19–24
Coastal Bermuda grass	25	35.7	6.4
Switch grass	45	31.4	12

Characterization of Biomass Feedstocks

Characterization of biomass feedstocks involves understanding their physical, chemical, and thermal properties [6]. Key physical characteristics include particle size, density, and moisture content, which influence processing. Additionally, chemical analysis examines elemental composition and energy content. Thermal properties, such as calorific value and ash content, are vital for evaluating energy potential and combustion behaviour.

Properties of Biomass Feedstocks

Moisture Content: The moisture content of biomass affects its combustion efficiency and storage stability. Elevated moisture content can result in reduced energy yield, as well as challenges in transportation and storage.

Calorific Value: It represents the energy content of biomass. Higher calorific values indicate greater energy potential, making biomass more suitable for bioenergy applications.

Ash Content: Ash content is an important factor in biomass combustion. A higher ash content can lead to issues, such as slagging and fouling, which negatively impact the efficiency of energy conversion processes.

Elemental Composition: The elemental composition, including carbon, hydrogen, nitrogen, sulfur, and oxygen content, influences the combustion characteristics and emissions during biomass utilization.

CATALYSTS IN BIOMASS CONVERSION

Biomass is a diverse and renewable resource that can be converted into valuable fuels and chemicals through a variety of thermochemical and biochemical processes. The use of catalysts in these conversion processes is critical for breaking down complex biomass structures and facilitating the formation of desired products. Functional catalysts, which play a pivotal role in enhancing reaction rates, selectivity, and overall process efficiency, enable the transformation of biomass-derived intermediates into various valuable end-products, including biofuels and chemicals (Fig. **1**).

Types of Functional Catalysts

Heterogeneous Catalysts

Catalysts that have distinct phases from the reactants are called heterogeneous catalysts.There are typically solid materials, including metal oxides, zeolites, and supported metal nanoparticles. Heterogeneous catalysts are classified as acid catalysts, base catalysts, metal catalysts, and so on.

Homogeneous Catalysts

Catalysts that exist in the same phase as the reactants are called homogeneous catalysts. Typically, these are complex organic or metal-containing compounds that are effective in specific reaction environments.

Biological Catalysts

Biological catalysts include enzymes and microorganisms that can facilitate certain biochemical pathways, providing high specificity and selectivity in the conversion of biomass.

Bifunctional Catalysts

Bifunctional catalysts are the ones that possess multiple active sites, thereby enabling a sequence of reactions in a single reactor. This approach enhances efficiency and reduces the need for multiple processing steps.

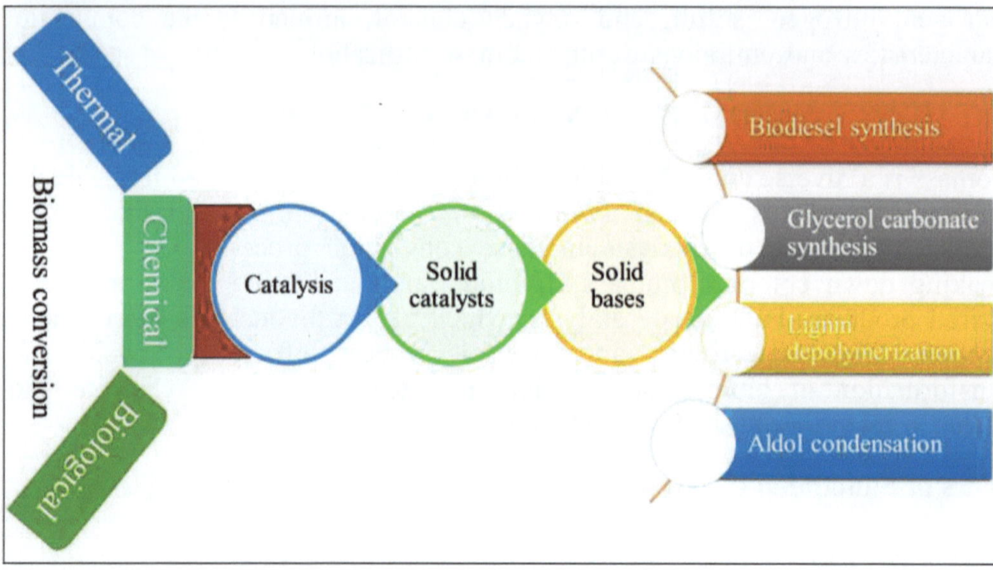

Fig. (1). Schematic catalytic transformation of biomass to value-added fuels and chemicals.

Characterization of Basic Sites of Solid Base Catalysts

Strong basic sites are responsible for the transesterification of oils and glycerols. The presence of strong basic sites in the catalysts drives the transesterification of glycerol to glycidol rather than the glycerol carbonate. Therefore, optimizing catalyst systems for the right balance of basic sites is crucial for producing selective products in biomass transformation using solid-base catalysts.

The basicity of solid bases is mostly measured using two methods: the Hammet Indicator method and the TPD of CO_2.

Hammet Indicator Method

The basicity of the solid bases is assessed utilizing the Tanabe and Yamaguchi titration method [7]. In this procedure, a solid catalyst is suspended in benzene and titrated with benzoic acid.

The calculation of catalyst basicity is based on the titer of 0.1 N benzoic acid required, as indicated in Equation (2):

$$Basicity = Vx\frac{N}{W}$$

Where:

V represents the volume of benzoic acid solution (mL)

N denotes the normality of the benzoic acid solution (mgeq.mL)

W represents the weight of the catalyst sample (g) [7 - 9]

Furthermore, the following Hammett indicators are used: bromothymol blue (H− = 7.2), phenolphthalein (H− = 9.3), and alizarine yellow (H− = 11.0). The quantitative determination of basicity is performed by the method of Hammett indicator-benzene carboxylic acid (0.01 mol/L anhydrous methanol solution) titration [10].

Higher basic strength is known to enhance glycerol conversion percentage, but excessively high basic strength can lead to the generation of glycidol in the reaction mixture, thereby reducing glycerol carbonate yield. Therefore, the optimal catalyst for glycerol carbonate synthesis exhibits a moderate basic strength within the range of $15.0 < H\_ < 18.4$.

The Hammett indicator titration method is employed to investigate the basic strength and total basicity of the synthesized $MgTiO_3$ catalyst [11]. The analysis reveals that the basic strength of the designed perovskite falls within the range of $15 < H\_ < 18.6$. Various indicators, such as 2,4-dinitroaniline ($H\_ = 15.6$), phenolphthalein ($H\_ = 9.6$), bromothymol blue ($H\_ = 7.8$), and 4-nitroaniline ($H\_ = 15.4$), are utilized to determine the basic strength of the synthesized catalyst. The total basicity of $MgTiO_3$ is notably high at approximately 11.46 mmol/g, highlighting its exceptional performance in the transesterification reaction of glycerol.

TPD of CO_2

The surface basicity of MgO is assessed utilizing the CO_2-TPD (Temperature-Programmed Desorption of Carbon Dioxide) method [12]. In the TPD spectrum, a higher desorption temperature is indicative of a stronger basic site, and a larger peak area signifies a greater number of basic sites.

To assess the basic site strength, a CO_2-TPD analysis is conducted on both bare Coal Fly Ash (CFA)and CFA loaded with 2% lithium, both calcined at 500 °C [13]. In CO_2-TPD, the basic strength of the catalyst is correlated with the temperature at which CO_2 desorbs. Specifically, desorption peaks occurring at < 200 °C correspond to weak basic sites, those in the 300–450 °C range represent moderate basic sites, and peaks above 450 °C indicate strong basic sites [14]. Bare CFA demonstrates a weak desorption peak, approximately within the 100–250 °C range, with a basic site concentration of 0.69 mmol/g. In contrast, the Li/CFA catalyst exhibits a strong desorption peak around 450–600 °C, indicative of a higher concentration of strong basic sites on the catalyst surface, measuring 1.36 mmol/g. Moreover, the basic strength of lithium-loaded CFA (falling within the range of $9.3 < H\_ < 15$) surpasses that of bare CFA.

Functional catalysts play a crucial role in the synthesis of fuels and chemicals from biomass. Therefore, continued research and development in this field are essential for advancing sustainable and economically viable pathways for biomass conversion. Particularly, solid base catalysts are important in the upgrading of biomass-derivedchemicals [15, 16].

APPLICATIONS OF SOLID BASE CATALYSTS

Transesterification of Oils

The transformation of biomass into biodiesel, an eco-friendly fuel derived from renewable sources and residual lipids, holds considerable potential for use in transportation (Fig. **2**). Among various methodologies employed for this purpose, transesterification stands out as a frequently utilized chemical process. This technique facilitates the conversion of animal fats, cooking oils, and vegetable oils into long-chain fatty acid alkyl esters (Fig. **3**). Typically, methanol is the preferred reagent for transesterification due to its cost-effectiveness and the resulting production of FAME (Fatty Acid Methyl Ester), a non-toxic, renewable, and biodegradable biodiesel.

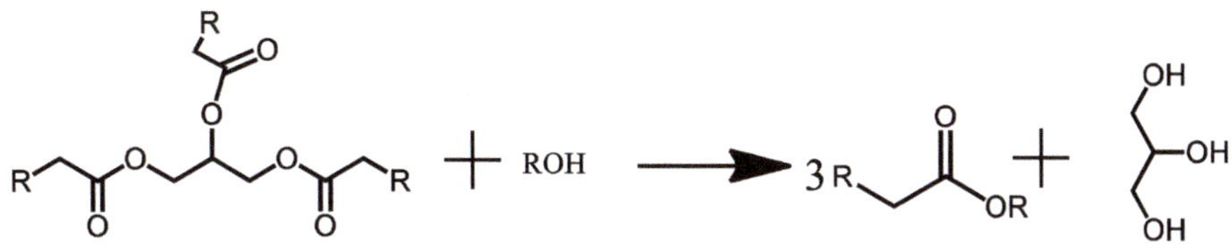

Fig. (2). Transesterification of oils using alcohol in the presence of catalysts.

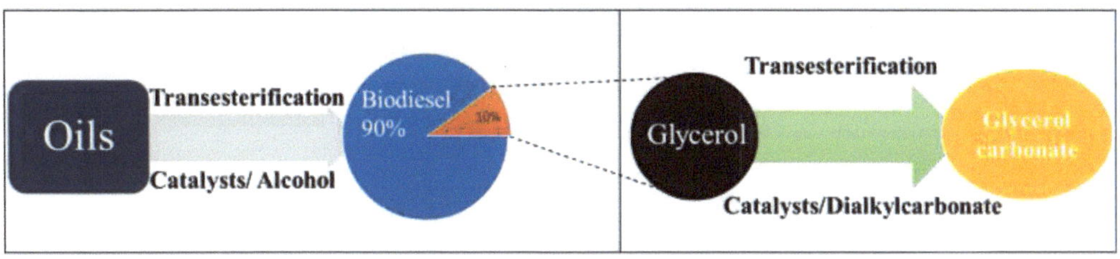

Fig. (3). Schematic catalytic transformation of oils to biodiesel and glycerol to glycerol carbonate by transesterification.

Under basic environments, a base, in general, abstracts a proton from alcohol, generating an alkoxide anion. This anion subsequently attacks the carbonyl carbon of the ester, resulting in the release of three moles of FAME and one mole of glycerol. Transesterification of biomass-derived oils, similar to aldol condensation, can be catalysed by homogeneous alkali metal catalysts. Nevertheless, the usage of such catalysts may lead to the production of water and soaps, reducing methyl ester yields and catalyst efficiency. This issue can be mitigated by employing heterogeneous bases, addressing post-reaction separation challenges and allowing for catalyst recycling, thus promoting a more environmentally friendly process.

Various solid bases have been utilized for transesterification reactions, namely supported alkali, alkaline earth metal oxides, metal hydroxides, *etc*. Calcium oxide (CaO) has emerged as a commonly used catalyst. Notably, a study by Kouzu *et al*. studied soybean oil transesterification using modified graphitic carbon nitrides enhanced with sodium (Na) [17]. The co-polymerized catalyst with optimized Na content demonstrated exceptional performance, yielding over 90% biodiesel across ten successive cycles. Robust bonds between sodium (Na) and nitrogen (N) were crucial in preventing leaching and sustaining catalytic activity.

In a study, a novel catalyst, NCSO, consisting of Na_2CaSiO_4 and CaO phases, exhibited an impressive 98.2% FAME yield within 60 minutes at 80°C, and even after 12 consecutive cycles, it maintained a respectable 57.3% FAME yield [18].

This silicate-strengthening strategy presents a promising approach for synthesizing robust and highly efficient solid base catalysts. Additionally, a porous solid base catalyst crafted from dewatered paper sludge demonstrated exceptional catalytic performance, yielding 91.6% biodiesel under optimal conditions and retaining catalytic activity after multiple cycles [19].

A novel class of heterogeneous base catalysts, Zeo-Geo, was engineered using a zeolite-geopolymer hybrid material as a substrate for potassium in the transesterification of soybean oil [20]. The 15K/Zeo-Geo catalyst, featuring a 97.64% biodiesel yield, demonstrated robust reusability over five cycles, emphasizing its potential for the production of biodiesel on a large scale and presenting a promising approach.

In line with the growing demand for environmentally friendly synthesis methods, this study investigates the potential usage of calcined ash derived from *Terminalia arjuna* bark (TABCA) as a versatile, renewable, and heterogeneous alkaline catalyst for the economicproduction of high-quality, fuel-grade FAME through the transesterification of Soybean Oil (SO) under microwave irradiation [21].

TABCA exhibits exceptional attributes, including a mesoporous structure, significant surface area, and remarkable basicity due to the presence of various metal oxides. These properties effectively ease the transformation of SO into FAME. Optimal conditions, such as a methanol-to-SO ratio of 24:1, an 8 wt% catalyst concentration, a temperature of 80°C, and a 60-minute reaction period under microwave irradiation, result in the highest biodiesel production and an impressive 99.2% (± 0.6%) yield with outstanding selectivity (100%). The produced soybean oil biodiesel meets ASTM D6751 and EN 14214 standards, ensuring compatibility with existing engines. The mesoporous TABCA catalyst shows strong stability over five transesterification reactions, consistently yielding 80% (± 0.8%) biodiesel without extra treatments.This underscores the substantial promise of TABCA as a sustainable and effective catalyst, meeting both environmental and technical standards.

In another approach, mechanical vapor recompression combines with the Organic Rankine Cycle (ORC) to refine crude biodiesel, which is produced through the transesterification method [22]. Compared to conventional purification processes, this approach results in a significant 55.40% reduction in the total annual cost and a substantial decrease in greenhouse gas emissions, including a 76.31% reduction in CO_2 emissions, a 76.41% reduction in SO_2 emissions, and a 76.42% reduction in NOx emissions. Moreover, the proposed process parameters have been optimized, achieving a coefficient of performance of 16.03 for mechanical vapor recompression and a thermal efficiency of 14.06% for the Organic Rankine Cycle (ORC), enhancing the overall efficiency of the system.

In another study, various catalysts derived from different parts of the *Musa champa* plant were employed in the production of biodiesel [23]. The synthesized *Musa champa* peel (CMCP-550) catalyst exhibited superior catalytic efficiency, boasting the highest potassium content (47.49 wt%) and the highest basicity (1.25 mmol g^{-1}). Under optimal reaction conditions (ORCs), CMCP-550 achieved a remarkable yield of biodiesel (98.27%). The catalyzed reaction displayed a favorable activation energy (54.256 kJ mol^{-1}) and a notable turnover frequency (TOF), *i.e.*, 14.15 h^{-1}.

Henbane seed oil, rich in oil content, was explored as a source for biodiesel production using fourteen supported solid base catalysts [24]. $K_2O/\gamma\text{-}Al_2O_3$ demonstrated exceptional performance, with optimized reaction conditions resulting in a yield of 90.25%, closely matching the actual measured value of 90.54%. The produced biodiesel met stringent European biodiesel standard EN14214.

Silica nanoparticles obtained from rice husk were coupled with modified Calcium Oxide (CaO) for efficient production of Fatty Acid Methyl Ester (FAME) [25]. The modified CaO yielded an impressive 97.8% FAME from WCO after 3 hours at 80°C, surpassing the 83.5% yield achieved with unmodified CaO under the same conditions. The modified CaO-nanoparticles proved to be resilient heterogeneous transesterification catalysts, remaining effective after undergoing fourteen cycles of use.

Furthermore, the transesterification of Parkia biglobosa oil, a bi-functional Clay-Na_2CO_3 heterogeneous catalyst, was performed, yielding an impressive maximum biodiesel yield of 94.7% under optimized conditions [26]. The catalyst exhibited potential for cost-effectiveness and superior catalytic activity, making it a promising candidate, and the resulting biodiesel also met the standards set by the Ghana Standard Authority.

Brahma *et al.* developed effective solid catalysts utilizing post-harvest materials from *Musa chinensis* [27]. These catalysts were successfully applied in biodiesel production from a blend of five different oils (soybean, canola, sunflower, pongamia, and jatropha oil). Characterization of the catalysts revealed their highly basic nature, micro-mesoporous structure, and polycrystalline composition. The study examined the role of catalyst load, plant material source, Molar Ratio of Methanol-to-oil (MRMO), and temperature. Notably, the calcined *M. chinensis* peel catalyst demonstrated exceptional effectiveness, containing 44.15 wt% potassium in carbonate and oxide forms. Under optimized conditions, including a

5 wt% dosage of catalyst, a 9:1 MRMO ratio, and 65°C temperature, this peel catalyst achieved an impressive 95.82% biodiesel yield in just 11 minutes. Additionally, the calcined peel catalyst exhibited the highest basicity at 0.93 mmol g^{-1}, surpassing the rhizome catalysts (0.2 mmol g^{-1}) and trunk (0.15 mmol g^{-1}).

Turnover frequencies were determined as 16.85 h^{-1} (peel), 2.317 h^{-1} (rhizome), and 7.025 h^{-1} (trunk). Importantly, the catalyst proved to be highly effective, reusable, and capable of producing high-quality biodiesel, meeting international standards, making it a promising renewable candidate for the biodiesel industry.

Prajapati *et al.* identified Karanja Seed Shell (KSS) ash, a byproduct resulting from the incineration of dried seed shells, as an effective green heterogeneous catalyst for the manufacturing of biodiesel [28]. Their study focused on the catalytic capabilities of KSS ash as a solid base catalyst in soybean oil methanolysis. Conversion of soybean oil into methyl ester was achieved under specific conditions determined through GC-MS and FT-IR analysis. This synthesized heterogeneous catalyst proves to be a highly efficient and environmentally friendly alternative, given its benign nature, reusability, sustainability, and wide availability.

To produce biodiesel from waste cooking oil with an acidic value of 1.86 mg KOH/g, a study was conducted with five distinct catalysts *via.*, CaO, Ba(OH)2, ZnO, MgO, and $AlCl_3$ [29]. To optimize reaction conditions, the influence of various process variables was studied. These included the amount of catalyst used, the ratio of methanol to oil, and the reaction time. Additionally, the study explored the impact of acetone as a cosolvent on the production of biodiesel along with an evaluation of the reusability. Statistical evaluation at a 95% confidence level revealed that among the catalysts, $Ba(OH)_2$ exhibited exceptional performance, yielding 92 wt% biodiesel, closely followed by CaO. Other catalysts did not meet the desired standards. Regardless of the catalyst type used, the addition of 20 vol% cosolvent increased the production of biodiesel. Furthermore, all the heterogeneous catalysts proved to be reusable.

The main goal of the investigation was to enhance the biodiesel yield from *Argemone mexicana* oil, utilizing CaO as a catalyst derived from chicken eggshells [30]. The study employed response surface methodology to optimize crucial parameters. This approach identified optimal conditions for achieving a remarkable 99.07% yield, including a 9.7:1 molar ratio of methanol-to-oil, 3.05 wt% loading of catalyst, and 3.01 hours of reaction time. The biodiesel produced was characterized through FTIR and 1H Nuclear Magnetic Resonance (NMR) techniques. In terms of catalyst reusability, it was found to exhibit significant stability and could be employed for more than four cycles.

A Calcium Oxide (CaO)-based catalyst, known as Calcined Filter Cake (CFC), was prepared using Waste Filter Cake (WFC) obtained from a sugar beet processing facility by calcination at 900°C in the air for two hours. The CFC catalyst facilitated the transesterification of rapeseed oil with methanol under mild conditions [31]. The low free fatty acid content (approximately 2.0 mg KOH/g) allowed for the direct utilization of the base CFC catalyst in rapeseed oil transesterification. Rapeseed oil was found to have a higher proportion of unsaturated fatty acids (around 93%), with oleic acid being the predominant one. A simplified model was developed to describe the kinetics of transesterification, incorporating the changes in the reaction mechanism and the limitation imposed by the mass transfer of triacylglycerols. The model demonstrated a good agreement with experimental results, exhibiting a mean relative percentage deviation of only ±7.43% for the conversion degree. High-quality Fatty Acid Methyl Esters (FAME) with a content exceeding 96.5% were obtained within a broad range of catalyst amounts (4% to 10%) and reaction times (approximately 45 to 70 minutes).

Bharti *et al.* employed a bifunctional catalyst consisting of tin (Sn) supported on CaO [18, 32]. The resulting Sn-CaO catalyst, proved to be highly effective for converting waste cooking oil into biodiesel in a one-step reaction process. The study utilized a Box-Behnken design to optimize the biodiesel conversion process. Key variables influencing the reaction included the molar ratio of methanol-to-oil, reaction time, temperature, and catalyst concentration. Optimized conditions resulted in an impressive biodiesel conversion rate of 97.39%. To validate the predicted model, three experiments were conducted under these conditions, yielding an experimental biodiesel conversion value of 96.72% ± 0.61%. The study also measured the kinematic viscosity and acid value of the biodiesel at 40°C. These values complied with the standards set by ASTM and EN. The study additionally described the mechanism of simultaneous esterification and transesterification reactions, offering an economically viable and environmentally friendly approach to biodiesel production.

To reduce biodiesel production costs and utilize steel slag waste, this study synthesized a CaO-CeO2/slag solid base catalyst using a two-step impregnation for efficient transesterification of palm oil into FAMEs [33]. The catalytic performance was highly dependent on the Ca-Ce ratio in the steel slag, with the $CaO-CeO_2$/slag-0.8 catalyst demonstrating exceptional efficiency. The study showed that the optimal conditions for obtaining FAMEs included a 7 wt.% catalyst loading, a 9:1 molar ratio of methanol to palm oil, a 3-hour time, and a temperature of 70°C, resulting in a 95.3% yield. Moreover, the catalyst remained effective for at least three cycles, indicating its effectiveness in biodiesel production.

A study focused on the transesterification of rapeseed oil with butanol utilized a catalyst consisting of dolomite particles with a size range of 0.315–0.1 mm, calcined at 850°C [34]. The research involved the development of an experimental plan and subsequent analysis of results using response surface methodology. The study comprehensively discussed the analysis of the interaction between four independent variables. Results highlighted that temperature and reaction duration exerted the most significant influence on ester yield. Following process optimization, it was determined that the yield of 94.55 wt% was achieved within 8 hours, using 5.24 wt% of dolomite as the catalyst at a temperature of 110°C and a butanol-to-oil molar ratio of 13.71:1.

Lani *et al.* [22] explored the CaO–ZSM-5 zeolite/Fe3O4 catalyst for the production of biodiesel from Used Cooking Oil (UCO) with a high free fatty acid content by direct transesterification. The results demonstrated that the catalyst exhibited exceptional catalytic activity, yielding 91% biodiesel with recyclability for up to four cycles. This impressive performance of the catalyst was attributed to its high surface area and its acid-base properties.

In another study [35], a magnetic solid base catalyst, $CaO@Fe_3O_4/SiO_2$, was used for the production of biodiesel from palm oil. The catalyst achieved a biodiesel conversion rate of over 92% under optimal conditions, employing a concentration of 5.0% by weight, a methanol-to-oil ratio of 5:1, and a reaction time of 3 hours at 80°C. The catalytic activity and recovery rate of $CaO@Fe_3O_4/SiO_2$ outperformed that of CaO catalyst. The properties of the synthesized biodiesel conformed to the standards set by the Department of Energy Business, highlighting its potential for efficient and high-quality biodiesel production.

To facilitate efficient biodiesel production, a catalyst was prepared by modifying acid-activated Clay (AC) with $Ca(OH)_2$ and NaOH in another study, resulting in CaNa/AC and Na/AC [36]. Both catalysts underwent thorough characterization using Hammett indicators, CO_2-Temperature-Programmed Desorption (CO_2-TPD), XRD, FT-IR, *etc*. The study examined the effects of various factors, including catalyst dosage, reaction temperature, time, methanol-to-oil molar ratio, and so on. The catalysts showed 97% oil conversion under optimized reaction conditions and maintained 80% activity after five cycles of usage. This study showed a method for large-scale biodiesel production, highlighting its potential for industrial application.

To enhance the sustainability of biodiesel production by utilizing cost-effective feedstocks, such as agricultural and food wastes, along with UCO, a solid base catalyst was synthesized [37]. The catalyst was derived from a mixture of ripe and unripe plantain peels, which underwent calcination at 500 °C for 4 hours. Before biodiesel production, WCO underwent a preliminary treatment involving an esterification reaction with 3% (v/v) H_2SO_4 to address its elevated acidic value, which subsequently transformed into biodiesel through a transesterification reaction. The complete procedure was precisely modelled and optimized with the help of the Taguchi L9 orthogonal array design method by taking reaction time, temperature, catalysts amount, and methanol WCO molar ratio as input variables. The analysis of the prepared catalyst indicated a predominant composition of potassium phases, accounting for 45.16 wt%. The morphology of the catalyst revealed a crystalline mesoporous nanocomposite structure. The acidic value after the esterification of WCO was reduced from 5 to 1 mg KOH/g. The optimal conditions for the transesterification process achieved a biodiesel yield of 97.96 wt%.

In this study, the UCO was transformed into biodiesel (FAME)) using a composite catalyst made from zeolite and eggshells (ZE/ES) [38]. The catalyst was fabricated through the wet impregnation method and subsequent calcination at diverse temperatures in the range of 700 to 900 °C. Optimization of the transesterification of UCO with methanol was conducted using a central composite design. The highest FAME content of 93.7% was attained under the optimal transesterification conditions. The remarkable performance of the 2ZE/ES-800 composite catalyst, prepared using 2 g of zeolite with 5 g of eggshell and calcining at 800 °C, contributed to its exceptional catalytic activity. This catalyst exhibited a high basic strength of 1181.1 μmol/g, with a predominance of CaO on its surface. Moreover, the reusability study demonstrated that the 2ZE/ES-800 catalyst can be effectively reused for up to five successive cycles without a substantial decline in its catalytic activity.

The study introduces an innovative approach using Ordered Mesoporous Carbon (OMC) encapsulating KF to create a novel KF@OMC-x, where x signifies the potassium weight ratio [39]. This involves the mechanical grinding of terephthalaldehyde, resorcinol, and a block copolymer template (Pluronic F127) followed by solidothermal treatment at 150 °C and calcination at 360 °C to remove the template. The resulting OMC exhibits several key features, including substantial BET surface areas (ranging from 351 to 445 m^2/g), well-defined mesoporosity, and even distribution of KF superbase sites within a highly stable structure. Consequently, the KF@OMC-x catalysts prove to be highly effective and reusable in base-catalyzed reactions, such as biodiesel and fine chemicals production. Notably, their performance surpasses that of widely used strong base catalysts like KF, Na_2CO_3, CaO, and hydrotalcite, which advance the development of robust and efficient solid base catalysts for large-scale biofuel production in demanding industrial processes.

Transesterification of Glycerol

The transesterification of glycerol leads to the production of a versatile product, glycerol carbonate, which can be synthesized through various procedures, as illustrated in Fig. (**3**). In a study, a range of catalysts, including Ni, Mg, K, La, Ce, K/Mg, and Li incorporated into MCM-41, were evaluated for glycerol transesterification with Dimethyl carbonate (Fig. **4**). Among these, the Li/MCM-41 catalyst exhibited the best catalytic performance, achieving 75% conversion of glycerol and 74% yield of glycerol carbonate [40]. The impact of calcination was also investigated, revealing that the catalyst calcined at 450°C was optimal for glycerol transesterification, supported by TGA analysis. Additionally, the study explored the influence of Li loading, with 4 wt% loading showing excellent performance (99% conversion of glycerol and 93% yield of glycerol carbonate) at 90°C in 3 hours. The E-factor for this reaction was calculated as 0.64, indicating process sustainability. Moreover, the catalytic system proved to be recyclable for up to 4 cycles without a substantial decrease in activity.

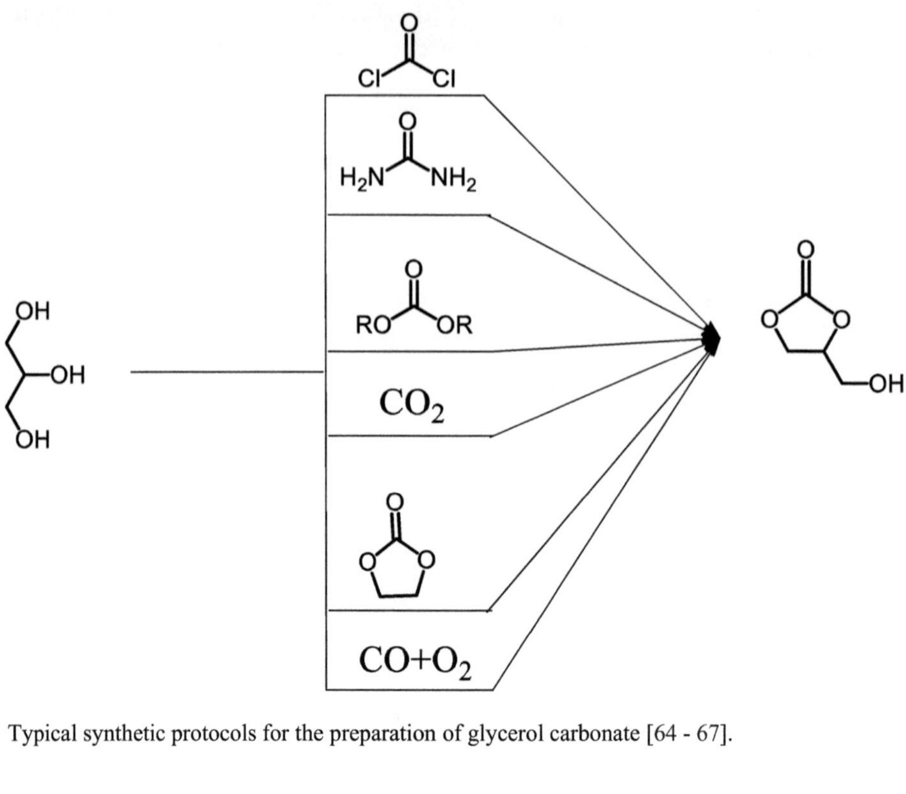

Fig. (4). Typical synthetic protocols for the preparation of glycerol carbonate [64 - 67].

In another investigation, Song *et al.* screened catalysts, including Li, K, and Cs-impregnated ZnO catalysts, for glycerol transesterification. The li/ZnO catalyst exhibited higher catalytic activity [41]. The study also explored the impact of supports, such as Al_2O_3 and ZrO_2, with ZnO identified as the best support. Furthermore, the influence of Li loading and calcination was studied, indicating that 1 wt% Li loading and calcination at 500°C provided the highest activity (97% conversion of glycerol and 96% yield of glycerol carbonate at 95°C for 4 hours). Although the catalyst system demonstrated reusability for up to 3 cycles, a drastic decrease in activity in the 4th cycle was observed as a result of the leaching of active species.

Arora *et al.* reported a Li-loaded coal fly ash-based catalyst for glycerol transesterification, revealing that 2 wt% Li loading and 500°C calcination resulted in 96% conversion of glycerol and 92% yield of glycerol carbonate. The catalyst exhibited sustainability for up to 4 cycles without a loss of activity [13].

Keogh *et al.* investigated Sodium Aluminate (SA)-supported catalysts for glycerol transesterification. Catalysts prepared by the impregnation method showed higher initial activity, and the study explored the influence of supports, revealing that MgO-supported SA yielded excellent catalytic activity [42]. The reusability of the catalytic system was evaluated, showing a marginal activity decrease in the impregnation method, while stable activity was observedwith the Al2O3-supported catalyst.

Das *et al.* investigated the use of *Mangifera indica* Peel Calcined Ash (MIPCA) as a solid base catalyst for the transesterification of glycerol with dimethyl carbonate [43]. MIPCA, derived from biowaste, demonstrated high basicity and stability for up to 5 cycles.

In order to prepare glycerol carbonate through the glycerol and CO2 interaction, various morphologies of MgO species were synthesised [44]. The cubic morphology of MgO with more (111) facets exhibited superior catalytic activity because of the existence of strong basic sites.

The catalyst derived from oil palm fuel ash was applied for glycerol transesterification, resulting in a 97% glycerol carbonate yield with about 95% glycerol conversion at 80°C. The catalyst showed recyclability for up to 5 cycles with a gradual decrease in activity [45].

Pradhan *et al.* investigated the valorization of glycerol using $MgCr_2O_4$ and MgV_2O_4 spinel catalysts [46]. $MgCr_2O_4$ proved to be an effective catalyst for glycerol-to-glycerol carbonate conversion. $MgTiO_3$ catalyst synthesized by the co-precipitation method also demonstrated good catalytic performance (93%

glycerol conversion with 88% selectivity towards glycerol carbonate) and reusability for up to 6 cycles [11].

Hameed and co-workers reported $Mg_{1.2}Ca_{0.8}O_2$ exhibited 100% yield of glycerol carbonate with a molar ratio of Mg/Ca 3:2[47]. Optimal reaction conditions include a glycerol to DMC molar ratio of 1:2, a reaction temperature of 70 °C, and a reaction duration of 90 minutes. The transesterification outcomes indicated that the strong catalytic performance was due to the synergistic interaction between calcium and magnesium components within the catalyst structure.

Hydrocalumites, a specific class of anionic clays, were explored for glycerol carbonate production [48]. The catalytic properties of Ca/Al-mixed oxide derived from hydrocalumites were highlighted for various catalytic reactions, demonstrating their versatility in catalysis [49 - 51].

A study by Àlvarez *et al.* used MgAl hydrotalcites supported on α- and γ-Al2O3 substrates to study the formation of glycerol carbonate [52]. The nature of the basic sites in the catalyst significantly influenced catalytic activity, with catalysts exhibiting predominant Brönsted basicity and showing superior performance.

Wu *et al.* synthesized Ce–NiO catalysts for glycerol transesterification using DEC under solvent-free conditions, achieving 94.14% conversion and 90.95% selectivity to glycerol carbonate [53]. $Ce_{1-x}Cd_xO$-mixed oxide systems were also explored, with the $Ce_{0.7}Cd_{0.3}O$ catalyst exhibiting the highest activity (97% conversion, 100% selectivity) under optimized conditions [54].

Mixed base oxide catalysts based on Mg, Zr, and Sr were synthesized and investigated for glycerol transesterification, with a molar ratio of Mg/Zr/Sr at 3:1:1 demonstrating superior catalytic activity [55]. Glycerol carbonate was decarbonylated by the catalysts, which made it easier to produce glycidol.

MgO-CeO_2 catalysts demonstrated notable competence in selectively producing glycerol carbonate through glycerol transesterification [56]. The performance of the catalysts was contingent upon the molar ratio of Mg and Ce, with a MgO to CeO_2 mole ratio of 3:1 exhibiting the highest activity.

Nanocrystalline mesoporous MgO-ZrO_2 solid solutions, synthesized using a one-pot synthesis method, demonstrated remarkable efficiency in facilitating glycerol transesterification with dimethyl carbonate, achieving a 96% conversion rate and an 88% yield [57].

A novel catalyst, MgO/g-C_3N_4, exhibited significantly higher catalytic performance when glycerol was transformed into glycerol carbonate through transesterification [58]. The catalyst with a specific MgO:g-C_3N_4 ratio of 6:1 displayed the highest yield at 95.47% and demonstrated good reusability. The presented catalysts showcased diverse approaches to glycerol transesterification, emphasizing their potential for various industrial applications.

The *Musa acuminata* Peel Ash (MAPA) as a solid base catalyst was employed for the synthesis of Glycerol Carbonate(GC) using glycerol and dimethyl carbonate *via* transesterification reaction [59]. The high basicity, substantial surface area, and mesoporous nature of the catalyst showed efficacy in the formation of glycerol carbonate. In this reaction, microwave irradiation greatly outperformed conventional heating (18% conversion and 98.5% selectivity) in terms of conversion (99%) and selectivity (99.5%) under optimized reaction conditions. The obtained GC was confirmed using gas chromatography, ^{1}H and ^{13}C NMR, and mass spectrometry. The reusability of the catalyst was assessed over six successive cycles, revealing excellent stability in facilitating the formation of GC.

Different MgO loadings were used to create MgO@ZIF-8 in the transesterification of glycerol to glycerol carbonate [60]. Interestingly, the 50 wt% MgO@ZIF-8 catalyst exhibited superior catalytic activity compared to individual MgO and ZIF-8 counterparts. The observed enhancement in catalytic activity was attributed to a proposed mechanism involving acid-base bifunctional sites on the surface of the catalyst. The outcomes revealed that MgO nanoparticles were effectively deposited onto the ZIF-8 surface using the wet-impregnation and calcination procedure, maintaining the ZIF-8 structure while achieving high atomic efficiency. The study shed light on the synergistic interplay between MgO and ZIF-8, contributing valuable insights into the design of efficient catalysts for transesterification reactions.

A wet impregnation approach was effectively used to create La_2O_3 catalysts doped with alkali and alkaline earth metals, and their effectiveness in converting glycerol to glycerol carbonate was evaluated [61]. Among the alkali and alkaline earth metals tested, Na emerged as the most effective dopant for enhancing catalytic performance. For two hours of reaction at 70 °C, the 25 mol% Na-doped La_2O_3 catalyst produced 60% of Glycerol Carbonate (GC). The ionic radius and valence state of the dopants had a major impact on the catalytic performance of La_2O_3. On the surface of La_2O_3, alkali metals were evenly distributed, whereas alkaline earth metals formed another phase and were accumulated there. On the surface of La_2O_3, doping alkali metals changed the distribution of electrons around La, while doping alkaline earth metals changed the activity of oxygen sites. A key element in improving the catalytic activity of La_2O_3 was found to be its comparable ionic radii with dopant. Dopants with ionic radii close to La demonstrated more substantial improvements compared to those with smaller or larger radii, and dopants with lower valence states exhibited superior enhancements. In contrast, basic sites on the La_2O_3 surface were crucial for the transesterification of glycerol and dimethyl carbonate (DMC). These findings regarding the impact of radius and charge of dopants on the catalytic activity of La_2O_3 provide valuable insights for designing more efficient and cost-effective catalysts for the value-added conversion of glycerol-to-glycerol carbonate.

In order to create an effective material for the synthesis of Glycerol Carbonate (GC), a CaO-based catalyst impregnated with chloride salts was tested [51]. LiCl/CaO, the catalyst, showed exceptional catalytic activity under mild reaction conditions. The study examined how calcination temperature, catalyst amount, and LiCl loadings affected catalytic activity. At a reaction temperature of 65 °C, ideal conditions were reached with CaO loaded with 10% LiCl, resulting in the maximum glycerol carbonate yield of 94.19% after an hour. Moreover, the catalyst exhibited robust stability across multiple reuses. The observed catalytic activity of LiCl-modified CaO was attributed to structural modifications and changes in the basicity of the catalyst. The formation of Li_2O_2 species, acting as strong basic sites due to the substitution of Ca2+ in the CaO lattice by Li+, was identified as a key factor contributing to the catalyst's efficacy in transesterification. This research sheds light on the role of LiCl modification in enhancing the catalytic performance of CaO, offering valuable insights into the structural aspects and basicity influencing the synthesis of glycerol carbonate.

Dalai et al studied the glycerol carbonate synthesis using various SBA-15 catalysts [62]. The results confirmed that the most effective catalyst for the conversion of glycerol to glycerol carbonate was Ti-SBA-15 and the lewis acidic nature of Ti was the main driving force in facilitating the reaction.

The incorporation of potassium into a three-dimensional mesoporous silica structure was revealed as a highly promising approach for the synthesis of glycerol carbonate [63]. Specifically, the 7% Cal/K-TUD-1 catalyst demonstrated exceptional performance with 91.53% yield at a reaction temperature of 90 °C, a DMC-to-glycerol molar ratio of 5, and a catalyst dosage of 6 wt.% within a 2.5 h reaction time. The synthesized catalyst holds commercial potential for valorizing glycerol derived from biodiesel into valuable glycerol carbonate, showcasing its significance in sustainable applications.

Depolymerization of lignin

The transformation of lignin into compounds with additional value is pivotal for advancing the sustainability of biorefineries [68]. Various types of lignins sourced from diverse botanical origins, including softwood, herbaceous, and hardwood lignins, are obtained through prominent industrial pulping techniques, such as soda and kraft processes.

The structural complexity of lignin poses important challenges to its valorisation. To overcome this, various solid base catalysts have been employed [16]. Research focuses on quantifying and characterizing two primary depolymerization fractions: lignin oil and lignin residue. The evaluation includes parameters, such as lignin conversion, lignin oil yield, selectivity towards phenolic monomers, and the production of lignin residue and char. Advanced spectroscopic and chromatographic techniques analyze the products, revealing lignin oil yields ranging from 20% to 31% by weight, with total monomer contents varying from 48% to 57% by weight [69].

A Ni-Fe-supported Magnesium Silicate ($MgSiO_3$) catalyst was synthesized and applied for the kraft lignin depolymerization, yielding valuable phenolic monomers [70]. The study explored the impact of temperature, time, and Ni-Fe ratio on $MgSiO_3$, showing that the 10Ni10Fe/MS catalyst achieved the highest lignin depolymerization products at 14.29 wt% under optimised conditions of 300°C for 1 hour, with primary products being 0.79 wt%phenol, 11.38 wt% catechol, and 1.51 wt% of guaiacol.

A series of CaO-supported Co metal catalysts were examined for lignin depolymerization at various loadings and conditions [71]. The study revealed that the 26.6% yield of bio-oil was accomplished at 160°C with water as the solvent. Optimal conditions for bio-oil production (60.2 wt%) were observed over 10 wt% Co/CaO catalyst in methanol at 160°C for 60 minutes, promoting the depolymerization of lignin macromolecules and increasing selectivity (58.7%) towards vanillin compounds.

Eco-friendly catalysts derived from eggshells were employed for lignin depolymerization, resulting in the largest output of bio-oil using the CaO/TiO_2-800 catalyst in ethanol solvent in the presence of a mixture of N_2 and H_2 atmospheres [72]. The catalytic depolymerization under a nitrogen environment primarily generated 68% of alkyl phenolic monomers and 24% of methoxy phenolic compounds, while hydrogen gas prompted deoxygenation and hydrogenation reactions, producing a bigger proportion of cyclohexane compounds (38.2%) and aromatic hydrocarbons (19%).

A variety of solid base catalysts were used to depolymerise alkali lignin, including CaO/Al_2O_3, CaO/CeO_2, and CaO/ZrO_2 [73]. The application of solid base catalysts significantly increased bio-oil yield, with the highest yield reaching 50.0 wt% using ethanol and methanol as solvents and CaO/ZrO_2 and CaO/CeO_2 catalysts at a temperature of 180°C.

A catalytic system consisting of a minimal amount of solid base $K_7HNb_6O_{19}$ (KNb6) in combination with Cu/C_3N_4 demonstrated improved catalytic efficacy for the oxidative cleavage of β-O-4 lignin linkages by means of molecular oxygen [74]. The synergistic interplay among KNb6 and Cu/ C_3N_4 allowed for the conversion of 96% of the β-O-4 ketone model under moderate reaction conditions, yielding phenol (90%) and other aromatic monomers (90%) with sustained catalytic activity over five cycles.

In a recent paper, microwave-assisted hydrothermal liquefaction was employed for the base-catalysed lignin depolymerization under mild conditions [75]. The approach, applied to various lignin types, yielded substantial quantities of low-molecular-weight bio-oil in just 1 minute at 220°C. Grass lignin exhibited exceptional efficiency in monomer production, with a remarkable monomer selectivity of 66% for 4-vinylphenol.

A work reported by Zhu *et al.* extensively studied the impact of reaction time on the product yield and distribution of the Base-Catalyzed Depolymerisation (BCD) process [76]. The research aimed to achieve good yield and extremely selective phenol-based products through a chain reaction mechanism. The findings confirmed that guaiacol underwent hydrolysis of the methoxy group, leading to the formation of catechol. The utilization of reaction time as the controlling factor for the secondary depolymerization of methoxy groups, without the need for critical metals, reduced operational costs and enhanced the yield and selectivity of phenol compounds.

Understanding the structural composition of lignin is crucial for comprehending its degradation behavior, as highlighted in a study characterizing lignin samples derived from different sources [77]. The compounds observed by GC-MS analysis were categorized into phenolics, heterocyclic compounds, aromatic hydrocarbons, acidic substances, and others. Pyrolysis of alkali lignin using CaO catalyst produced 93.5% of phenolics. Ethanone, mequinol, 1-(2-hydroxy-5-methyl-phenyl), benzaldehyde,2-methoxy-5-methylphenol, and 3-hydroxy-4-methoxy were recognized as key compounds that were formed with a high degree of selectivity.

Na2CO3, NaOH, and CaO were used to depolymerize sorghum stalks in order to produce a methanol-soluble fraction [78]. GC-MS analysis revealed predominant products of phenols and esters, with Na_2CO_3-catalyzed sorghum stalk depolymerization primarily yielding methoxylphenols, whereas alkylphenols predominated in depolymerization catalyzed by NaOH and CaO.

In another study, lignin samples from different origins were subjected to depolymerization at 386 °C and 260 bar pressure in alkaline supercritical water

[79]. The study aimed to generate compounds with low molecular weights and high values, and the resulting products were analyzed for yield, monomer selectivity, molecular weight distribution, and chemical structure. The method was found to be effective, with technical wood-based lignins showing higher suitability for monomer production compared to non-wood biorefinery lignins.

Using a Pt/HAP-3 catalyst [80], the hydrogenolysis of lignin to provide valuable arene monomers was investigated. With a monomer output of 15.1%, the metal nanoparticles supported by an acidic carrier demonstrated excellent catalytic performance.

A novel hydrogenation catalyst combining tungsten (W) modification with an acid-base catalyst was established for the efficient conversion of lignin into liquid fuels [81]. The catalyst comprised Lewis acid sites (TiO_2 and WO_x) and base sites (MgO), facilitating the cleavage of C–C and C-O bonds. This offers valuable insights for developing an effective catalyst for the hydrogenolysis of lignin in future applications.

Tian *et al.* investigated the pivotal role of Sm_2O_3 structure in catalyzing the depolymerization of lignin [82]. The study found that different morphologies of Sm_2O_3 exerted distinct influences on the conversion of lignin and the yield of catalytic products under identical reaction conditions. The catalytic performance ranking of different Sm_2O_3 morphologies was determined as follows: r- Sm_2O_3 > m- Sm_2O_3 > o- Sm_2O_3 > t- Sm_2O_3 > i- Sm_2O_3. Specifically, rod-like nano samarium oxide (r-Sm_2O_3) showed optimal catalytic activity at 220°C, 1 MPa O_2, and 14 h, achieving an impressive conversion of 93.9%. The primary products included benzaldehyde dimethyl acetal, dimethyl butyrate, and methyl benzoate, with yields of 23.7%, 29.7%, and 58.4%, respectively. Various characterization analyses revealed that the superior catalytic activity of r- Sm_2O_3 was primarily attributed to its abundance of oxygen vacancies. The crystal surface {222} of r-Sm_2O_3 played a crucial role in facilitating oxygen transition from the bulk to the surface through a vacancy hopping mechanism, activating oxygen and generating reactive oxygen species, which is an essential factor in enhancing catalytic activity.

The distribution of catalytic products indicated that r- Sm_2O_3 simultaneously broke C-O (β-O-4), $C\alpha$-C_β, and Csp2-Csp2 bonds in the aryl ring. These results highlight the unique catalytic properties exhibited by r- Sm_2O_3. Compared to precious metal catalysts, Sm_2O_3 offers advantages in terms of cost-effectiveness and abundant reserves, making it a valuable candidate for research and broad applicability. This study opens new avenues for designing and implementing heterogeneous catalysts in lignin processing, providing innovative insights and practical approaches.

Aldol Condensation

The catalytic conversion of ketones and aldehydes into high-value chemicals is crucial for sustainable development. The development of active and selective catalytic materials is essential for biomass conversion into chemicals and fuels. Arumugam *et al.* conducted an aldol condensation reaction between aldehydes (furfural) and ketones (acetone) using MgAl mixed oxide, leading to the formation of important products, FAc and F2Ac. These products undergo hydrodeoxygenation to generate diesel and jet fuels, addressing current fuel market challenges [83].

In another study, K-modified MgAl-SBA-15 was used for the condensation reaction between furfural and acetone, emphasizing the significance of K in producing FAc and F2Ac for biofuel production [84]. Chottiratanachote *et al.* utilized a Mg-Al mixed oxide acid-base catalyst for the aldol condensation of furfural and 2-butanone, producing branched-chain C_9 products [85]. The catalyst with a Mg:Al atomic ratio (3:1) and high surface area demonstrated high effectiveness under optimal conditions.

Researchers [86] explored the manufacture of aviation fuel using common carbonyl compounds over aldol condensation and subsequent hydrodeoxygenation using a bifunctional catalyst, *i.e.*, Ni/Mg-Al-O/Ac. This process achieved a notable 79.23% yield of jet fuel intermediates within the carbon number range of C7 to C14.

Ghosh *et al.* conducted cross-aldol and self-aldol reactions using Cs+ single sites confined with β-zeolite channels, demonstrating the superior activity of Cs+/β-zeolite catalysts in aldol reactions [87]. Despite lacking significant acid-base characteristics, the Cs+ single site/β catalyst displayed unprecedented efficiency and reusability.

Furfural was transformed into a biofuel precursor through aldol condensation with 2-butanone using mesoporous silica (SBA-15)-supported Sr and Ti mixed oxide [88]. The optimized catalyst, 20STO(2:1)/SBA15, achieved over 99% furfural conversion and a 78.3% yield of α,β-unsaturated C9 ketones (C9B).

Deshpande *et al.* explored the influence of micropores on amino silica catalytic activity, revealing that materials with minimal micropores exhibited greater activity in aldol reactions [89]. Jet fuel generation offers a promising yet difficult effort in the area of biomass consumption. One approach to producing highly branched alkanes involves two steps. Methyl Isobutyl Ketone (MIBK) undergoes a self-aldol condensation first, followed by a one-step hydrodeoxygenation (HDO) employing cyclohexane as the solvent and MgAl-O in the self-aldol condensation. The study found that the self-aldol condensation of MIBK is greatly aided by basic sites with moderate to severe acidity.

Structured Hydroxyapatite Composites (SHCs) are highly efficient reusable solid base catalysts for a diversity of condensation reactions. Butyraldehyde to 2-ethylhexenal self-aldol condensation reaction at mild reaction conditions is one example [82]. According to Jose T. *et al.*, SHC catalysts were more efficient than traditional solid base catalysts, namely., MgO, TiO2, $CaCO_3$, and hydroxyapatites. After the synthesis of SHC catalysts, they were characterized using a variety of electron microscopy, spectroscopic, and surface analysis techniques, and it was discovered that their exceptional performance was mostly due to a balanced acid/base ratio and a high BET surface area [83].

The functionalized aminosilanes were thoroughly characterized with the help of standard techniques. The findings revealed that catalysts with minimal or no micropores exhibited greater activity in the aldol reaction, both for furfuraldehyde (FA) combined with acetone and hydroxymethylfurfural (HMF) combined with acetone. Moreover, in recycling experiments, the NMP catalyst (the method that minimizes or eliminates micropores) demonstrated a higher level of retained activity for four cycles when compared to the catalyst prepared by the conventional method. These results have significant implications for the design of materials aimed at converting biomass into valuable chemicals and fuels [90].

A study on hydrotalcites (HTCs) demonstrated their capability as solid base materials for advanced biofuel production. The nature of the M3+ element in HTCs impacted crystallinity and basicity, affecting furfural conversion in aldol condensation [91].

Microwave-assisted aldol condensation of 5-hydroxymethyl furfural (HMF) with acetone using hydrotalcite-based catalysts showed excellent conversions in shorter reaction times. Various heterogeneous catalysts resulting from the same parent hydrotalcite were explored, and effective catalyst-recycling strategies were identified [92].

Korolova *et al.* reported the significant impact of variations in reaction parameters, specifically pH value and co-precipitation temperature, on the physicochemical features and catalytic activity of MgAl hydrotalcite with Mg/Al = 3 [93]. The XRD data confirmed that all synthesized materials were pure in phase and exhibited high crystallinity as Hydrotalcite-Like Compounds (HTCs), but their crystallite size varied considerably within the range of 80–230 Å (d003) due to changes in synthesis conditions. Upon calcination, the resulting materials led to structural defects and extra-framework Al-rich species. The subsequent rehydration procedure was effective in restoring the original size of HTC platelets for small-size precursors, while the restoration degree diminished as the crystallite size of the precursor increased from 80 to 230 Å (d003). The catalytic performance of calcined and shortly rehydrated catalysts was linked with the initial crystallite size of the precursors, where a smaller crystallite size resulted in higher furfural conversion. The approachability of catalytically active sites was impacted by extra-framework species and/or external Al-rich species.

On the other hand, all evaluated samples achieved 99% conversion of furfural in 180 minutes, demonstrating that the catalytic performance of rehydrated catalysts for 120 minutes did not correlate with the crystallite size of their as-prepared counterparts. SEM pictures demonstrating the establishment of secondary platelets in addition to the surface of primary platelets demonstrated that extended rehydration promoted the participation of extra-framework and/or partially framework species in the recrystallization of the reinstated HTC structure. Furthermore, the improved furfural conversion suggested that this occurrence led to better access to catalytically active regions for reactants. As a result of variations in pH and coprecipitation temperature during HTC synthesis, materials with diverse particle sizes were produced. These variations affected the physicochemical characteristics of materials, degree of restoration, capacity for rehydration, and, ultimately, their catalytic efficiency in aldol condensation.

Three different waste seashells were used for the synthesis of CaO catalysts and applied for the self-aldol condensation of cyclopentanone [94]. Notably, the S-shell-750 catalyst demonstrated superior catalytic activity and selectivity, achieving a dimer yield of 92.1% and up to 100% selectivity. This performance was comparable to that of commercial CaO (dimer yield: 88.2%). Moreover, multiple characterizations revealed that the outstanding catalytic efficacy of S-shell catalysts could be attributed to their relatively higher CaO content, low hydrophilicity, and stronger basicity/acidity. The findings suggest that the S-shel--750 catalyst can directly substitute commercial CaO in this reaction. This innovation not only highlights the viability of waste seashells as a cost-effective source of bio-based catalysts but also presents a new pathway for utilizing such catalysts in the transformation of biomass.

CONCLUSION

Base catalysis is a widely employed technique in diverse reactions, including the transesterification of oils, glycerol transesterification, lignin depolymerization, and aldol condensation. In particular, base-catalyzed aldol condensation plays a vital role in the transformation of biomolecules into larger functionalized oxygenates through C-C bond formation. The synthesized compounds have applications in the fine chemical industry, contributing to the manufacture of hydrocarbon fuels and lubricants. One extensively studied process involves the transesterification of vegetable oils, not only generating environmentally friendly fuel but also yielding a significant amount of glycerol, which serves as a starting material for bio-based fine chemical production.

In general, heterogeneous catalysts offer distinct advantages over homogeneous catalysts in these transformations, including the potential for the reuse and recovery of solid base catalysts. These materials can be tailored by appropriate basicity or by combining acid–base pairs to meet the specific requirements of reactions. However, a persistent challenge is the design of selective, active, and stable catalysts that prevent deactivation and leaching processes. A promising avenue for fine chemical manufacturing involves the progress of cascade processes, where multiple reaction stages arise in a one-pot process, eliminating the need for separation and purification operations. This approach can lead to significant economic and environmental benefits.

Lastly, achieving this goal may require the creation of tailor-made catalytic structures with numerous active sites, such as combining basic sites with metal, redox, and acid sites. This integration is crucial for advancing the field. Consequently, ongoing research focused on process development and catalyst improvement remains critical for the continued progress of this field.

REFERENCES

[1] Sudarsanam, P.; Zhong, R.; Van den Bosch, S.; Coman, S.M.; Parvulescu, V.I.; Sels, B.F. Functionalised heterogeneous catalysts for sustainable biomass valorisation. *Chem. Soc. Rev.*, **2018**, *47*(22), 8349-8402.
[http://dx.doi.org/10.1039/C8CS00410B] [PMID: 30226518]

[2] Ono, Y. Solid base catalysts for the synthesis of fine chemicals. *J. Catal.*, **2003**, *216*(1-2), 406-415.https://doi.org/https://doi.org/10.1016/S0021-9517(02)00120-3
[http://dx.doi.org/10.1016/S0021-9517(02)00120-3]

[3] Jambhulkar, D.K.; Ugwekar, R.P.; Bhanvase, B.A.; Barai, D.P. A review on solid base heterogeneous catalysts: preparation, characterization and applications. *Chem. Eng. Commun.*, **2022**, *209*(4), 433-484.
[http://dx.doi.org/10.1080/00986445.2020.1864623]

[4] Dhyani, V.; Bhaskar, T. A comprehensive review on the pyrolysis of lignocellulosic biomass. *Renew. Energy*, **2018**, *129*, 695-716.
[http://dx.doi.org/10.1016/j.renene.2017.04.035]

[5] Gupta, V. K.; Tuohy, M. G. *Biofuel Technologies: Recent Developments*, **2014**.
[http://dx.doi.org/10.1007/978-3-642-34519-7]

[6] Gundekari, S.; Mitra, J.; Varkolu, M. *Classification, Characterization, and Properties of Edible and Non-Edible Biomass Feedstocks*; Elsevier Inc., **2020**.
[http://dx.doi.org/10.1016/B978-0-12-820236-4.00004-0]

[7] Tanabe, K. Catalysis by solid bases and related subjects. *Stud. Surf. Sci. Catal.*, **1985**, *20*, 1-14.
[http://dx.doi.org/10.1016/S0167-2991(09)60151-8]

[8] Buchori, L.; Diponegoro, U.; Istadi, I.; Diponegoro, U.; Purwanto, P.; Diponegoro, U. Synthesis of biodiesel on a hybrid catalytic-plasma reactor over K□O/CaO-ZnO catalyst Luqman Buchori, Istadi Istadi *, Purwanto Purwanto Diponegoro University, Faculty of Engineering. *Department of Chemical.,* **2021**, *2017*(January), 302-318.

[9] Take, J.; Kikuchi, N.; Yoneda, Y. Base-strength distribution studies of solid-base surfaces. *J. Catal.,* **1971**, *21*(2), 164-170.
[http://dx.doi.org/10.1016/0021-9517(71)90134-5]

[10] Miladinović, M.R.; Zdujić, M.V.; Veljović, D.N.; Krstić, J.B.; Banković-Ilić, I.B.; Veljković, V.B.; Stamenković, O.S. Valorization of walnut shell ash as a catalyst for biodiesel production. *Renew. Energy,* **2020**, *147*, 1033-1043.
[http://dx.doi.org/10.1016/j.renene.2019.09.056]

[11] Pradhan, G.; Maurya, S.; Pradhan, S.; Sharma, Y.C. An accelerated route for synthesis of Glycerol carbonate using MgTiO3 perovskite as greener and cheaper catalyst. *Molecular Catalysis,* **2023**, *545*(May), 113162.
[http://dx.doi.org/10.1016/j.mcat.2023.113162]

[12] Ruhaimi. A.H; Aziz.M.A.A ; Jalil. A.A; Magnesium oxide-based adsorbents for carbon dioxide capture: Current progress and future opportunities, HYPERLINK "https://www.sciencedirect.com/journal/journal-of-co2-utilization" *Journal of CO_2 Utilization*, **2021**, *43*, 101357.
[http://dx.doi.org/10.1016/j.jcou.2020.101357]

[13] Arora, S.; Gosu, V.; Subbaramaiah, V.; Hameed, B.H. Lithium loaded coal fly ash as sustainable and effective catalyst for the synthesis of glycerol carbonate from glycerol. *J. Environ. Chem. Eng.,* **2021**, *9*(5), 105999.
[http://dx.doi.org/10.1016/j.jece.2021.105999]

[14] Wang, S.; Hao, P.; Li, S.; Zhang, A.; Guan, Y.; Zhang, L. Synthesis of glycerol carbonate from glycerol and dimethyl carbonate catalyzed by calcined silicates. *Appl. Catal. A Gen.,* **2017**, *542*, 174-181.https://doi.org/https://doi.org/10.1016/j.apcata.2017.05.021
[http://dx.doi.org/10.1016/j.apcata.2017.05.021]

[15] Zhang, Z.H.; Sun, Z.; Yuan, T.Q. Recent advances in the catalytic upgrading of biomass platform chemicals *via* hydrotalcite-derived metal catalysts. *Transactions of Tianjin University,* **2022**, *28*(2), 89-111.
[http://dx.doi.org/10.1007/s12209-021-00307-6]

[16] Schutyser, W.; Renders, T.; Van den Bosch, S.; Koelewijn, S.F.; Beckham, G.T.; Sels, B.F. Chemicals from lignin: an interplay of lignocellulose fractionation, depolymerisation, and upgrading. *Chem. Soc. Rev.,* **2018**, *47*(3), 852-908.
[http://dx.doi.org/10.1039/C7CS00566K] [PMID: 29318245]

[17] Eun Kim, S.; Hu Kim, J.; Keun Kim, D.; Chul Ham, H.; Lee, K.Y.; Joo Kim, H. Na-modified carbon nitride as a leach-resistant and cost-effective solid base catalyst for biodiesel production. *Fuel,* **2023**, *341*(341), 127548.
[http://dx.doi.org/10.1016/j.fuel.2023.127548]

[18] Zhang, S.; Fu, J.; Xing, S.; Li, M.; Liu, X.; Yang, L.; Lv, P. Sodium Silicates Modified Calcium Oxide as a High-Performance Solid Base Catalyst for Biodiesel Production. *Catalysts,* **2023**, *13*(4), 775.
[http://dx.doi.org/10.3390/catal13040775]

[19] Zhou, G.; Liang, Y.; Zheng, Z.; Ju, L. Application of dewatered paper sludge-derived porous solid base catalyst for biodiesel production: Physicochemical properties, reaction kinetics and thermodynamic studies. *Environ. Technol.,* **2023**, *44*(21), 3263-3280.
[http://dx.doi.org/10.1080/09593330.2022.2056087] [PMID: 35306978]

[20] Supamathanon, N.; Boonserm, K.; Osakoo, N.; Wittayakun, J.; Prayoonpokarach, S.; Chanlek, N.; Dungkaew, W. Potassium supported on zeolite-geopolymer hybrid materials as a new solid base catalyst for transesterification of soybean oil. *Renew. Energy,* **2023**, *202*,

1460-1469.https://doi.org/https://doi.org/10.1016/j.renene.2022.12.018
[http://dx.doi.org/10.1016/j.renene.2022.12.018]

[21] Das, A.; Li, H.; Kataki, R.; Agrawal, P.S.; Moyon, N.S.; Gurunathan, B.; Rokhum, S.L. Terminalia arjuna bark – A highly efficient renewable heterogeneous base catalyst for biodiesel production. *Renew. Energy,* **2023**, *212*, 185-196.https://doi.org/https://doi.org/10.1016/j.renene.2023.05.066
[http://dx.doi.org/10.1016/j.renene.2023.05.066]

[22] Sun, D.; Gao, L.; Wei, R.; Pan, X.; Xiao, G. Mechanical vapor recompression coupling organic rankine cycle process for purification of crude biodiesel obtained by solid base-catalyzed transesterification. *Energy,* **2023**, *266*, 126499.https://doi.org/https://doi.org/10.1016/j.energy.2022.126499
[http://dx.doi.org/10.1016/j.energy.2022.126499]

[23] Basumatary, B.; Brahma, S.; Nath, B.; Basumatary, S.F.; Das, B.; Basumatary, S. Post-harvest waste to value-added materials: Musa champa plant as renewable and highly effective base catalyst for Jatropha curcas oil-based biodiesel production. *Bioresour. Technol. Rep.,* **2023**, *21*(21), 101338.
[http://dx.doi.org/10.1016/j.biteb.2023.101338]

[24] Chen, Q.; Wang, A.; Quan, W.; Gong, W. Efficient synthesis of biodiesel from Hyoscyamus niger L. seed oil by base catalysis. *Fuel Process. Technol.,* **2023**, *241*, 107630.https://doi.org/https://doi.org/10.1016/j.fuproc.2022.107630
[http://dx.doi.org/10.1016/j.fuproc.2022.107630]

[25] Ozor, P.A.; Aigbodion, V.S.; Sukdeo, N.I. Modified calcium oxide nanoparticles derived from oyster shells for biodiesel production from waste cooking oil. *Fuel Communications,* **2023**, *14*(14), 100085.
[http://dx.doi.org/10.1016/j.jfueco.2023.100085]

[26] Takase, M.; Kipkoech, R.; Miller, D.L.; Buami, E.K. Optimisation of the reaction conditions for biodiesel from Parkia biglobosa oil *via* transesterification with heterogeneous clay base catalyst. *Fuel Communications,* **2023**, *15*(May), 100089.
[http://dx.doi.org/10.1016/j.jfueco.2023.100089]

[27] Brahma, S.; Basumatary, B.; Basumatary, S.F.; Das, B.; Brahma, S.; Rokhum, S.L.; Basumatary, S. Biodiesel production from quinary oil mixture using highly efficient Musa chinensis based heterogeneous catalyst. *Fuel,* **2023**, *336*, 127150. https://doi.org/https://doi.org/10.1016/j.fuel.2022.127150
[http://dx.doi.org/10.1016/j.fuel.2022.127150]

[28] Prajapati, P.; Shrivastava, S.; Sharma, V.; Srivastava, P.; Shankhwar, V.; Sharma, A.; Srivastava, S.K.; Agarwal, D.D. Karanja seed shell ash: A sustainable green heterogeneous catalyst for biodiesel production. *Results in Engineering,* **2023**, *18*(March), 101063.
[http://dx.doi.org/10.1016/j.rineng.2023.101063]

[29] Miyuranga, K.A.V.; Arachchige, U.S.P.R.; Marso, T.M.M.; Samarakoon, G. Biodiesel production through the transesterification of waste cooking oil over typical heterogeneous base or acid catalysts. *Catalysts,* **2023**, *13*(3), 546.
[http://dx.doi.org/10.3390/catal13030546]

[30] Ashine, F.; Kiflie, Z.; Prabhu, S.V.; Tizazu, B.Z.; Varadharajan, V.; Rajasimman, M.; Joo, S.W.; Vasseghian, Y.; Jayakumar, M. Biodiesel production from Argemone mexicana oil using chicken eggshell derived CaO catalyst. *Fuel,* **2023**, *332*, 126166. https://doi.org/https://doi.og/ 10.1016/j.fuel.2022.126166
[http://dx.doi.org/10.1016/j.fuel.2022.126166]

[31] Nježić, Z.B.; Kostić, M.D.; Marić, B.D.; Stamenković, O.S.; Šimurina, O.D.; Krstić, J.; Veljković, V.B. Kinetics and optimization of biodiesel production from rapeseed oil over calcined waste filter cake from sugar beet processing plant. *Fuel,* **2023**, *334*, 126581. https://doi.org/https://doi.org/10.1016/j.fuel.2022.126581
[http://dx.doi.org/10.1016/j.fuel.2022.126581]

[32] Bharti, R.; Singh, B.; Oraon, R. Synthesis of Sn-CaO as a bifunctional catalyst and its application for biodiesel production from waste cooking oil. *Biofuels,* **2023**, *14*(6), 607-617. [http://dx.doi.org/10.1080/17597269.2022.2161128]

[33] Sun, J.; Yu, H.; Zhang, P.; Qi, G.; Chen, X.; Liang, X.; Si, H. Steel slag decorated with calcium oxide and cerium oxide as a solid base for effective transesterification of palm oil. *Processes (Basel),* **2023**, *11*(6), 1810. [http://dx.doi.org/10.3390/pr11061810]

[34] Gaide, I.; Makareviciene, V.; Sendzikiene, E.; Gumbyte, M. Application of dolomite as solid base catalyst for transesterification of rapeseed oil with butanol. *Sustain. Energy Technol. Assess.,* **2022**, *52*, 102278.https://doi.org/https://doi.org/10.1016/j.seta.2022.102278 [http://dx.doi.org/10.1016/j.seta.2022.102278]

[35] Singhanatkaisi, N.; Kandiah, M.; Zheng, A.L.T.; Ando, Y.; Boonyuen, S. Biodiesel production from used palm oil using magnetic solid base catalyst. *J. Phys. Conf. Ser.,* **2022**, *2175*(1), 012025. [http://dx.doi.org/10.1088/1742-6596/2175/1/012025]

[36] Wang, Y.; Muhammad, Y.; Yu, S.; Fu, T.; Liu, K.; Tong, Z.; Hu, X.; Zhang, H. Preparation of Ca- and Na-modified activated clay as a promising heterogeneous catalyst for biodiesel production *via* transesterification. *Appl. Sci. (Basel),* **2022**, *12*(9), 4667. [http://dx.doi.org/10.3390/app12094667]

[37] Falowo, O.A.; Oladipo, B.; Taiwo, A.E.; Olaiya, A.T.; Oyekola, O.O.; Betiku, E. Green heterogeneous base catalyst from ripe and unripe plantain peels mixture for the transesterification of waste cooking oil. *Chemical Engineering Journal Advances,* **2022**, *10*, 100293. https://doi.org/https://doi.org/10.1016/j.ceja.2022.100293 [http://dx.doi.org/10.1016/j.ceja.2022.100293]

[38] Yusuff, A.S.; Gbadamosi, A.O.; Atray, N. Development of a zeolite supported CaO derived from chicken eggshell as active base catalyst for used cooking oil biodiesel production. *Renew. Energy,* **2022**, *197*, 1151-1162.https://doi.org/https://doi.org/10.1016/j.renene.2022.08.032 [http://dx.doi.org/10.1016/j.renene.2022.08.032]

[39] Zheng, W.; Kan, X.; Liu, F.; Li, H.; Song, F.; Xia, G.; Liu, J.; Liu, F. Ordered mesoporous carbon encapsulating KF: Efficient and stable solid base for biodiesel and fine chemical catalytic synthesis. *ACS Sustain. Chem.& Eng.,* **2022**, *10*(11), 3477-3487. [http://dx.doi.org/10.1021/acssuschemeng.1c07157]

[40] Arora, S.; Gosu, V.; Kumar, U.K.A.; Subbaramaiah, V. Valorization of glycerol into glycerol carbonate using the stable heterogeneous catalyst of Li/MCM-41. *J. Clean. Prod.,* **2021**, *295*, 126437. [http://dx.doi.org/10.1016/j.jclepro.2021.126437]

[41] Song, X.; Wu, Y.; Cai, F.; Pan, D.; Xiao, G. High-efficiency and low-cost Li/ZnO catalysts for synthesis of glycerol carbonate from glycerol transesterification: The role of Li and ZnO interaction. *Appl. Catal. A Gen.,* **2017**, *532*, 77-85. [http://dx.doi.org/10.1016/j.apcata.2016.12.019]

[42] Keogh, J.; Deshmukh, G.; Manyar, H. Green synthesis of glycerol carbonate *via* transesterification of glycerol using mechanochemically prepared sodium aluminate catalysts *Fuel,* **2022**, *310*, 122484. [http://dx.doi.org/10.1016/j.fuel.2021.122484]

[43] Das, A.; Shi, D.; Halder, G.; Lalthazuala Rokhum, S. Microwave-assisted synthesis of glycerol carbonate by transesterification of glycerol using Mangifera indica peel calcined ash as catalyst. *Fuel,* **2022**, *330*(April), 125511. [http://dx.doi.org/10.1016/j.fuel.2022.125511]

[44] Shen, Q.; Yan, H.; Yuan, X.; Li, R.; Kong, D.; Zhang, W.; Zhang, H.; Liu, Y.; Chen, X.; Feng, X.; Zhou, X.; Yang, C. Tailoring morphology of MgO catalyst for the enhanced coupling reaction of CO2 and glycerol to glycerol carbonate. *Fuel,* **2023**, *335*(335), 126972. [http://dx.doi.org/10.1016/j.fuel.2022.126972]

[45] Okoye, P.U.; Wang, S.; Khanday, W.A.; Li, S.; Tang, T.; Zhang, L. Box-Behnken optimization of glycerol transesterification reaction to glycerol carbonate over calcined oil palm fuel ash derived catalyst. *Renew. Energy,* **2020**, *146*, 2676-2687.
[http://dx.doi.org/10.1016/j.renene.2019.08.072]

[46] Pradhan, G.; Jaiswal, S.; Sharma, Y.C. Exploring the promotional effect of transition metals (Cr and V) on the catalytic activity of MgO for glycerol carbonate synthesis. *Molecular Catalysis,* **2022**, *526*(April), 112332.
[http://dx.doi.org/10.1016/j.mcat.2022.112332]

[47] Khayoon. M.S.; Hameed. B.H. Mg1+xCa1−xO2 as reusable and efficient heterogeneous catalyst for the synthesis of glycerol carbonate *via* the transesterification of glycerol with dimethyl carbonate. *Appl. Catal. A Gen.* **2013**, *466*, 272–281
[http://dx.doi.org/10.1016/j.apcata.2013.06.044]

[48] Souza Júnior, R.L.; Rossi, T.M.; Detoni, C.; Souza, M.M.V.M. Glycerol carbonate production from transesterification of glycerol with diethyl carbonate catalyzed by Ca/Al-mixed oxides derived from hydrocalumite. *Biomass Convers. Biorefin.,* **2023**, *13*(2), 661-673.
[http://dx.doi.org/10.1007/s13399-020-01110-4]

[49] Tian, J.; Guo, Q. Thermal decomposition of hydrocalumite over a temperature range of 400–1500°c and its structure reconstruction in water. *J. Chem.,* **2014**, *2014*, 1-8.
[http://dx.doi.org/10.1155/2014/454098]

[50] Rossi, T.M.; Campos, J.C.; Souza, M.M.V.M. An evaluation of calcined hydrocalumite as carbon dioxide adsorbent using thermogravimetric analysis. *Appl. Clay Sci.,* **2019**, *182*(February), 105252.
[http://dx.doi.org/10.1016/j.clay.2019.105252]

[51] Granados-Reyes, J.; Salagre, P.; Cesteros, Y. Boosted selectivity towards glycerol carbonate using microwaves *vs* conventional heating for the catalytic transesterification of glycerol. *Appl. Clay Sci.,* **2018**, *156*(156), 110-115.
[http://dx.doi.org/10.1016/j.clay.2018.01.025]

[52] Álvarez, M.G.; Plíšková, M.; Segarra, A.M.; Medina, F.; Figueras, F. Synthesis of glycerol carbonates by transesterification of glycerol in a continuous system using supported hydrotalcites as catalysts. *Appl. Catal. B,* **2012**, *113-114*, 212-220.
[http://dx.doi.org/10.1016/j.apcatb.2011.11.040]

[53] Ochoa-Gómez, J.R.; Gómez-Jiménez-Aberasturi, O.; Maestro-Madurga, B.; Pesquera-Rodríguez, A.; Ramírez-López, C.; Lorenzo-Ibarreta, L.; Torrecilla-Soria, J.; Villarán-Velasco, M.C. Synthesis of glycerol carbonate from glycerol and dimethyl carbonate by transesterification: Catalyst screening and reaction optimization. *Appl. Catal. A Gen.,* **2009**, *366*(2), 315-324.
[http://dx.doi.org/10.1016/j.apcata.2009.07.020]

[54] Wu, Y.; Song, X.; Zhang, J.; Li, S.; Yang, X.; Wang, H.; Wei, R.; Gao, L.; Zhang, J.; Xiao, G. Synthesis of glycerol carbonate from glycerol and diethyl carbonate over CeO_2 -CdO catalyst: The role of Ce 4+ doped into CdO lattice. *J. Taiwan Inst. Chem. Eng.,* **2018**, *87*, 131-139.
[http://dx.doi.org/10.1016/j.jtice.2018.03.023]

[55] Parameswaram, G.; Srinivas, M.; Hari Babu, B.; Sai Prasad, P.S.; Lingaiah, N. Transesterification of glycerol with dimethyl carbonate for the synthesis of glycerol carbonate over Mg/Zr/Sr mixed oxide base catalysts. *Catal. Sci. Technol.,* **2013**, *3*(12), 3242-3249.
[http://dx.doi.org/10.1039/c3cy00532a]

[56] Parameswaram, G.; Rao, P.S.N.; Srivani, A.; Rao, G.N.; Lingaiah, N. Magnesia-ceria mixed oxide catalysts for the selective transesterification of glycerol to glycerol carbonate. *Molecular Catalysis,* **2018**, *451*, 135-142.
[http://dx.doi.org/10.1016/j.mcat.2017.12.006]

[57] Varkolu, M.; Burri, D.R.; Kamaraju, S.R.R.; Jonnalagadda, S.B.; Van Zyl, W.E. Transesterification of glycerol with dimethyl carbonate over nanocrystalline ordered mesoporous MgO–ZrO_2 solid base

catalyst. *J. Porous Mater.,* **2016**, *23*(1), 185-193.
[http://dx.doi.org/10.1007/s10934-015-0069-8]

[58] Chermahini, B. R.; A. N., *Modification G-C3N4 by MgO and its application for glycerol carbonate,* **2023**,

[59] Changmai, B.; Laskar, I.B.; Rokhum, S.L. Microwave-assisted synthesis of glycerol carbonate by the transesterification of glycerol with dimethyl carbonate using Musa acuminata peel ash catalyst. *J. Taiwan Inst. Chem. Eng.,* **2019**, *102*(102), 276-282.
[http://dx.doi.org/10.1016/j.jtice.2019.06.014]

[60] Chang, C.W.; Gong, Z.J.; Huang, N.C.; Wang, C.Y.; Yu, W.Y. MgO nanoparticles confined in ZIF-8 as acid-base bifunctional catalysts for enhanced glycerol carbonate production from transesterification of glycerol and dimethyl carbonate. *Catal. Today,* **2020**, *351*(351), 21-29.
[http://dx.doi.org/10.1016/j.cattod.2019.03.007]

[61] Yu, J.; Wang, K.; Shao, S.; Li, W.; Du, S.; Chen, X.; Chao, C.; Fan, X. Effect of ionic radius and valence state of alkali and alkaline earth metals on promoting the catalytic performance of La_2O_3 catalysts for glycerol carbonate production. *Chem. Eng. J.,* **2023**, *458*(January), 141486.
[http://dx.doi.org/10.1016/j.cej.2023.141486]

[62] Devi P; Das U; Dalai. A. K; Production of glycerol carbonate using a novel Ti-SBA-15 catalyst, "https://www.sciencedirect.com/journal/chemical-engineering-journal" *Chemical Engineering Journal*, **2018**, *346*, 477-488.
[http://dx.doi.org/10.1016/j.cej.2018.04.030]

[63] Arora, S.; Gosu, V.; Subbaramaiah, V. One-pot synthesis of glycerol carbonate from glycerol using three-dimensional mesoporous silicates of K/TUD-1 under environmentally benign conditions. *Molecular Catalysis,* **2020**, *496*(May), 111188.
[http://dx.doi.org/10.1016/j.mcat.2020.111188]

[64] Sonnati, M.O.; Amigoni, S.; Taffin de Givenchy, E.P.; Darmanin, T.; Choulet, O.; Guittard, F. Glycerol carbonate as a versatile building block for tomorrow: synthesis, reactivity, properties and applications. *Green Chem.,* **2013**, *15*(2), 283-306.
[http://dx.doi.org/10.1039/C2GC36525A]

[65] de Caro, P.; Bandres, M.; Urrutigoïty, M.; Cecutti, C.; Thiebaud-Roux, S. Recent progress in synthesis of glycerol carbonate and evaluation of its plasticizing properties. *Front Chem.,* **2019**, *7*(MAY), 308.
[http://dx.doi.org/10.3389/fchem.2019.00308] [PMID: 31179264]

[66] Van Mileghem, S.; De Borggraeve, W.M.; Baxendale, I.R. A robust and scalable continuous flow process for glycerol carbonate. *Chem. Eng. Technol.,* **2018**, *41*(10), 2014-2023.
[http://dx.doi.org/10.1002/ceat.201800012]

[67] Sahani, S.; Jaiswal, S.; Mishra, S.; Sharma, Y.C.; Han, S.S. Recent advances in bio-glycerol valorization to glycerol carbonate by heterogenous base-catalyzed transesterification. *Molecular Catalysis,* **2023**, *550*(August), 113508.
[http://dx.doi.org/10.1016/j.mcat.2023.113508]

[68] Gundekari, S.; Biswas, B.; Varkolu, M.; Srinivasan, K.; Bhaskar, T. Catalytic approaches for the selective preparation of cyclohexanone from lignin-based methoxyphenols/phenols In: *Biofuels, Biochemicals*; A. B. T.-B., Eds.; Elsevier, **2021**; pp. 301-327.
[http://dx.doi.org/10.1016/B978-0-12-820294-4.00013-2]

[69] Pazhavelikkakath Purushothaman, R.K.; van Erven, G.; van Es, D.S.; Rohrbach, L.; Frissen, A.E.; van Haveren, J.; Gosselink, R.J.A. New insights into the base catalyzed depolymerization of technical lignins: A systematic comparison. *RSC Advances,* **2023**, *13*(8), 4898-4909.
[http://dx.doi.org/10.1039/D2RA06998A] [PMID: 36762076]

[70] Laobuthee, A.; Khankhuean, A.; Panith, P.; Veranitisagul, C.; Laosiripojana, N. Ni–Fe Cocatalysts on magnesium silicate supports for the depolymerization of kraft lignin. *ACS Omega,* **2023**, *8*(9), 8675-8682.

[http://dx.doi.org/10.1021/acsomega.2c08045] [PMID: 36910962]

[71] Biswas, B.; Kumar, A.; Krishna, B.B.; Baltrusaitis, J.; Adhikari, S.; Bhaskar, T. Catalytic depolymerization of lignin for the selective production of phenolic monomers over cobalt-supported calcium catalysts. *Energy Fuels,* **2023**, *37*(5), 3813-3824. [http://dx.doi.org/10.1021/acs.energyfuels.2c04193]

[72] Ma, Y.; Li, H.; Yang, H.; Zhu, Y.; Zhao, L.; Li, M. Calcium oxide derived from eggshells supported on titanium oxide green catalysts for lignin hydrogenolysis under supercritical alcohol solvents. *Journal of the Energy Institute,* **2022**, *101*, 160-167. [http://dx.doi.org/10.1016/j.joei.2021.09.005]

[73] Biswas, B.; Kumar, A.; Krishna, B.B.; Bhaskar, T. Effects of solid base catalysts on depolymerization of alkali lignin for the production of phenolic monomer compounds. *Renew. Energy,* **2021**, *175*, 270-280.https://doi.org/https://doi.org/10.1016/j.renene.2021.04.039 [http://dx.doi.org/10.1016/j.renene.2021.04.039]

[74] Li, J.; Li, Z.; Dong, J.; Fang, R.; Chi, Y.; Hu, C. Hexaniobate as a recyclable solid base catalyst to activate C–H bonds in lignin linkage boosting the production of aromatic monomers. *ACS Catal.,* **2023**, *13*(8), 5272-5284. [http://dx.doi.org/10.1021/acscatal.2c06206]

[75] Cao, Y.; Gao, J.; Zhang, C.; Tsang, D.C.W.; Fan, J.; Clark, J.H.; Luo, G.; Zhu, X.; Zhang, S. Fast and selective production of aromatics *via* efficient lignin depolymerization: Critical factors and mechanism studies. *ACS Sustain. Chem.& Eng.,* **2022**, *10*(46), 15273-15283. [http://dx.doi.org/10.1021/acssuschemeng.2c05018]

[76] Zhu, H.; Du, B.; Bai, Y.; Pan, Z.; Sun, Y.; Wang, X.; Zhou, J. Base-catalyzed depolymerization of lignin into phenols: methoxy groups' secondary reactions triggered phenol regulation and repolymerization. *Biomass Convers. Biorefin.,* **2022**, *13*(14), 13009-13021. [http://dx.doi.org/10.1007/s13399-021-02190-6]

[77] Kumar, A.; Biswas, B.; Kaur, R.; Rani, R.; Krishna, B.B.; Bhaskar, T. Structure elucidation of prot, alkali and dealkaline lignin(s) by NMR, FT-IR and Py-GC/MS: effect of solid acid and base catalysts. *Sustain. Energy Fuels,* **2023**, *7*(8), 1942-1954. [http://dx.doi.org/10.1039/D3SE00045A]

[78] Li, Z.K.; Li, X.L.; Yan, H.L.; Yan, J.C.; Lei, Z.P.; Ren, S.B.; Wang, Z.C.; Kang, S.G.; Shui, H.F. Two-step separation for enriching phenols and esters from base-catalyzed depolymerization of sorghum stalk. *Fuel,* **2022**, *314*, 122739. https://doi.org/https://doi.org/10.1016/j.fuel.2021.122739 [http://dx.doi.org/10.1016/j.fuel.2021.122739]

[79] Pérez, E.; Abad-Fernández, N.; Lourençon, T.; Balakshin, M.; Sixta, H.; Cocero, M.J. Base-catalysed depolymerization of lignins in supercritical water: Influence of lignin nature and valorisation of pulping and biorefinery by-products. *Biomass Bioenergy,* **2022**, *163*(July), 106536. [http://dx.doi.org/10.1016/j.biombioe.2022.106536]

[80] Zhang, J.; Su, Z.; Wu, Z.; Wang, P.; Xiao, F.S. Basic carrier promoted Pt-catalyzed hydrogenolysis of alkaline lignin. *Catal. Today,* **2021**, *365*, 193-198. https://doi.org/https://doi.org/10.1016/j.cattod.2020.06.027 [http://dx.doi.org/10.1016/j.cattod.2020.06.027]

[81] Zhang, X.; Li, W.; Wang, J.; Zhang, B.; Guo, G.; Shen, C.; Jiang, Y. Depolymerization of Kraft lignin into liquid fuels over a WO3 modified acid-base coupled hydrogenation catalyst. *Fuel,* **2022**, *323*, 124428. https://doi.org/https://doi.org/10.1016/j.fuel.2022.124428 [http://dx.doi.org/10.1016/j.fuel.2022.124428]

[82] Tian, Z.; Zeng, Y.; Zhao, H.; Yang, J.; Zhang, H. A new approach to explore the catalytic depolymerization of lignin *via* samarium oxide. *Catal. Commun.,* **2023**, *177*(January), 106665. [http://dx.doi.org/10.1016/j.catcom.2023.106665]

[83] Tišler, Z.; Vondrová, P.; Peroutková, K.; Šimek, J.; Skuhrovcová, L.; Strejcová, K.; Svobodová, E.;

Gholami, Z. Influence of Water on the Production of Liquid Fuel Intermediates from Furfural *via* Aldol Condensation over MgAl Catalyst. *Processes (Basel),* **2023**, *11*(1), 261. [http://dx.doi.org/10.3390/pr11010261]

[84] Arumugam, M.; Kikhtyanin, O.; Osatiashtiani, A.; Kyselová, V.; Fila, V.; Paterova, I.; Wong, K.L.; Kubička, D. Potassium-modified bifunctional MgAl-SBA-15 for aldol condensation of furfural and acetone. *Sustain. Energy Fuels,* **2023**, *7*(13), 3047-3059. [http://dx.doi.org/10.1039/D3SE00444A]

[85] Chottiratanachote, A.; Suttipong, M.; Rashid, U.; Parasuk, V.; Kondo, J.N.; Yokoi, T.; Alsalme, A.; Ngamcharussrivichai, C. Selective synthesis of renewable bio-jet fuel precursors from furfural and 2-butanone *via* heterogeneously catalyzed aldol condensation. *Catalysts,* **2023**, *13*(2), 242. [http://dx.doi.org/10.3390/catal13020242]

[86] Li, Z.; Shao, S.; Hu, X.; Li, X.; Cai, Y. Insight into the production of aviation fuel by aldol condensation of biomass-derived aldehydes and ketones followed by hydrogenation. In: *Biomass Convers. Biorefinery*; , **2022**. [http://dx.doi.org/10.1007/s13399-022-03083-y]

[87] Ghosh, S.; Acharyya, S.S.; Yoshida, Y.; Kaneko, T.; Iwasawa, Y.; Sasaki, T. Nontraditional aldol condensation performance of highly efficient and reusable cs $^{+}$ single sites in β-zeolite channels. *ACS Appl. Mater. Interfaces,* **2022**, *14*(16), 18464-18475. [http://dx.doi.org/10.1021/acsami.2c01312] [PMID: 35426658]

[88] Meemanah, T.; Chotirattanachote, A.; Ahmad, J.; Rashid, U.; Ngamcharussrivichai, C. Bifunctional acid–base strontium–titanium mixed oxides supported on SBA-15 for selective synthesis of renewable branched bio-jet fuel precursor. *Fuel,* **2023**, *351*, 128895. https://doi.org/https://doi.org/10.1016/j.fuel.2023.128895 [http://dx.doi.org/10.1016/j.fuel.2023.128895]

[89] Deshpande, N.; Chen, J.Y.; Kobayashi, T.; Cho, E.H.; Pineault, H.; Lin, L.C.; Brunelli, N.A. Investigating the impact of micropore volume of aminosilica functionalized SBA-15 on catalytic activity for amine-catalyzed reactions. *J. Catal.,* **2022**, *414*, 356-364. https://doi.org/https://doi.org/10.1016/j.jcat.2022.09.016 [http://dx.doi.org/10.1016/j.jcat.2022.09.016]

[90] Chen, J.Y.; Brunelli, N.A. Investigating the impact of microporosity of aminosilica catalysts in aldol condensation reactions for biomass upgrading of 5-hydroxymethylfurfural and furfuraldehyde to fuels. *Energy Fuels,* **2021**, *35*(18), 14885-14893. [http://dx.doi.org/10.1021/acs.energyfuels.1c01934]

[91] Korolova, V.; Kikhtyanin, O.; Veselý, M.; Vrtiška, D.; Paterová, I.; Fíla, V.; Čapek, L.; Kubička, D. On the effect of the m3+ origin on the properties and aldol condensation performance of mgm3+ hydrotalcites and mixed oxides. *Catalysts,* **2021**, *11*(8), 992. [http://dx.doi.org/10.3390/catal11080992]

[92] Tampieri, A.; Russo, C.; Marotta, R.; Constantí, M.; Contreras, S.; Medina, F. Microwave-assisted condensation of bio-based hydroxymethylfurfural and acetone over recyclable hydrotalcite-related materials. *Appl. Catal. B,* **2021**, *282*, 119599. https://doi.org/https://doi.org/10.1016/j.apcatb.2020.119599 [http://dx.doi.org/10.1016/j.apcatb.2020.119599]

[93] Veselý, M.; Lhotka, M.; Kikhtyanin, O.; Kubi, D. On the influence of synthetic parameters on the properties and aldol condensation performance of the mgal mixed oxides and rehydrated hydrotalcites. *Appl. Clay Sci.,* **2023**, *2024*, 249. [http://dx.doi.org/10.1016/j.clay.2024.107263]

[94] Sheng, X.; Xu, Q.; Wang, X.; Li, N.; Jia, H.; Shi, H.; Niu, M.; Zhang, J.; Ping, Q. Waste seashells as a highly active catalyst for cyclopentanone self-aldol condensation. *Catalysts,* **2019**, *9*(8), 661. [http://dx.doi.org/10.3390/catal9080661]

CHAPTER 9

Solid-Supported Base Catalysis for Carbon-Carbon Bond Formation

Trapti Aggarwal[1,*], **Tarnija Midha**[1] and **Simer Kaur**[1]

[1] *Department of Chemistry, University of Delhi, Delhi, India*

Abstract: Solid-Phase Organic Synthesis (SPOS) has gained prominence after the pioneering work of Merrifield for the development of peptide synthesis using solid-phase catalysis was published in the 1960s. Since then, chemists have shown tremendous interest in expanding this field to develop environment-friendly, cost-effective, and sustainable protocols for synthesizing diverse compounds used in material synthesis. The major emphasis of solid base synthesis is on the growth and emergence of sustainable and green chemistry. This approach largely relies on converting solution-phase reactions into solid-phase processes. Key green chemistry principles that support solid-phase synthesis include reduced use of toxic solvents, fewer reaction steps, improved energy efficiency, and high atom economy. However, the solid-phase approach has high efficiency, excellent selectivity, easy work-up, and good dispersion of active reagent sites. Solid-supported basic catalysis has advanced significantly over the last few decades, and it is widely used in the synthesis of organic molecules, bench-top catalysts, and fine chemicals. An overview of solid basic catalysts and their application in the synthesis of important organic molecules was disclosed in this chapter. The initial section discussed the importance and synthesis of a variety of supported solid base catalysts. The second part discusses the application of solid base catalysts in various organic reactions and evaluates their catalytic performance.

Keywords: Aldol condensation, Coupling reactions, Hydrogenation, Knoevenagel condensation, Michael addition, Solid-supported catalyst.

INTRODUCTION

Catalysis attracts attention due to its wide applicability in numerous chemical production from small scale in research laboratories to large industrial scale [1]. Catalysis has high importance in the manufacturing of our daily life essentials like medicines, polymers, fuels, automotive and so on [2]. Homogeneous catalysts usually dominate over heterogeneous catalysts because of their high solubility and

* **Corresponding author Trapti Aggarwal:** Department of Chemistry, University of Delhi, Delhi, India; E-mail: trapti0685@gmail.com

Ravi Tomar, Vivek Mishra, Kapil Arya & Garima Pandey (Eds.)

reactivity with reactants [3]. In this regard, the main challenge faced by the chemist is to purify the desired product, which is difficult to separate from the catalysts. To overcome this difficulty, an alternative approach has been adopted by designing novel catalysts to replace homogeneous catalysts. Chemists have turned to heterogeneous catalysts, which are easier to separate and recycle. In addition to this, heterogeneous catalysts are easily removed from the product, and their reaction requires a minimum quantity of toxic organic solvents [4]. Thus, approaches using this catalyst are environmentally benign for small-scale synthesis as well as for industrial-level synthesis of organic compounds. The increasing demand for sustainable and green chemistry also led scientists to design new strategies using heterogeneous catalysts instead of conventional homogeneous catalysts.

Among various heterogeneous catalysts, solid base catalysts have attracted more attention due to their valuable properties over their liquid counterparts. Importantly, the solid bases are much less corrosive and cause fewer disposal problems. Solid base-catalyzed reactions generally show higher selectivity towards the targeted product. Further, solid bases are easier to separate from the product and make the recovery of products from reaction systems faster, and subsequently it become facile to recycle the catalysts. Therefore, solid bases provide an environmentally benign and economical pathway for the synthesis of organic compounds [5]. Consequently, the major concern of chemists is to develop heterogeneous catalyst-based methodologies and design new solid base catalysts by altering the solid support along with the structural and morphological characteristics of the catalysts during the reaction.

Pioneering work on solid base catalysts was reported by Pines and Haag in 1958. They designed the new catalyst system by the dispersion of metal Na on Al_2O_3 and evaluated its efficiency in various organic reactions [6]. The results of this catalyst system were amazing enough to encourage the organic chemist to expand new ventures in this field. During the last two decades, solid-phase synthesis has gained tremendous interest among chemists due to its wide applicability [7]. Now, sustainable chemistry has been in demand for the synthesis of a myriad of chemicals through the environment-benign approach, which was used in material chemistry and catalysis. The novel solid-supported basic catalysts, which range from single to multicomponent metal hydroxides as well as oxides (*e.g.*, hydrotalcite), have vast applications in various chemical processes. Till now, the size, texture, morphology, category, and surface area of the particles have been the determining factors for the research on solid base catalysts [8]. The structure and characteristics of alkaline sites on the surface of the developed solid base catalyst have proved its catalytic behavior [9].

The chapter provides a concise outline of the progression in the construction and designing of novel solid basic catalysts as well as their use in a variety of organic reactions such as Knoevenagel reaction, Aldol condensation, Michael addition, double bond migration, transesterification, and hydrogenation (Fig. **1**). This section provides a detailed description of the structure and design of solid base catalysts, along with their controlled synthesis and broad application in various organic reactions. The content described in this chapter will influence the synthetic chemist towards the development of novel solid basic catalysts and it is anticipated that this will open new ventures in the field of sustainable chemistry.

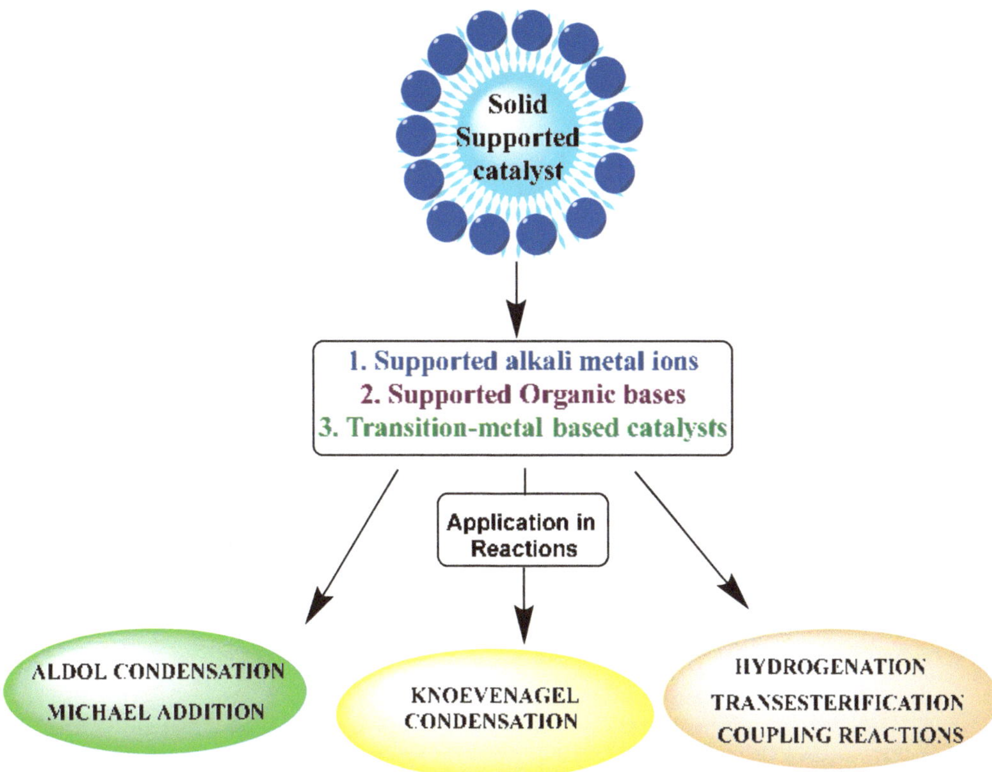

Fig. (1). General overview.

SUPPORTED SOLID BASIC CATALYST

In heterogeneous base catalysis reactions, supported solid basic catalysts play a crucial role. Alkali and alkali earth metals have some basic functional groups incorporated in them, which behave as potent bases and nucleophiles. These metals can also work as solid catalysts or as initiators of anionic chain reactions. In a solid basic catalyst, the basic strength of a solid surface is defined as the ability of the surface to convert an adsorbed neutral acid to its conjugate base

[10]. Furthermore, the solid texture of heterogeneous catalysts and the intrinsic catalytic activity of their counterparts have a significant impact on their effectiveness. While performing the organic reactions, controlling the pore size of the solid base is highly important as each reaction required a different pore size of catalyst and by simple tuning of its pore sizes, the selectivity of product was observed. The alumina pores in alumina-based catalysts can be found inside individual particles, inside pellets, in agglomerates, and the gaps between particles [11]. In 2017, the Kobayashi group created a stereoselective technique for creating glutamic acid derivatives by 1,4-addition of glycine derivatives to β-unsaturated esters utilising CsF/Al_2O_3 as a reusable solid base catalyst [12]. They stated that the catalyst CsF/Al_2O_3 was used to obtain great selectivity towards the 1,4-addition process. The catalyst's primary function was to inhibit the undesirable production of pyrrolidine derivatives through [3 + 2] cycloaddition. Furthermore, an outstanding yield of 1,4-adduct with anti-diastereoselectivity was provided by this reaction. A thorough preliminary analysis employing a variety of spectroscopic techniques verified that the CsF/Al_2O_3 catalyst's basicity was most likely caused by interactions between the O atom in the support and Cs^+ in Cs_3AlF_6. Strong basic F-containing sites were produced when the negative charge was localised on F, in addition to the basicity associated with the oxygen species. In addition to the aforementioned conclusion, the research also demonstrated that the most likely active basic site for the 1,4-addition reaction was the highly basic F in Cs_3AlF_6. Metal oxides are the most common solid base catalysts because they offer high thermal and chemical stability to the solid-supported catalyst system. Additionally, they are easily prepared and functionalized, enhance the activity of this catalyst, and become the most attractive catalyst in heterogeneous catalysis. Hydroxides, such as $Mg(OH)_2$, hydrotalcite, or anhydrous hydrotalcite, are also frequently employed as solid catalysts in addition to oxides as heterogeneous catalytic systems for organic processes [13] Among various oxides, hydrotalcite, due to its high surface area and abundant hydroxyl groups, and rehydrated hydrotalcite have demonstrated exceptional catalytic activity in a range of organic reactions such as condensation and coupling [14]. The multicomponent transesterification of edible and non-edible oils utilising metal hydroxides as a solid basic catalyst was described by the Shanbhag group in 2016. This reaction system leads to the conversion of fatty acids to methyl esters [15] $CaSn(OH)._6$ was used as the catalyst, featuring a cuboid shape with particle sizes ranging from 6 to 10 μm. When compared to other oxides and hydrotalcite solid bases, the proposed catalyst produced a higher yield of 94% for sunflower oil in the biodiesel synthesis. The fascinating fact about this catalyst system is that it can withstand temperatures up to 250 °C and has low hygroscopicity.

Recently, Hu and co-workers designed the C-H bond activation in lignin using a novel recyclable solid base catalyst containing hexaniobate, which helps in increasing the production of aromatic monomers [16]. This catalyst was important for enabling the targeted cleavage of lignin to synthesize aromatic monomers. The catalytic system contains a small amount of solid base $K_7HNb_6O_{19}$ (KNb_6), which is coupled with copper-modified graphitic carbon nitride (Cu/C_3N_4) to enhance the catalytic performance for the oxidative cleavage of β-O-4 lignin linkages using molecular oxygen. Due to the synergetic effect between Cu/C_3N_4, phenol, and KNb_6, other aromatic monomers were obtained by 96% conversion of the β-O-4 ketone model under relatively mild conditions. Interestingly, along with its high reactivity, the KNb_6–Cu/ C_3N_4 catalytic system recycled up to five catalytic cycles. Various studies, including control experiments with spectroscopic analysis and theoretical studies, including kinetic and density functional theory calculations, revealed the facts about the catalytic system. The basic O atoms of KNb_6 activate the Cβ–H bond of β-O-4 ketone to facilitate the insertion of molecular oxygen, with the concomitant cleavage of the Cα–Cβ bond selectively through and Cu/C_3N_4 catalysis system. The oxidative cleavage of other lignin models was also activated through this catalytic system.

ORGANIC REACTIONS CATALYZED BY SUPPORTED BASES

Solid base catalysts are well known for displaying unique catalytic activity across different C–C bond-forming reactions. It is well known from the literature that the acidic and basic sites in combination are more efficient for the catalytic activity for various condensation reactions, including Aldol, Michael, and Knoevenagel reactions. The most active catalysts, such as hydroxyapatite, have basic sites with moderate strength, and it is important to mention that the probability of basic site accommodation rises with pore volume. The strength and ratio of basic and acid sites should be prominently balanced for the condensation reaction to have a high efficiency and yield. Similarly, other reactions, such as hydrogenation, have been achieved using solid-supported catalysts based on metal oxides through methods like doping, strong metal support interactions, or confinement of active metal nanoparticles within the micro or mesopores of the supports. Due to the stable catalytic activity, solid-supported catalysts also show great potential for various catalytic reactions. Further, the details of all these reactions, along with the catalytic system, have been discussed.

Aldol Condensation

Earlier, Malinowski and Basinki [17] reported the condensation of propionaldehyde in the presence of $NaOH/SiO_2$. Later, other catalysts, like $Ca(OH)_2$, Lithium phosphate, *etc.*, were employed for condensation reactions [18]. This catalyst also shows high efficiency for the Cannizzaro reaction [19]. An ion exchange resin, *i.e.*, Amberlite IRA-400, also catalyzes the formation of aldol product from acetaldehyde and acetone. Other hydroxides and oxides are also well-known for this condensation reaction. Among them, $Ba(OH)_2$ was extensively used to catalyze the formation of diacetone alcohol by aldol addition of acetone [20]. along with alkali and alkaline, inner transition metal oxides like lanthanum oxide (La_2O_3) and ZrO_2 were also reactive for this condensation reaction [21]. Interestingly, MgO, along with a small amount of water, increases the efficiency of the catalyst, suggesting that the active sites for this reaction are basic OH^- ions together with O^{2-} ions [22].

Recently, Ishihara group designed a porous boron nitride base catalyst and confirmed the efficacy of this new solid basic catalyst on aldol reaction [23]. The catalyst was synthesized through pyrolysis by varying a mixture of boric acid and urea. The prepared boron nitride had high surface areas with both microporous and mesoporous structures. They have reported that the edge X-ray absorption fine structure spectroscopy of Boron-K, revealed that the surface structure consisted of BN_3 sites along with BN_2O, BNO_2, and BO_3 sites. They have also performed the FTIR which shows porous boron nitride has very weak base sites of strength $+7.2 > H\_ \geq +6.3$. Interestingly, they have examined the catalyst reactivity with nitro-aldol condensation and found it shows high selectivity for nitroalkene. This suggests that the catalytic activity of boron nitride correlates well with its porous structure (Scheme **1**). The reaction was supposed to proceed *via* two plausible mechanisms: the base mechanism and the acid-base cooperative mechanism. The base-catalyzed mechanism involves a proton abstraction of nitromethane by base sites (**6**), amino groups on the boron nitride, while in other, a nucleophilic attack of the deprotonated nitromethane to benzaldehyde (**5**) and subsequently dehydration led to nitro alkene **3**. This boron nitride catalyst is much superior to the other conventional solid base catalysts such as CaO and hydrotalcite.

Later, Matsuhashi designed a catalyst from Zinc(II) Acetylacetonate with ZnO/MgO and explored its reactivity with retro aldol reaction [24]. The catalyst was prepared using solid base ZnO/MgO, having MgO modified with 10 mol% ZnO deposited on zinc(II) acetylacetonate ($Zn(acac)_2$) (in polar and nonpolar organic solvents) on the surface of MgO and $Mg(OH)_2$. Then, the catalyst was placed for 3h under a thermal decomposition process at 773K. The application of this catalyst for retro aldol reaction of the formation of acetone from the decomposition of diacetone alcohol (4-hydroxy-4-methyl-2-pentanone) showed higher yield and reactivity as compared with previously used unmodified MgO. The X-ray diffraction studies showed the presence of highly intense peaks assigned to the ZnO units present in the sample with high catalytic activity. The catalyst with high activity has ZnO in the form of large particles deposited on the surface of MgO, as shown in (Scheme **2**). The decomposition of $Zn(acac)_2$ was done by hydroxyl groups on $Mg(OH)_2$, which led to the formation of free acetylacetone from Zn and resulted in the surface bonding and Mg–O–Zn bond formation.

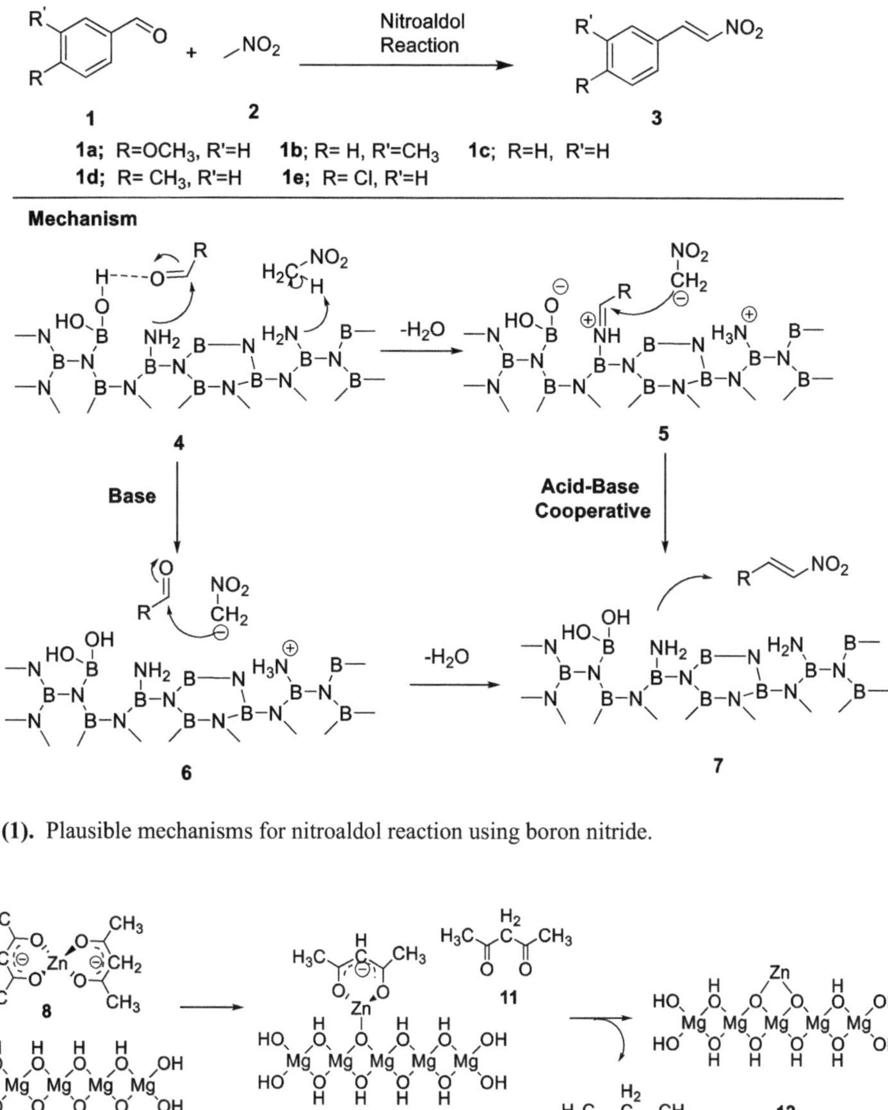

Scheme (1). Plausible mechanisms for nitroaldol reaction using boron nitride.

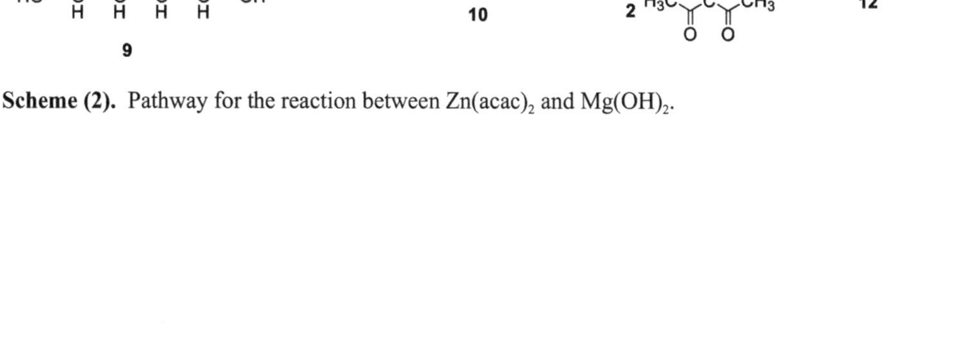

Scheme (2). Pathway for the reaction between $Zn(acac)_2$ and $Mg(OH)_2$.

Michael Addition

There are various scientific groups that have examined solid base catalysts for Michael's addition reaction. In 2018, the Qiao group designed a redox approach for Zeolite NAa-supported K_2O and used it as a catalyst for the Michael addition reaction of ethyl acrylate [25]. Previously, supported K_2O was employed for this reaction, but it required a high temperature. Thus, they have designed the Zeolite-based catalyst and used a greener approach for the Michael addition reaction. They have studied thermogravimetry−mass spectrometry to investigate the solvent-assisted redox decomposition process of KNO_3 at elevated temperatures, which shows the decomposition of organic solvent to make carbon after reducing a minor amount of KNO_3 at 400 °C. This carbon then promotes the reduction of KNO_3 to produce the zeolite NaA with strong basicity at 600°C. In comparison with previous reports for Michael's addition reaction, the K_2O/NaA-S shows better catalytic activity. To our interest, this catalyst can withstand at least four catalytic cycles without any loss in activity. In (Scheme **3**), we have presented the comparative study using different catalytic systems for Michael addition.

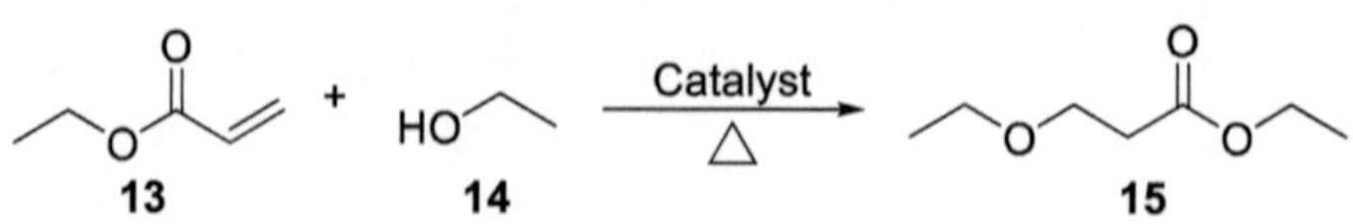

Entry	Catalyst	Conversion (%)	Selectivity (%)
13a	K_2O/NaA-I	52	99
13b	K_2O/NaA-II	54	98
13c	K_2O/NaA-S	95	100
13d	K_2O/Al_2O_3	54	99
13e	PPh_3	31	33
13f	C_2H_5OK	97	99

Scheme (3). Reactivity of catalyst for Michael's addition.

Knovevenagel Condensation

In 2009, Gupta and co-workers designed a greener approach using hydroxyapatite-supported calcium carbonate as a new recyclable solid base catalyst for the Knoevenagel condensation [26]. The hydroxyapatite was prepared by the co-precipitation approach that is required for the preparation of the hydroxyapatite-supported caesium carbonate catalyst. The hydroxyapatite and Cs_2CO_3 mixture were stirred in Double Distilled Water (DDW) for four hours to create HAP–Cs_2CO_3. The solid was filtered, washed with DDW, and dried for one hour at 700–800 °C in a muffle furnace after being dried overnight at 100 °C. The efficacy of the catalyst was tested using a variety of substituted aldehydes having electron-donating as well as withdrawing groups (Scheme **4**). Heteroaromatic aldehydes were also employed with the designed catalyst and provided a good yield of CN substituted product.

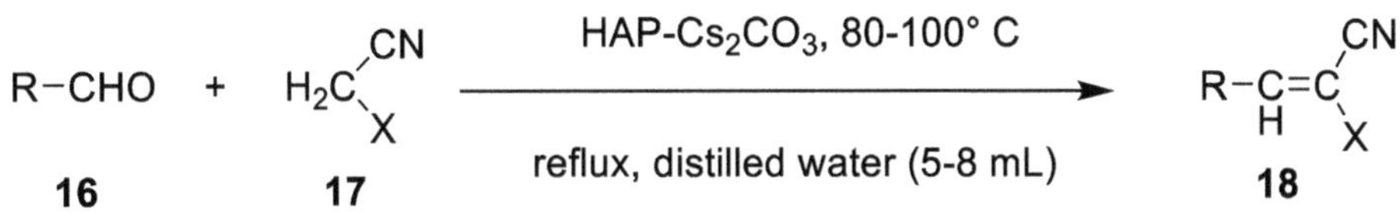

Scheme (4). Knovevenagel Condensation using HAP-Cs_2CO_3.

Ogura group in 2011 reported the synthesis of Methylated N-substituted SBA-15 as a solid-base catalyst [27]. A class of high specific surface area nitrogen-substituted solid-base materials using MCM-41 and SBA-15 (SBA) was developed by scientists [28]. The presence of nitrogen atoms in these material's frameworks gives rise to their basic sites. These nitrogen species are thought to be as strongly basic as traditional solid-base catalysts like mesoporous silica materials impregnated with basic salts, alkaline-ion-exchanged zeolites, or oxides of alkali or alkaline earth metals. Since the catalytic sites of these conventional base catalysts tightly adsorb ambient water or carbon dioxide, pretreatment at high temperatures (>873 K) is necessary before use. Here, this research group has designed a new catalytic system by a simple modification on mesoporous silica SBA-15, which enhances the solid-base catalyst's reactivity. In the silica framework, the nitrogen atom receives an electron donation from the methyl group. The rate of Knoevenagel condensation using benzaldehyde and diethyl malonate increased when this catalyst was used (Scheme **5**).

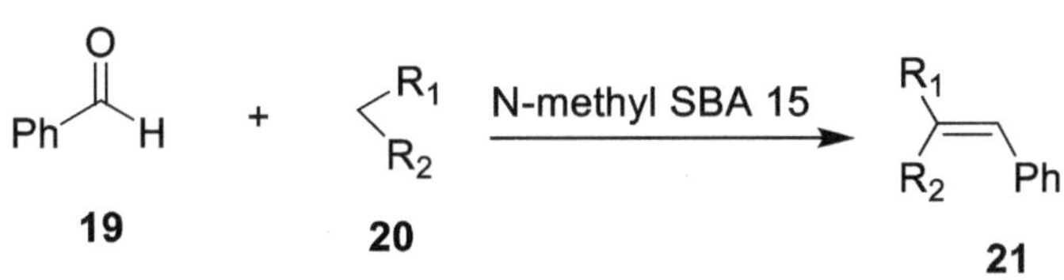

Scheme (5). Knoevenagel condensation using N-methyl SBA-15.

When it comes to mediating chemical reactions, metal-organic frameworks (MOFs) have become a novel class of porous crystalline solids. Here, Corma and co-workers have documented the use of organic alkylaryl spacers, and aluminium chains joined in a single layer to create a mesoscopic metal-organic structure [29]. The catalytic material may be recycled and utilised again because of its mesoscopic phase, which makes it easier for the products to diffuse. They have created Al-ITQ-HB, a new metal-organic material based on organised individual sheets of aluminium clusters. This material mesoscopic structure, hydrophobicity, and chemical stability in water make it an appealing supramolecular catalyst for enabling (asymmetric) conversions under more benign circumstances. A mesoscopic and lamellar metal-organic substance serves as a host catalyst for the condensation processes in water when the circumstances are neutral. The demonstrated activity appears to mimic how enzymes function in natural systems, where the hydrophobic pocket and non-covalent interactions account for the increased efficiency. The mesoscopic host material can also provide shelter for a chiral organocatalyst and two reacting components, allowing for asymmetric transformations under more ecologically friendly circumstances, including room temperature, reduced catalyst loading, and the use of water as a green solvent. Although the produced lipophilic compounds prefer the hydrophobic environment of the catalyst system to the aqueous solvent, the products are readily removed, and the material can be recycled and utilised again. The high yield of the condensation product **24** was produced when aldehyde reacted with Meldrum's acid in the presence of Al-ITQ-HB (Scheme **6**).

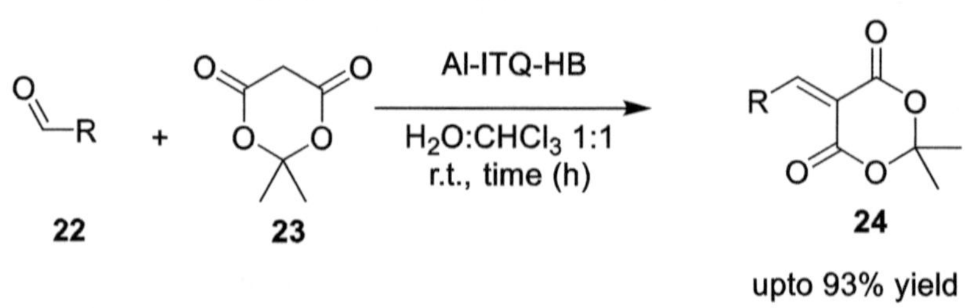

Scheme (6). Knoevenagel condensation using Meldrum's acid in presence of Al-ITQ-HB.

Recently, Huang and co-workers [30] developed the pyrolysis of a bipyridyl metal-organic framework, MOF-253, which results in N-doped porous carbons (Cz-MOF-253). This catalytic system showed excellent activity and performed better than other nitrogen-containing MOF-derived carbons for the Knoevenagel condensation reaction (Scheme 7). More significantly, Pd nanoparticles supported by Cz-MOF-253 (Pd/CzMOF-253-800) showed strong Lewis basicity and porous nature features, making them ideal for one-pot sequential Knoevenagel condensation-hydrogenation reactions. Through a solvothermal reaction between 2,2′-bipyridine-5,5′-dicarboxylic acid and $AlCl_3 \cdot 6H_2O$, the catalyst MOF-253 was synthesized [31]. MOF-253 possesses a large surface area of 1940 m^2/g and major micro-pores at ~11 Å, according to the N_2 sorption and pore size distribution analysis as reported in the paper.

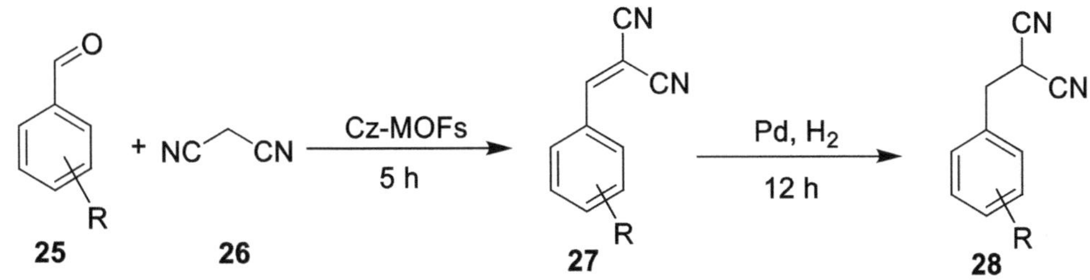

Scheme (7). Knoevenagel condensation reaction using Cz-MOF-253.

Transesterification

Transesterification is also one of the well-known applications of solid base catalysts. Potassium carbonate supported by activated carbon was used as a catalyst for the transesterification of dimethyl carbonate with alcohol to synthesize dipropyl carbonate (DPC) at atmospheric pressure, as reported by Fan group in 2007 [32]. The optimum catalyst dosage for catalytic activity was found to be about 5%, and the suitable K_2CO_3-loading was about 15% for 5h. The importance of this catalytic system was its reusability. It was recycled and used three times with a high catalytic activity. In (scheme **8**), the transesterification reaction over supported K_2CO_3 base catalyst for the synthesis of DPC from DMC and propyl alcohol was explained with a plausible mechanism for alcoholysis [33].

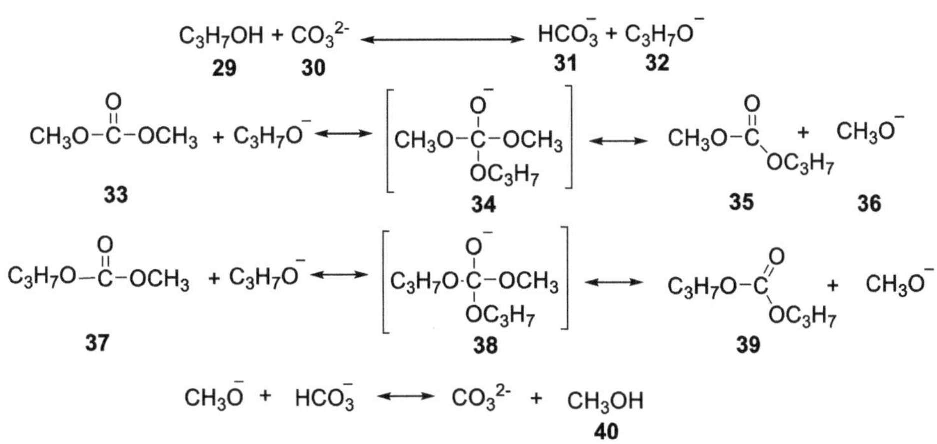

Scheme (8). Alcoholysis of dipropyl carbonate.

Liu's group, in 2008, proposed calcium ethoxide-based transesterification of soybean oil to biodiesel with methanol and ethanol [34]. Firstly, a reaction of calcium and ethanol was used to synthesize calcium ethoxide. Then, using instrumental techniques like particle size distribution measurements, scanning electron micrographs, and Brunauer-Emmett-Teller surface area measurements, its physical and chemical characteristics were determined. To optimize the reaction conditions, the effects of the mass ratio of catalyst to oil, the molar ratio of methanol to oil, and the reaction temperature were investigated. According to the experimental results, the optimum conditions for the reaction were 65°C temp with 3% $Ca(OCH_2CH_3)_2$ catalyst addition at mild pressure and a molar ratio of methanol to oil of 12:1. These conditions afforded 95.0% biodiesel yield in 1.5 hours, with an activation energy of 54149 J/mol. It also catalyzed soybean oil to biodiesel with ethanol, yielding 91.8% biodiesel. The moderate surface area, a relatively broader particle size distribution, and better low solubility in methanol and ethanol made this catalyst highly efficient for the extraction of biodiesel from alcohol.

Various other reactions were also performed using a solid base catalyst for higher yields. Polyethylene glycol embedded with potassium bromide [K^+{PEG}Br^-] was used as an initiator for the synthesis of DMC with 12.8% yield and 98% selectivity from the reaction of methanol and carbon dioxide (CO_2) (Scheme **9**) [35]. Polyethylene glycol enfolded potassium bromide, or [K^+{PEG}Br^-], was readily obtained by stirring KBr and PEG_{400}(**49**) at room temperature in water. Dichloromethane was used to extract the resulting mixture, and it was subjected to the usual workup to yield [K^+ {PEG}Br^-] (**50**) as a viscous brownish-yellow oil in quantitative yield. UV−Vis and ESI-MS analyses confirmed the successful formation of [K^+{PEG}Br^-] [36].

After the completion of the reaction, the reaction products were isolated by distillation, and the potassium carbonate (K_2CO_3) and potassium bromide (KBr)-containing PEG layer that was produced was effectively reused for the next six runs without causing a significant loss in DMC yield. While in the absence of [K^+{PEG}Br^-], only 1−2% of DMC was produced using only K_2CO_3 as a catalyst.

Hydrogenation Reaction

Radhakrishan group in 2011, reported potassium triphosphate as a solid catalyst for the catalytic transfer hydrogenation of carbonyl compounds [37]. To obtain the active catalytic system, a pretreatment of base at 600°C is essential. After treatment for 5 hours, sites of lower basicity were created, which was indicated by Temperature-Programmed Desorption (TPD) of CO_2. Then, this catalyst was used for the reduction of aromatic aldehydes to the corresponding alcohols using 2-

propanol as the hydride source (Scheme **10**). When compared to unsubstituted benzaldehyde, the rate of reaction is higher when electron-withdrawing chloro- and nitro-substituents are present at the benzene ring; however, the opposite is true when electron-donating methyl or 2-propyl groups are present. Compared to aldehydes, ketones were reduced at a slower rate of reaction. Preliminary studies with competing substrates, along with the absence of product formation during controlled reactions without the solid base, highlight the important role of K_3PO_4 as a heterogeneous catalyst.

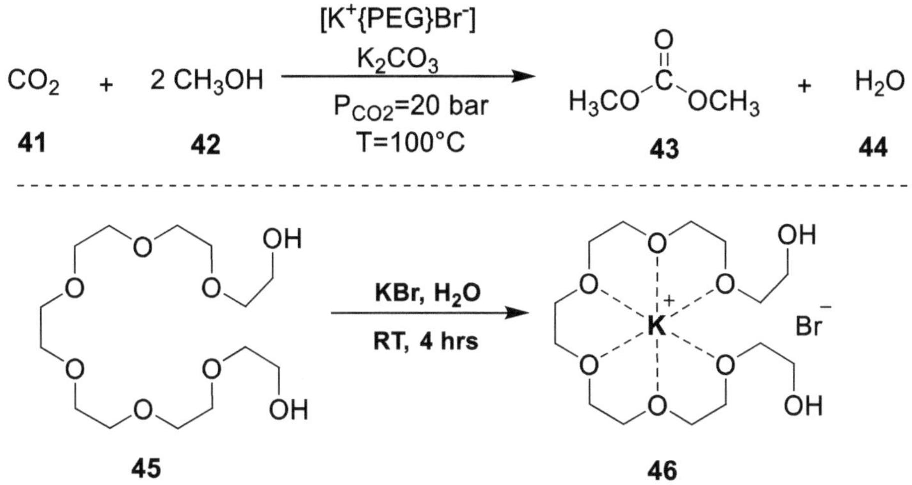

Scheme (9). The reaction of methanol and carbon dioxide (CO_2).

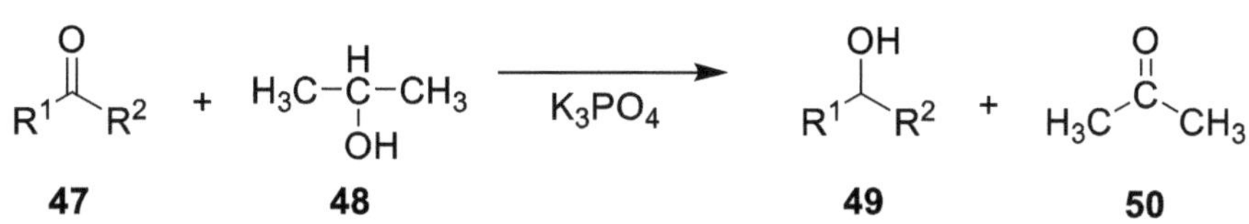

Scheme (10). Reduction of carbonyl compound.

C-C Coupling Reactions

An essential class of conjugated polyaromatic compounds, biaryls, and higher homologues are unique building blocks that originate from benzene [38]. Furthermore, a wide range of pharmaceuticals, agrochemicals, and biologically active natural products often contain biphenyls and polyaromatics as subunits. Basu and co-workers successfully designed a palladium-catalyzed cross-coupling reaction of polyhaloaromatics with arylboronic acids utilising KF/alumina under microwave conditions (Scheme **11**) [39]. The synthesis of a wide range of polyaromatics has been made possible by this catalyst system, which has broadened the extent of coupling reactions using solvent-free inorganic surfaces.

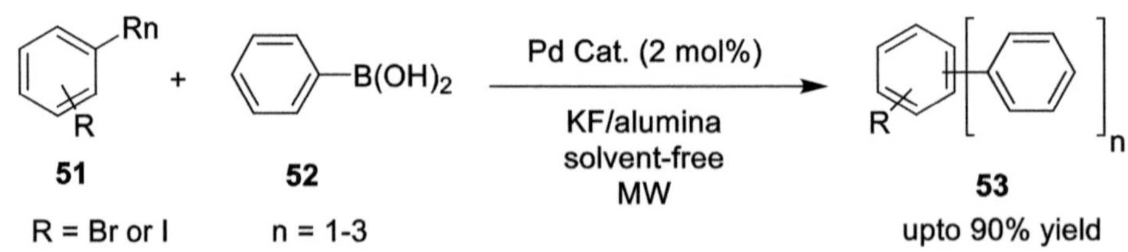

Scheme (11). Talladium-catalyzed cross-coupling reaction of polyhaloaromatics with arylboronic acids using KF/alumina.

Ion-exchange resins were also employed in an effort to find suitable polymeric supports. It was found that commercially available, inexpensive polyionic resins, such as Amberlite IRA-420 anion (chloride form) exchange resins, could quantitatively and readily exchange the anion with formate anion (CH_3COO^-), resulting in the formation of Amberlite Resin Formate (anion), designated as ARF. In the presence of ARF-Pd, the Suzuki-Miyaura reaction (Scheme **12**) was carried out using 4-bromo-2-methyl anisole, phenylboronic acid, and Na_2CO_3 at 110 °C under an N_2 atmosphere for 5 h.

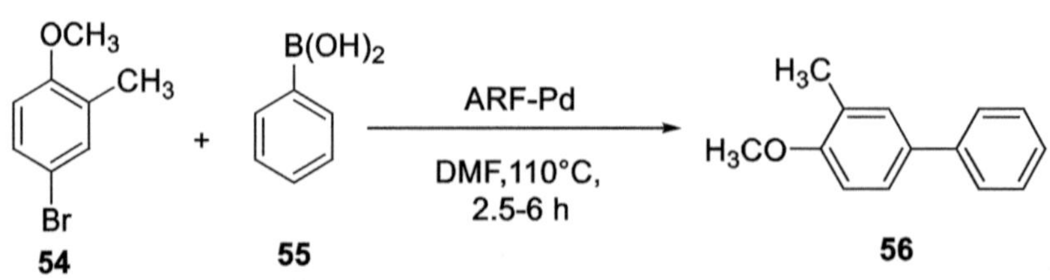

Scheme (12). The Suzuki-Miyaura reaction using ARF-Pd.

The removal of ARF-Pd was done by simple filtration and chromatographic purification of the residual mixture, and thus, 82% of the desired unsymmetrical biphenyl was obtained [40]. The efficiency of this catalytic system was also explored using acrylate for Heck coupling and alkynes for the Sonogashira coupling reaction (Scheme **13**).

In several organic transformations, the resin-capture-release method, which produced polymer-supported reactive species, has emerged as a unique, greener approach for the synthesis of several compounds. Polymer-bound boronic acids were first studied in 1994 by Frenette and Friesen for the Suzuki coupling reaction [41]. The possibility for additional use of polyionic resin-bound reagents and/or catalysts has been established since the formation of an ion-exchange resin-supported organoborate species as a heterogeneous phenylating agent [42]. Amberlite ion-exchange resins (chloride form) are subsequently exchanged with tetraphenylborate anion (Ph_4B^-) through continuous washing with an aqueous solution of $NaBPh_4$ until the solution gave negative response to chloride anion (monitored with $AgNO_3$ solution and then aqueous ammonia addition). The Suzuki coupling reaction has been performed directly in the presence of the polyionic resins which are embedded with the tetraphenylborate counter anion (Ph_4B^-). The biphenyls with different substituents were obtained in 90% yield according to initial studies on coupling with 3-iodotoluene in the presence of palladium catalyst [$Pd(OAc)_2$; 2 mol% and Na_2CO_3; 1 equiv] (Scheme **14**). Interestingly, the absence of a base has no effect on the coupled product's yield. These base-free conditions, with the organoborate ion bound on the solid-phase polymeric frameworks, provided an important advantage for Suzuki-Miyaura coupling reactions.

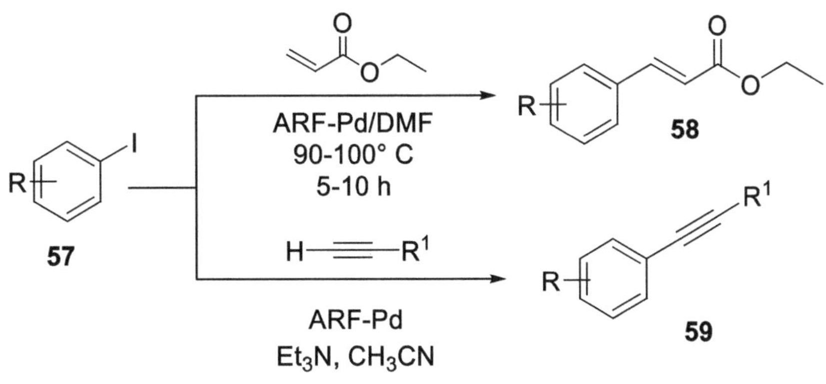

Scheme (13). Sonogashira coupling reaction.

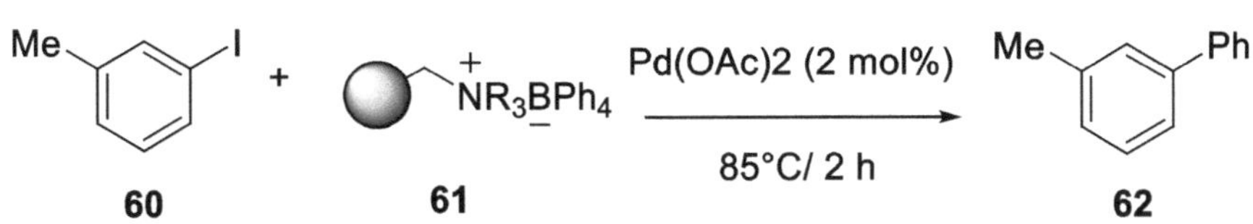

Scheme (14). Coupling with 3-iodotoluene in the presence of palladium catalyst.

Due to their exceptional mechanical qualities and exceptionally high heat resistance, both microporous and mesoporous materials have been used extensively as solid support catalysts and will function as polymeric phenolic resins in a variety of organic transformations [43]. Deng *et al.* created a new hyper-cross-linked phenolic resin (PFN-P) direct polycondensation method in 2018. This reaction produced a polymeric phenolic resin-supported palladium catalyst (Pd@PFN-P) that produced the coupling products in high yield by reacting a wide range of aryl halides and phenylboronic acid in an aqueous medium with palladium metal immobilised on it [44]. Five consecutive reaction cycles yield no change in the conversions. Due to the catalyst's excellent hydrophilicity and wide range of pore sizes, the reactants are easily absorbed onto the catalytic sites, thereby enhancing the reaction rate. Later, Gu *et al.* developed Pd@PPP-1 and Pd@PPP-2, two extremely stable porous porphyrin-based Pd nanoparticles (PdNPs) supported by poly-imide polymer (PPP). Electron microscopy (TEM) demonstrated the layered-sheet packing generated by the π–π stacking, whereas Scanning Electron Microscopy (SEM) depicted regular spheres of aggregation [45]. On PPP supports, uniform dispersion of PdNPs was also observed. Using a range of substituted phenylboronic acids and benzyl halides, Pd@PPP-1 and Pd@PPP-2 catalysts were tested for catalytic activity in coupling reactions. Under ideal reaction conditions, both catalysts produced yields of up to 94% and 97%, respectively. Since the reactivity did not alter after five cycles in a row, the catalysts' stability was verified [46]. Three catalysts' catalytic activity was evaluated for the chemoselective Suzuki-Miyaura coupling reaction between 4-methoxyphenylboronic acid and 2,6-dibromopyridine (Scheme **15**). Yamamoto *et al.* observed that PS-PdMC's colour changed from yellow to dark yellow after the second cycle when only bisarylpyridine was generated and that there was a decrease in chemoselectivity towards the synthesis of mono-arylpyridine. They came to a conclusion after a second consecutive run that the colour change and lack of chemoselectivity might have been caused by the removal of Pd atoms from the ligand, which reduced the Pd(II) complex to Pd(0) atoms [47]. Additionally, at ambient temperature, Lim and colleagues created a photocatalyst, PdNPs supported by MoS2 nanosheets (Pd-MoS2), for the visible-light-driven Suzuki-Miyaura cross-coupling that has a yield of >99% and selectivity for the synthesis of biphenyl coupled products [48]. In another study, the catalytic activity of Pd^0 and Pd^{2+} photocatalysts immobilised on Graphene Oxide (GO) and reduced-GO (r-GO) were used to study the Suzuki-Miyaura C-C coupling.

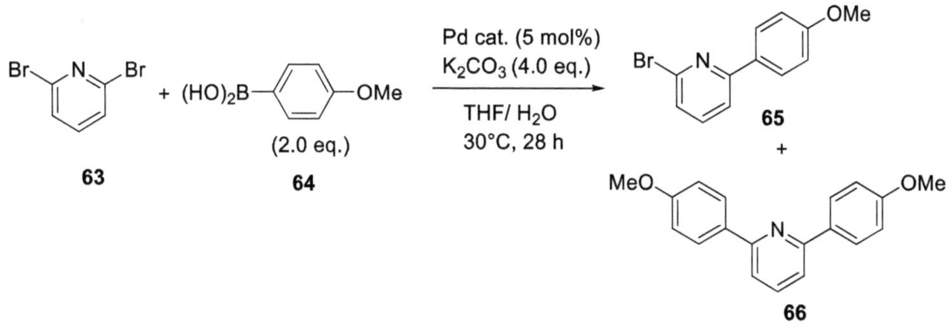

Entry	Pd cat.	Yield of 65 (%)	Yield of 66 (%)
63	BnO-PdMC	66	4
63	PS-PdMC (1st)	76	8
63	PS-PdMC (2nd)	77	7

Scheme (15). Suzuki-Miyaura coupling reaction between 4-methoxyphenylboronic acid and 2,6-dibromopyridine.

CONCLUSION

Based on the extensive literature review, the alkaline site structure has a great influence on catalytic performance, and it is still under exploration. Therefore, altering the microstructure of alkaline species to increase the catalyst reactivity remains a challenge for the chemist. In this chapter we have discussed various reactions based on supported solid catalysts, which exhibit good selectivity and high yield of targeted product. Mild reaction conditions, highly selective target molecule synthesis, and the solid-supported catalyst's propensity for reuse are the work's highlights. It is anticipated that the catalyst system and reactions discussed in this chapter open new areas in the field of solid-supported catalysts for the development of sustainable protocols and encourage the chemist to push forward the emerging area of solid catalysis by designing new catalysts in the future.

REFERENCES

[1] a)Trost, B.M. Atom Economy—a challenge for organic synthesis: homogeneous catalysis leads the way. *Angew. Chem. Int. Ed. Engl.,* **1995**, *34*(3), 259-281.
[http://dx.doi.org/10.1002/anie.199502591]
b) Parshall, G.W. Industrial applications of homogeneous catalysis. A review. *J. Mol. Catal.,* **1978**, *4*(4), 243-270.
[http://dx.doi.org/10.1016/0304-5102(78)85023-8]
c) Cole-Hamilton, D.J. Homogeneous catalysis--new approaches to catalyst separation, recovery, and recycling. *Science,* **2003**, *299*(5613), 1702-1706.

[http://dx.doi.org/10.1126/science.1081881] [PMID: 12637737]

[2] a) Kadokawa, J.; Kobayashi, S. Polymer synthesis by enzymatic catalysis. *Curr. Opin. Chem. Biol.*, **2010**, *14*(2), 145-153.
[http://dx.doi.org/10.1016/j.cbpa.2009.11.020] [PMID: 20022287]
b) Ouchi, M.; Terashima, T.; Sawamoto, M. Transition metal-catalyzed living radical polymerization: toward perfection in catalysis and precision polymer synthesis. *Chem. Rev.*, **2009**, *109*(11), 4963-5050.
[http://dx.doi.org/10.1021/cr900234b] [PMID: 19788190]
c) Ethiraj, J.; Wagh, D.; Manyar, H. Advances in upgrading biomass to biofuels and oxygenated fuel additives using metal oxide catalysts. *Energy Fuels,* **2022**, *36*(3), 1189-1204.
[http://dx.doi.org/10.1021/acs.energyfuels.1c03346]
d) Taylor, K.C. Nitric oxide catalysis in automotive exhaust systems. *Catal. Rev., Sci. Eng.,* **1993**, *35*(4), 457-481.
[http://dx.doi.org/10.1080/01614949308013915]

[3] Bergbreiter, D.E.; Tian, J.; Hongfa, C. Using soluble polymer supports to facilitate homogeneous catalysis. *Chem. Rev.*, **2009**, *109*(2), 530-582.
[http://dx.doi.org/10.1021/cr8004235] [PMID: 19209941]

[4] a) Blaser, H.U. Heterogeneous catalysis for fine chemicals production. *Catal. Today,* **2000**, *60*(3-4), 161-165.
[http://dx.doi.org/10.1016/S0920-5861(00)00332-1]
b) Sápi, A.; Rajkumar, T.; Kiss, J.; Kukovecz, Á.; Kónya, Z.; Somorjai, G.A. Metallic nanoparticles in heterogeneous catalysis. *Catal. Lett.,* **2021**, *151*(8), 2153-2175.
[http://dx.doi.org/10.1007/s10562-020-03477-5]

[5] a) Bing, W.; Wei, M. Recent advances for solid basic catalysts: Structure design and catalytic performance. *J. Solid State Chem.,* **2019**, *269*, 184-194.
[http://dx.doi.org/10.1016/j.jssc.2018.09.023]
b) Basu, B.; Paul, S. Solid-Phase organic synthesis and catalysis: Some recent strategies using alumina, silica, and polyionic resins. *J. Catal.,* **2013**, *2013*, 1-20.
[http://dx.doi.org/10.1155/2013/614829]

[6] Pines, H.; Haag, W. Communications - stereoselectivity in the carbanion-catalyzed isomerization of 1-butene. *J. Org. Chem.,* **1958**, *23*(2), 328-329.
[http://dx.doi.org/10.1021/jo01096a627]

[7] a) Varnava, K.G.; Sarojini, V. Making solid-phase peptide synthesis greener: A review of the literature. *Chem. Asian J.,* **2019**, *14*(8), 1088-1097.
[http://dx.doi.org/10.1002/asia.201801807] [PMID: 30681290]
b) Cankařová, N.; Schütznerová, E.; Krchňák, V. Traceless solid-phase organic synthesis. *Chem. Rev.,* **2019**, *119*(24), 12089-12207.
[http://dx.doi.org/10.1021/acs.chemrev.9b00465] [PMID: 31747265]

[8] a) Shang, J.; Liu, S.; Lu, L.; Ma, X.; He, Y.; Deng, Y. Efficient synthesis of 2-oxazolidinone from epoxides and carbamates with binary Mg Fe oxides as a magnetic solid base catalyst. *Catal. Commun.,* **2012**, *28*, 13-17.
[http://dx.doi.org/10.1016/j.catcom.2012.08.006]
b) Quesada, J.; Faba, L.; Díaz, E.; Bennici, S.; Auroux, A.; Ordóñez, S. Role of surface intermediates in the deactivation of Mg Zr mixed oxides in acetone self-condensation: A combined DRIFT and *ex situ* characterization approach. *J. Catal.,* **2015**, *329*, 1-9.
[http://dx.doi.org/10.1016/j.jcat.2015.04.029]
c) Lauwaert, J.; Moschetta, e.g.; Van Der Voort, P.; Thybaut, J.W.; Jones, C.W.; Marin, G.B. Spatial arrangement and acid strength effects on acid–base cooperatively catalyzed aldol condensation on aminosilica materials. *J. Catal.,* **2015**, *325*, 19-25.
[http://dx.doi.org/10.1016/j.jcat.2015.02.011]

[9] a) Ge, J.; Zeng, Z.; Liao, F.; Zheng, W.; Hong, X.; Tsang, S.C.E. Palladium on iron oxide nanoparticles: The morphological effect of the support in glycerol hydrogenolysis. *Green Chem.,*

2013, *15*(8), 2064-2069.
[http://dx.doi.org/10.1039/c3gc40712h]
b) Lei, X.; Lü, Z.; Guo, X.; Zhang, F. MgAl-layered double hydroxide films on muscovite mica: Epitaxial growth and catalytic activity in acetone self-condensation. *Ind. Eng. Chem. Res.*, **2012**, *51*(3), 1275-1280.
[http://dx.doi.org/10.1021/ie202566x]
c) Frey, A.M.; Bitter, J.H.; de Jong, K.P. Lanthanum oxide supported on carbon nanofibers as solid base catalysts. *ChemCatChem*, **2011**, *3*(7), 1193-1199.
[http://dx.doi.org/10.1002/cctc.201100001]

[10] a) Sun, G.; Li, Y.; Cai, Z.; Teng, Y.; Wang, Y.; Reaney, M.J.T. K_2 CO_3 -loaded hydrotalcite: A promising heterogeneous solid base catalyst for biolubricant base oil production from waste cooking oils. *Appl. Catal. B*, **2017**, *209*, 118-127.
[http://dx.doi.org/10.1016/j.apcatb.2017.02.078]
b) Sano, N.; Yamada, K.; Tsunauchi, S.; Tamon, H. A novel solid base catalyst for transesterification of triglycerides toward biodiesel production: Carbon nanohorn dispersed with calcium ferrite. *Chem. Eng. J.*, **2017**, *307*, 135-142.
[http://dx.doi.org/10.1016/j.cej.2016.08.010]
c) Zhang, P.; Han, Q.; Fan, M.; Jiang, P. Magnetic solid base catalyst $CaO/CoFe_2O_4$ for biodiesel production: Influence of basicity and wettability of the catalyst in catalytic performance. *Appl. Surf. Sci.*, **2014**, *317*, 1125-1130.
[http://dx.doi.org/10.1016/j.apsusc.2014.09.043]
d) Zhu, F.; Yang, D.; Zhang, F.; Li, H. Amine-bridged periodic mesoporous organosilica nanospheres as an active and reusable solid base-catalyst for water-medium and solvent-free organic reactions. *J. Mol. Catal. Chem.*, **2012**, *363-364*, 387-397.
[http://dx.doi.org/10.1016/j.molcata.2012.07.015]

[11] Peri, J.B. Infrared and gravimetric study of the surface hydration of γ-alumina. *J. Phys. Chem.*, **1965**, *69*(1), 211-219.
[http://dx.doi.org/10.1021/j100885a032]

[12] Borah, P.; Yamashita, Y.; Kobayashi, S. *Angew. Chem. Ind. Ed*, **2017**, *56*, 10330.

[13] a) Shirotori, M.; Nishimura, S.; Ebitani, K. *J. Mater. Chem. A Mater. Energy Sustain.*, **2017**, 5.
b) Keller, T.C.; Desai, K.; Mitchell, S.; Pérez-Ramírez, J. Design of base zeolite catalysts by alkali-metal grafting in alcoholic media. *ACS Catal.*, **2015**, *5*(9), 5388-5396.
[http://dx.doi.org/10.1021/acscatal.5b00761]
c) Lei, X.; Zhang, F.; Yang, L.; Guo, X.; Tian, Y.; Fu, S.; Li, F.; Evans, D.G.; Duan, X. Highly crystalline activated layered double hydroxides as solid acid-base catalysts. *AIChE J.*, **2007**, *53*(4), 932-940.
[http://dx.doi.org/10.1002/aic.11118]
d) Kajihara, K.; Tezuka, N.; Shoji, M.; Wakasugi, J.; Munakata, H.; Kanamura, K. $Li_4B_4M_3O_{12}Cl$ (M= Al, Ga): An electrochemically stable, lithium-ion-conducting cubic boracite with substituted boron sites. *Bull. Chem. Soc. Jpn.*, **2017**, *90*(12), 1279-1286.
[http://dx.doi.org/10.1246/bcsj.20170242]
e) Frey, A.M.; Yang, J.; Feche, C.; Essayem, N.; Stellwagen, D.R.; Figueras, F.; de Jong, K.P.; Bitter, J.H. Influence of base strength on the catalytic performance of nano-sized alkaline earth metal oxides supported on carbon nanofibers. *J. Catal.*, **2013**, *305*, 1-6.
[http://dx.doi.org/10.1016/j.jcat.2013.04.019]

[14] a) Kikhtyanin, O.; Lesnik, E.; Kubička, D. The occurrence of Cannizzaro reaction over Mg-Al hydrotalcites. *Appl. Catal. A Gen.*, **2016**, *525*, 215-225.
[http://dx.doi.org/10.1016/j.apcata.2016.08.007]
b) Delidovich, I.; Palkovits, R. Structure–performance correlations of Mg–Al hydrotalcite catalysts for the isomerization of glucose into fructose. *J. Catal.*, **2015**, *327*, 1-9.
[http://dx.doi.org/10.1016/j.jcat.2015.04.012]

[15] Sandesh, S.; Kristachar, P.K.R.; Manjunathan, P.; Halgeri, A.B.; Shanbhag, G.V. Synthesis of

biodiesel and acetins by transesterification reactions using novel $CaSn(OH)_6$ heterogeneous base catalyst. *Appl. Catal. A Gen.,* **2016**, *523*, 1-11.
[http://dx.doi.org/10.1016/j.apcata.2016.05.006]

[16] Li, J.; Li, Z.; Dong, J.; Fang, R.; Chi, Y.; Hu, C. *CS Catal.,* **2023**, *13*, 5272-5284.

[17] Malinowski, S.; Basinski, S. Kinetics of aldolic reactions in the gaseous phase on solid catalysts of basic character. *J. Catal.,* **1963**, *2*(3), 203-207.
[http://dx.doi.org/10.1016/0021-9517(63)90044-7]

[18] Scheidt, F. Vapor-phase aldol condensations over heterogeneous catalysts. *J. Catal.,* **1964**, *3*(4), 372-378.
[http://dx.doi.org/10.1016/0021-9517(64)90042-9]

[19] a) Concepcion, J.J.; Jurss, J.W.; Templeton, J.L.; Meyer, T.J. One site is enough. catalytic water oxidation by $[ru(tpy)(bpm)(oh_2)]^{2+}$ and $[Ru(typ)(bpz)(OH_2)]^{2+}$. *J. Am. Chem. Soc.,* **2008**, *130*(49), 16462-16463.
[http://dx.doi.org/10.1021/ja8059649] [PMID: 19554681]
b) Cape, J.L.; Lymar, S.V.; Lightbody, T.; Hurst, J.K. Characterization of intermediary redox states of the water oxidation catalyst, [Ru(bpy)(2)(OH(2))](2)O(4+). *Inorg. Chem.,* **2009**, *48*(10), 4400-4410.
[http://dx.doi.org/10.1021/ic9001219] [PMID: 19371067]

[20] aHarriman, A.; Thomas, J.M.; Millward, G.R. *New J. Chem.,* **1987**, *11*, 757-762.bMills, A. Heterogeneous redox catalysts for oxygen and chlorine evolution. *Chem. Soc. Rev.,* **1989**, *18*, 285-316.
[http://dx.doi.org/10.1039/cs9891800285]

[21] Zhang, G.; Hattori, H.; Tanabe, K. Aldol addition of acetone, catalyzed by solid base catalysts: Magnesium oxide, calcium oxide, strontium oxide, barium oxide, lanthanum (iii) oxide and zirconium oxide. *Appl. Catal.,* **1988**, *36*, 189-197.
[http://dx.doi.org/10.1016/S0166-9834(00)80114-1]

[22] Zhang, G.; Hattori, H.; Tanabe, K. Aldol condensation of acetone/acetone d6 over magnesium oxide and lanthanum oxide. *Appl. Catal.,* **1988**, *40*, 183-190.
[http://dx.doi.org/10.1016/S0166-9834(00)80436-4]

[23] Nakamura, S.; Takagaki, A.; Watanabe, M.; Yamada, K.; Yoshida, M.; Ishihara, T. Porous boron nitride as a weak solid base catalyst. *ChemCatChem,* **2020**, *12*(23), 6033-6039.
[http://dx.doi.org/10.1002/cctc.202001435]

[24] Matsuhashi, H. Synthesis of a solid base catalyst formed from Zinc(II) acetylacetonate and consisting of MgO modified with ZnO. *Catalysis Research,* **2022**, *2*(2), 1-14.
[http://dx.doi.org/10.21926/cr.2202015]

[25] Zhang, Z.; Hu, M.; Lv, B.; Kang, J.; Tang, J.; Fei, Z.; Chen, X.; Liu, Q.; Cui, M.; Qiao, X. Solvent-assisted stepwise redox approach to generate zeolite NaA-supported K_2O as strong base catalyst for michael addition of ethyl acrylate with ethanol. *ACS Omega,* **2018**, *3*(8), 10188-10197.
[http://dx.doi.org/10.1021/acsomega.8b00704] [PMID: 31459147]

[26] Gupta, M. R. Gupta M. Anand. *Beilstein J. Org. Chem.,* **2009**, *5*, 68.
[PMID: 20300501]

[27] Sugino, K.; Oya, N.; Yoshie, N.; Ogura, M. A simple modification creates a great difference: new solid-base catalyst using methylated N-substituted SBA-15. *J. Am. Chem. Soc.,* **2011**, *133*(50), 20030-20032.
[http://dx.doi.org/10.1021/ja2070522] [PMID: 22103799]

[28] a) Lednor, P.W. Synthesis, stability, and catalytic properties of high surface area silicon oxynitride and silicon carbide. *Catal. Today,* **1992**, *15*(2), 243-261.
[http://dx.doi.org/10.1016/0920-5861(92)80178-P]
b) Xia, Y.; Mokaya, R. Highly ordered mesoporous silicon oxynitride materials as base catalysts. *Angew. Chem. Int. Ed.,* **2003**, *42*(23), 2639-2644.
[http://dx.doi.org/10.1002/anie.200350978]

c) Xia, Y.; Mokaya, R. Ordered mesoporous MCM-41 silicon oxynitride solid base materials with high nitrogen content: synthesis, characterisation and catalytic evaluation. *J. Mater. Chem.,* **2004**, *14*(16), 2507-2515.
[http://dx.doi.org/10.1039/b404949g]
d) Chino, N.; Okubo, T. Nitridation mechanism of mesoporous silica: SBA-15. *Microporous Mesoporous Mater.,* **2005**, *87*(1), 15-22.
[http://dx.doi.org/10.1016/j.micromeso.2005.07.034]

[29] P, Garcı'a-Garcı'a, J. M. Moreno, U. Dı'az1, M. Bruix, A. Corma. *Nat. Commun.,* **2016**, *7*, 10835.

[30] Li, X.; Zhang, B.; Fang, Y.; Sun, W.; Qi, Z.; Pei, Y.; Qi, S.; Yuan, P.; Luan, X.; Goh, T.W.; Huang, W. Metal-organic-framework-derived carbons: applications as solid-base catalyst and support for pd nanoparticles in tandem catalysis. *Chemistry,* **2017**, *23*(18), 4266-4270.
[http://dx.doi.org/10.1002/chem.201605852] [PMID: 28188655]

[31] Zeeland, R.V.; Li, X.; Huang, W.; Stanley, L.M. *RSC advances,* **2016**, *6*, 56330-56334.
[http://dx.doi.org/10.1039/C6RA12746K]

[32] Fan, M.; Zhang, P. Activated carbon supported K_2 CO_3 catalysts for transesterification of dimethyl carbonate with propyl alcohol. *Energy Fuels,* **2007**, *21*(2), 633-635.
[http://dx.doi.org/10.1021/ef060492m]

[33] Gryglewicz, S.; Oko, F.A.; Gryglewicz, G. Synthesis of Modern synthetic oils based on dialkyl carbonates. *Ind. Eng. Chem. Res.,* **2003**, *42*(21), 5007-5010.
[http://dx.doi.org/10.1021/ie030322m]

[34] Liu, X.; Piao, X.; Wang, Y.; Zhu, S. Calcium ethoxide as a solid base catalyst for the transesterification of soybean oil to biodiesel. *Energy Fuels,* **2008**, *22*(2), 1313-1317.
[http://dx.doi.org/10.1021/ef700518h]

[35] Radhakrishan, R.; Do, D.M.; Jaenicke, S.; Sasson, Y.; Chuah, G.K. Potassium phosphate as a solid base catalyst for the catalytic transfer hydrogenation of aldehydes and ketones. *ACS Catal.,* **2011**, *1*(11), 1631-1636.
[http://dx.doi.org/10.1021/cs200299v]

[36] Kumar, S.; Jain, S.L. Polyethylene glycol enfolded kbr assisted base catalyzed synthesis of dimethyl carbonate from methanol and carbon dioxide. *Ind. Eng. Chem. Res.,* **2014**, *53*(41), 15798-15801.
[http://dx.doi.org/10.1021/ie502579m]

[37] Kumar, S.; Jain, S.L. Polyethylene glycol wrapped potassium bromide assisted chemical fixation of carbon dioxide. *Ind. Eng. Chem. Res.,* **2014**, *53*(2), 541-546.
[http://dx.doi.org/10.1021/ie4033439]

[38] Basu, B.; Paul, S. *Hindawi publishing corporation journal of catalysts,* **2013**, *2013*, 1-20.

[39] Basu, B.; Das, P.; Bhuiyan, M.M.H.; Jha, S. Microwave-assisted Suzuki coupling on a KF–alumina surface: synthesis of polyaryls. *Tetrahedron Lett.,* **2003**, *44*(19), 3817-3820.
[http://dx.doi.org/10.1016/S0040-4039(03)00731-7]

[40] Basu, D.; Das, S.; Das, P.; Mandal, B.; Banerjee, D.; Almqvist, F. *Synthesis,* **2007**, *7*, 1137-1146.

[41] Frenette, R.; Friesen, R.W. Biaryl synthesis *via* suzuki coupling on a solid support. *Tetrahedron Lett.,* **1994**, *35*(49), 9177-9180.
[http://dx.doi.org/10.1016/0040-4039(94)88458-7]

[42] Basu, B.; Das, S.; Kundu, S.; Mandal, B. Polyionic heterogeneous phenylating agent for base-free suzuki-miyaura coupling reaction. *Synlett,* **2008**, *2008*(2), 255-259.
[http://dx.doi.org/10.1055/s-2008-1000874]

[43] Jia, S.H.; Ding, X.; Yu, H.T.; Han, B.H. Multi-hydroxyl-containing porous organic polymers based on phenol formaldehyde resin chemistry with high carbon dioxide capture capacity. *RSC Advances,* **2015**, *5*(87), 71095-71101.
[http://dx.doi.org/10.1039/C5RA13405F]

[44] Nafiu, S.A.; Ajeebi, A.M.; Alghamdi, H.S.; Aziz, M.A.; Shaikh, M.N. *Asian J. Org. Chem.,* **2023**, *12*, e202300051.
[http://dx.doi.org/10.1002/ajoc.202300051]

[45] Deng, G.; Wang, Z. Hierarchical porous phenolic resin and its supported pd-catalyst for suzuki–miyaura reactions in water medium. *Macromol. Rapid Commun.,* **2018**, *39*(3), 1700618.
[http://dx.doi.org/10.1002/marc.201700618]

[46] Zhu, W.; Wang, X.; Li, T.; Shen, R.; Hao, S.J.; Li, Y.; Wang, Q.; Li, Z.; Gu, Z.G. Porphyrin-based porous polyimide polymer/Pd nanoparticle composites as efficient catalysts for Suzuki–Miyaura coupling reactions. *Polym. Chem.,* **2018**, *9*(12), 1430-1438.
[http://dx.doi.org/10.1039/C8PY00092A]

[47] Yamamoto, K.; Nameki, R.; Sogawa, H.; Takata, T. Synthesis of polystyrene-supported Pd(II)-containing macrocyclic complex as a reusable catalyst for chemoselective Suzuki–Miyaura coupling reaction. *Tetrahedron Lett.,* **2020**, *61*(21), 151870.
[http://dx.doi.org/10.1016/j.tetlet.2020.151870]

[48] aShin, H.H.; Kang, E.; Park, H.; Han, T.; Lee, C.H.; Lim, D.K. Pd-nanodot decorated MoS_2 nanosheets as a highly efficient photocatalyst for the visible-light-induced Suzuki–Miyaura coupling reaction. *J. Mater. Chem. A Mater. Energy Sustain.,* **2017**, *5*(47), 24965-24971.
[http://dx.doi.org/10.1039/C7TA08441B] bShin, H.H.; Han, T.; Yang, W.; Lim, D.K. Pd2+-conjugated reduced graphene oxide as a highly efficient visible-light-induced photocatalytic Suzuki-Miyaura coupling. *Carbon,* **2019**, *143*, 362-370.
[http://dx.doi.org/10.1016/j.carbon.2018.11.039]

CHAPTER 10

Recent Advances in Solid Base Catalysis for Reduction Reactions

Vijay Bahadur[1,*], **Neha Yadav**[2] and **Ajam Chand Shaikh**[1]

[1] *Department of Science, Alliance University, Central Campus, Chikkahadage Cross, Chandapura- Anekal, Main Road, Bengaluru, Karnataka-562106, India*

[2] *Legal Affairs, IP, SABIC, Bengaluru, Karnataka, India*

Abstract: The demand for solid base catalysts has led to the development of green catalytic processes, owing to their advantages such as easy separation, low corrosion, and eco-friendliness. Over the last decade, substantial advancements have occurred in crafting these catalysts, amplifying their utility in organic synthesis, fine chemical production, and environmental catalysis. Consequently, propelling the progress of solid basic catalysts is crucial for both academic exploration and industrial implementation. The book chapter delves into the pivotal role of solid bases in organic reactions, particularly emphasizing their increasing usage in organic synthesis due to their superiority over liquid bases. Solid bases offer easier disposal, simplified separation and retrieval of products, solvents and catalysts, along with being non-corrosive. Moreover, they enable base-catalyzed reactions without the necessity of solvents and even in gas-phase conditions, thereby expanding the scope for discovering innovative reaction systems. Through numerous illustrative examples, this chapter delineates the significant impact of solid base catalysis in reduction reactions. Additionally, it sheds light on recent advancements and future prospects concerning solid base catalysts in reduction reactions.

Keywords: Catalyst, Eco-friendly, Effortless separation, Non-corrosive, Reduction, Solid base.

INTRODUCTION

While extensive research has focused on solid acids, relatively less attention has been directed towards heterogeneous basic catalysts, also known as solid base catalysts. The first pioneering investigation into heterogeneous base catalysts was conducted by Pines and Haag [1].

* **Corresponding author Vijay Bahadur:** Department of Science, Alliance University, Central Campus, Chikkahadage Cross, Chandapura- Anekal, Main Road, Bengaluru, Karnataka-562106, India; E-mail: vij85cy@gmail.com; vijay.bahadur@alliance.edu.in)

Ravi Tomar, Vivek Mishra, Kapil Arya & Garima Pandey (Eds.)

They demonstrated that sodium metal dispersed on alumina effectively catalyzed the isomerization of double bonds in alkenes.

In 1972, Tanabe and his team discovered that calcium oxide and magnesium oxide showed a marked increase in catalytic activity for the isomerization of 1-butene, especially after undergoing vacuum pretreatment. This study emphasized the critical role of preparation techniques and pretreatment procedures for base catalysts [2, 3].

Additionally, various techniques were developed to assess the basic characteristics of solid surfaces and to identify reaction intermediates. In 1989, Tanabe and his team released "New Solid Acids and Bases," summarizing the progress made in this field during the 1970s and 1980s [4].

Carbanions are essential intermediates in many organic reactions, such as isomerization, additions and condensations. They generally form when a base abstracts a proton from a C-H bond in an organic molecule, typically in homogeneous phases.

These organic transformations frequently necessitate a stoichiometric quantity of liquid bases to facilitate the generation of carbanions. However, this process also produces a stoichiometric amount of metal salts as by-products.

There is no necessity for employing a stoichiometric quantity of base compounds. Another notable advantage of solid bases compared to homogeneous bases is the reduced need for solvents. In homogeneous phase reactions, the requirement for both the base catalyst and reactants to be soluble in the same solvent limits the range of suitable solvents. Consequently, this often leads to the use of hazardous solvents like methyl chloride and dimethyl sulfoxide (DMSO) [5].

In reactions catalyzed by solid bases, solvents should only dissolve reactants (and products), allowing for a much wider range of solvent choices. The effects of solvents in solid-base catalyzed reactions differ from those in homogeneous reactions. No waste is produced from solvents like aqueous solutions with organics or organic solvents with alkaline water. Additionally, simplified separation processes reduce energy consumption. Moreover, solid bases can be reused after separation, improving atom economy in catalyst preparation. The absence of solvent selection constraints also increases the potential for discovering new reactions. Bases such as sodium hydroxide create corrosive solutions, necessitating corrosion-resistant materials for reactors and disposal systems. In contrast, this problem is typically absent in systems that use solid base catalysts.

The conversion of homogeneous processes to heterogeneous ones using solid-base (or solid-acid) catalysts is especially significant in the synthesis of fine chemicals and pharmaceuticals. This is because these sectors often produce more waste per unit mass of the desired product when employing stoichiometric or homogeneous catalytic processes. The advantages described earlier make this transformation essential in improving environmental sustainability and efficiency in these industries.

Moreover, solid base catalysts often achieve high activities and selectivity across a variety of reactions. Unlike acid-catalyzed reactions, reactions over basic catalysts typically avoid issues like cooking, which can be a significant problem. Furthermore, functionalized organic compounds like amines interact more with basic sites than with acidic sites. Consequently, these molecules are less likely to be adsorbed onto the surface and can desorb more readily. Consequently, reactions involving these molecules tend to proceed more smoothly over solid bases.

Organometallic compounds, such as Grignard reagents and alkyl lithiums, donate carbanion-like species, facilitating the formation of Si-C bonds through various reaction mechanisms.

$$R_3SiX + R'MgX \longrightarrow R'\text{-}SiR_3 + MgX_2$$

$$R_3SiX + R'Li \longrightarrow R'\text{-}SiR_3 + LiX$$

Once again, the use of organometallic reagents in stoichiometric quantities is necessary. However, solid bases provide alternative pathways to produce the same products in a more straightforward and environmentally friendly manner.

$$H_2SiEt_2 + {}^tBuC{\equiv}CH \longrightarrow {}^tBuC{\equiv}C\text{-}SiEt_2H + H_2$$

$$H_2SiEt_2 + C_6H_5CH_3 \longrightarrow C_6H_5CH_2\text{-}SiEt_2H + H_2$$

These two reactions proceed smoothly in the presence of catalytic systems such as KF/Al_2O_3 and KNH_2/Al_2O_3, respectively [6, 7].

The following are four key reasons certain materials are recognized as solid base catalysts:

1. Surface characterizations employing various methods, including changes in acid-base indicators, adsorption behavior towards acidic molecules, and spectroscopic techniques (such as UV, IR, XPS, ESR), consistently reveal the

presence of basic sites on the catalyst surfaces.

2. Catalytic activity demonstrates a strong correlation with the quantity or strength of these basic sites. Furthermore, the activity of these sites is hindered by the presence of acidic molecules such as HCl, H_2O, and CO, indicating their crucial role in the catalytic process.
3. These materials exhibit catalytic behavior similar to that of homogeneous base catalysts, suggesting that their surface mechanisms closely mirror those in homogeneous basic solutions. Mechanistic studies, product distributions, and spectroscopic analyses of surface species all suggest that anionic intermediates play a significant role in the reactions, underscoring the significance of basic sites in catalysis.

Role of Solid Base and Basic Sites as a Catalyst [8]

Definition of acid and base

The definitions provided by Brønsted-Lowry and Lewis are paramount in understanding the acid-base chemistry of solid surfaces. According to the Brønsted-Lowry definition, an acid (AH) donates a proton, while a base (B^-) accepts a proton.

$$AH + B^- \longrightarrow A^- + BH$$

According to Lewis's definition, a base (:B) donates a lone pair, whereas an acid (A) accepts a lone pair.

$$A + :B \longrightarrow A:B$$

Solids possessing sites that function as either a Brønsted acid or a Lewis acid are termed solid acids, while solids with sites that act as either a Brønsted base or a Lewis base are referred to as solid bases.

Abstraction of protons

On the surface of solid bases, specific sites or centers serve as basic sites. These sites extract protons from reactant molecules (AH), resulting in the formation of carbanions (A^-).

$$AH + B^- \longrightarrow A^- + BH$$

Here, the basic site B^- on the solid surface functions as a Brønsted base. More potent bases have the capability to abstract a proton from molecules with higher pKa values.

Activation of reactants without proton abstraction

Reactants like ketones and aldehydes are frequently activated by bases through processes that do not involve proton transfer, as exemplified by the equation below.

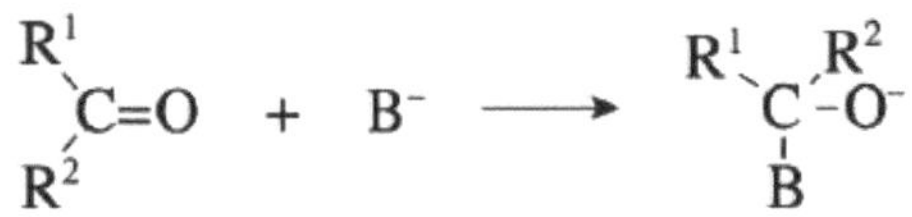

According to the Brønsted-Lowry definition, a base B^- accepts a proton from an adsorbate. Additionally, the same B^- site may act as a Lewis base by donating an electron pair to a different adsorbate. Thus, a single surface site can function as both a Brønsted base (proton acceptor) and a Lewis base (electron-pair donor), depending on the specific adsorbate involved.

Classification of Base Catalysts

Various materials exhibit solid base properties. The most significant classes of solid bases are categorized in Table **1**.

Generation of basic sites

The materials listed in Table **1** have surfaces that are typically covered with carbon dioxide, water, and sometimes oxygen due to exposure to air during handling. To expose basic sites on these surfaces, pre-treatment at high temperatures is necessary. For instance, in the case of magnesium oxide, during the evacuation of the magnesium hydroxide catalyst precursor, water and carbon dioxide start to evolve at temperatures around 700 K [9].

At this temperature, catalytic activities for various reactions become noticeable and generally increase with higher pre-treatment temperatures, peaking at certain temperatures. Beyond this peak, further increases in the evacuation temperature lead to a decrease in activity. High-temperature pre-treatment reveals basic sites on oxide surfaces. Nevertheless, surface and bulk atom rearrangements also take place during pre-treatment, potentially altering the number and characteristics of surface basic sites as the pre-treatment temperature increases.

Application to organic reactions

In recent years, the utilization of solid base catalysts in organic synthesis has been on the rise, albeit the number of applications remains significantly lower compared to solid acid catalysts [10]. However, certain solid base catalysts find applications in industrial processes [11].

Table 1. Classification of solid bases

Metal oxides	MgO, CaO, Al_2O_3, ZrO_2, La_2O_3, Rb_2O
Mixed oxides	SiO_2-MgO, SiO_2-CaO, MgO-La_2O_3, MgO-Al_2O_3 (calcined hydrotalcite)
Alkali or alkaline earth oxides on support	Na_2O/SiO_2, MgO/SiO_2, Cs oxides supported on zeolites
Metal oxynitrides and metal nitrides	AlPON, partially nitrided zeolites and mesoporous silica
Alkali compounds on support	KF/ Al_2O_3, K_2CO_3/ Al_2O_3 KNO_3/ Al_2O_3, NaOH/ Al_2O_3, KOH/ Al_2O_3
Amides, imines on support	KNH_2/ Al_2O_3, K, Y, Eu supported on zeolites from the ammoniacal solution
Alkali metals on support	Na/ Al_2O_3, K/ Al_2O_3, K/MgO, Na/zeolite
Anion exchangers	Anion exchange resins Hydrotalcite and modified hydrotalcite
Zeolites	K, Rb, Cs-exchanged X, Y-zeolites, ETS-10
Clays	Sepiolite, Talc
Phosphates	Hydroxyapatite, metal phosphates, natural phosphates
Amines or ammonium ions tethered to a support	Aminopropyl group /silica, MCM-41, SBA-15 Alkylammonium group/MCM-41

The applications of solid base catalysts extend to various reactions, including:

1. Isomerization of double bonds
2. Hydrogenation
3. Amination
4. Dehydro-cyclodimerization
5. Aldol addition
6. Nitroaldol reaction
7. Michael addition
8. Conjugate addition of alcohols
9. Cyanoethylation
10. Tishchenko reaction

Among the various reactions in organic chemistry, this chapter focuses exclusively on reduction reactions.

Hydrogenation

On heterogeneous basic catalysts, hydrogen molecules are adsorbed and undergo heterolytic dissociation to form H^+ and H^- ions. This mechanism is commonly employed in base-catalyzed hydrogenation reactions. The hydrogenation process on solid base catalysts possesses distinct characteristic features that differentiate them from conventional metallic hydrogenation catalysts [12, 13].

Hydrogenation of Alkenes

Studies have shown that sodium (Na) and lithium (Li) metals dispersed on alumina or silica exhibit activity for the hydrogenation of ethylene and propene, even at temperatures below room temperature [14]. Additionally, the hydrogenation of ethylene, propene, and 1-butene was investigated using alkaline earth oxides. The catalytic activities were found to be dependent on the evacuation temperature of the catalysts [15]. These oxides showed catalytic activity when evacuated at temperatures above 873 K, with peak activity occurring around 1273 K. Additionally, at 523 K, the catalytic activity followed the order: MgO < CaO < BaO < SrO.

Minachev *et al.* conducted research on the hydrogenation of ethylene using a range of rare earth oxides [16]. The catalytic activities were observed to vary based on the pretreatment temperature.

The most active among the investigated oxides are the sesqui-oxides of Pr, La, and Nd. Remarkably, La_2O_3 demonstrates activity even at temperatures as low as 157 K. Catalytic activity is strongly linked to the basicity of the catalysts, with acidic oxides like CeO_2, PrO_2, TbO_2, ZrO_2, and TiO_2 showing minimal activity. Additionally, acidic substances such as NO_2, CO_2, and water act as catalytic poisons. In ethylene hydrogenation with D_2 over Dy_2O_3 at 218 K, only $C_2H_4D_2$ is produced, indicating no H-D exchange between the hydrocarbon and deuterium.

Kokes and his team examined the hydrogenation of ethylene over ZnO [17]. They found that when ethylene is hydrogenated with D_2, the main product is >95% 1,2-dideuteroethane. Similarly, in the hydrogenation of ethylene, propene, and 3,3-dimethylbutene over ZrO_2, they observed a straightforward addition of two D atoms across the double bond. This is quite different from hydrogenation over metals, where a mixture of C_2H_6–xDx is typically obtained, suggesting that the hydrogenation process is irreversible.

$$C_2H_4(a) + H(a) \longrightarrow C_2H_5(a)$$

$$C_2H_5(a) + H(a) \longrightarrow C_2H_6$$

When the hydrogenation is performed using a 1:1 mixture of H_2 and D_2, the primary products observed at room temperature are C_2H_6 and $C_2H_4D_2$. This indicates that the hydrogen incorporated into ethane maintains its original molecular identity [18].

Dent and Kokes, through adsorption, kinetic, and tracer studies, identified two distinct types of hydrogen adsorption on ZnO [19]. Type I adsorption is characterized by rapid and reversible behavior. In contrast, Type II adsorption is initially rapid but becomes slower in later stages, and it is irreversible. Type I hydrogen adsorption gives rise to observed OH and ZnH bands in the infrared spectrum and primarily contributes to the hydrogenation of ethylene at room temperature. On the other hand, Type II adsorption does not actively participate in the hydrogenation process at room temperature but instead modifies the catalyst and enhances its activity.

Hydrogenation of Butadiene

The hydrogenation activity of alkaline earth oxides is notably higher for 1,3-butadiene compared to simpler alkenes like ethylene, propene, and butene [20]. Effective hydrogenation of simple alkenes typically necessitates reaction temperatures exceeding 520 K, whereas hydrogenation of 1,3-butadiene occurs readily even at 273 K.

Over CaO, SrO, and BaO catalysts, an equilibrium mixture of butene is formed, with no production of butane even after the disappearance of 1,3-butadiene. The higher reactivity observed for 1,3-butadiene compared to simple alkenes suggests the involvement of π-allyl carbanion species as intermediates. When hydrogenating 1,3-butadiene with D_2 over MgO, pretreated at 1100 K, cis--butene predominates, containing deuterium atoms at the 1 and 4 positions at 273 K [21].

Hydrogenation of 1,3-butadiene is also achievable over ThO_2 [22], ZrO_2 [23] and La_2O_3 [24]. In each instance, the primary products of hydrogenation with D_2 are 2-butene-d_2, resulting from the 1,4-addition of deuterium atoms. The molecular identity of H_2 (or D_2) remains unchanged, and butane is not generated. Unlike hydrogenation over MgO, where cis-2-butene is predominantly produced, hydrogenation over these oxides mainly yields trans-2-butene.

"The mechanism of hydrogenation (or deuteriation) as proposed by Hattori and colleagues is depicted as follows in Fig. (**1**)."

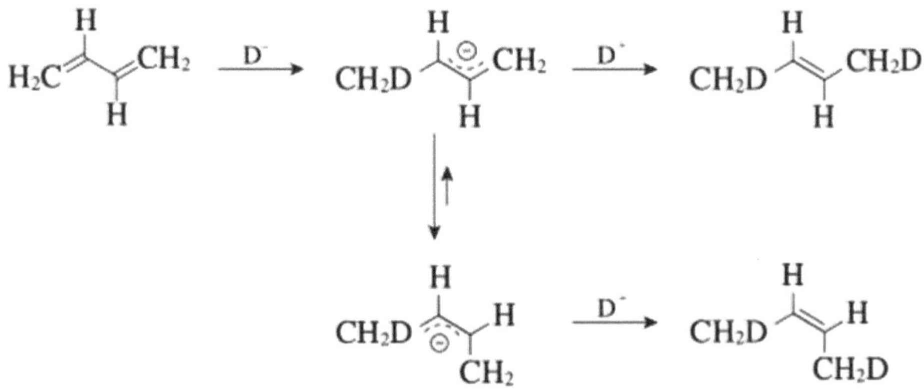

Fig. (1). Mechanism of Hydrogenation of 1,3-butadiene.

In the gaseous phase, 1,3-butadiene comprises 93% of the s-trans conformer and 7% of the s-cis conformer. It is presumed that this conformation in the gas phase persists in the allylic anion intermediates, which then undergo either interconversion to generate cis-allylic anions or react with D^+ to produce butene.

During the hydrogenation process of 1,3-butadiene with D_2 over ZnO, the addition of D atoms exclusively takes place at the 1,2-positions.

Hydrogenation of benzoic Acid

Mitsubishi Kasei Corp. (now Mitsubishi Chemical Corp.) successfully commercialized the hydrogenation of benzoic acid into benzaldehyde (Scheme **1**). This process is conducted within the temperature range of 350 to 400 °C under a hydrogen pressure ranging from 0 to 3 kg per square centimeter [25].

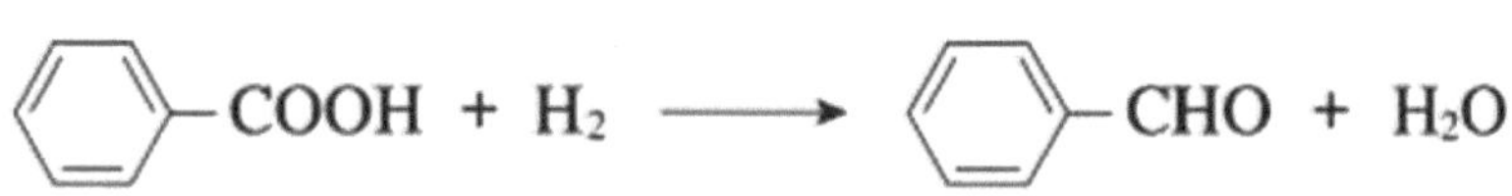

Scheme (1). Hydrogenation of benzoic acid into benzaldehyde.

Zirconium dioxide (ZrO_2) exhibited outstanding selectivity for benzaldehyde among the tested metal oxides, albeit with low activity. However, through modification with Pb^{2+}, In^{3+}, Cr^{3+}, or Mn^{2+}, the activity of ZrO_2 was significantly boosted.

The catalysts were useful for the hydrogenation of various acids, as listed in Table **2** below.

Table 2. Hydrogenation of acids/ester with their selectivity and conversion%

Carboxylic acids or esters	Conversion of acid/%	Selectivity to aldehyde/%
Benzoic acid	98	96
α-Metylbenzoic acid	98	97
m-Phenoxybenzoic acid	97	96
Dimethyle terephthalate	64	73
m-Chlorobenzoic acid	82	77
Trimethylacetic acid	97	99
Methyl *n*-hexanonate	50	70
Cyclohexanecarboxylic acid	95	98
Methyl nicotinate	86	83
4-Methyl-5-carbomethoxythiazole	74	80
3-Furoic acid	62	52

Infrared spectroscopy revealed the presence of bidentate carboxylate upon the adsorption of benzoic acid [26]. These findings suggest that the carboxylate serves as the intermediate, undergoing hydrogenation facilitated by dissociative adsorbed hydrogen. The presence of both weakly acidic and weakly basic sites is crucial for the reaction. Strongly acidic sites promote decarboxylation, whereas strongly basic sites form a salt with benzoic acid, inhibiting further reaction under the hydrogenation conditions.

Sakata *et al.* investigated the catalytic efficacy of various metal oxides in the conversion of benzoic acid under hydrogen, determining ZrO_2 and ZnO to be the most active and selective for benzaldehyde production. They proposed that a surface oxygen vacancy on these oxides captures one oxygen atom from benzoic acid [27].

Transfer Hydrogenation Reactions (Meerwein- Ponndorf-Verley Reduction)

The hydrogen transfer reaction known as the Meerwein-Ponndorf-Verley (MPV) reduction, employing aluminum isopropoxide, is a widely recognized process for reducing ketones or aldehydes. Its mechanism entails the transfer of a hydrogen atom from the alkoxide to the carbonyl carbon of the ketone (Fig. **2**).

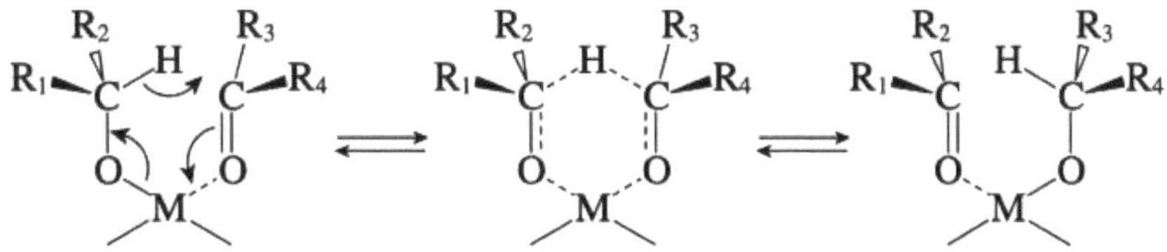

Fig. (2). Mechanism of meerwein- ponndorf-verley reduction.

The MPV reaction is effective for reducing α, β-unsaturated carbonyl compounds due to its high selectivity for the C=O double bond, leaving the C=C bond unaffected. However, to ensure efficiency, the MPV reaction necessitates an alkoxide excess of at least 100–200%, followed by neutralization of the remaining alkoxide in the solution with a strong acid.

Niiyama and Echigoya [28] investigated the catalytic properties of MgO and $MgO\text{-}SiO_2$ composite oxides for the vapor-phase transfer hydrogenation involving acetone and alcohols. The catalytic activities in the ethanol-acetone reaction at 473 K were found to be closely correlated with the number of basic sites, whereas for the 2-butanol-acetone reaction at 413 K, there was no significant correlation observed with either the amounts of basic sites or acidic sites. The authors concluded that in the ethanol-acetone reaction, the activation of ethanol through basic sites plays a more crucial role than the activation of ketone by acidic sites, whereas, in the butanol-acetone reaction, both acidic and basic sites contribute equally.

Ivanov *et al.* explored the vapor-phase MPV reaction between ethanol and acetone across a range of metal oxides with diverse acid-base characteristics. It was observed that Brønsted acid catalysts exhibited negligible activity in facilitating the MPV reaction [29].

Okamoto *et al.* documented the hydrogen transfer process between different alcohols and Methyl Ethyl Ketone (MEK) using MgO as the catalyst [30]. The reactivity of alcohols was found to be contingent upon the delocalization ability of the hydrogen bonded to the α-carbon of the alcohol.

The authors suggest a reaction mechanism involving the heterolytic dissociation of alcohol, succeeded by the direct transfer of the α-hydrogen (as H^-) of the alkoxide to a ketone molecule that is adsorbed, as depicted in Fig. (**3**).

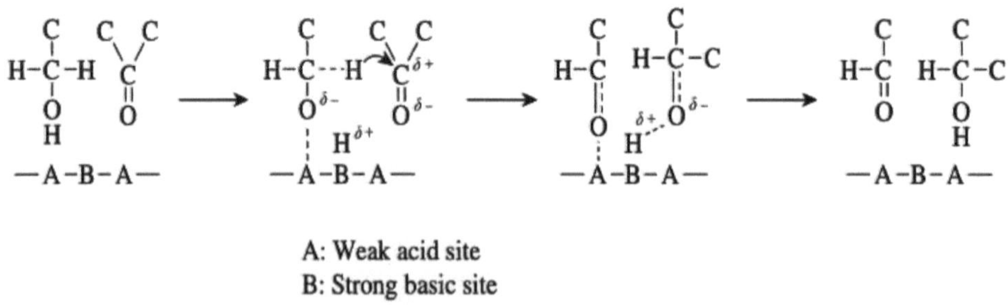

Fig. (3). Mechanism of MPV *via* hydrogen transfer.

Hydrogenation of Ketones with Alcohols

Posner *et al.* documented that 2-propanol, when utilized with alumina, achieves the reduction of structurally diverse aldehydes at ambient temperature. Specifically, α, β-unsaturated aldehydes undergo selective conversion to allylic alcohols. A significant excess of alumina relative to the reactants was employed in the process [31].

Ramana and Pillai documented that Al_2O_3, enriched with approximately 2.2 wt% of Na^+, exhibited notable efficacy as a catalyst for the reduction of various aliphatic and aromatic aldehydes and ketones using 2-propanol in the vapor phase at 573 K [32].

Kijenski *et al.* examined the vapor-phase reduction of various ketones by 2-propanol over MgO. In the reduction of methyl alkyl ketones with the general formula CH_3COR, where R = Et, Pr, ^{i}Pr, Bu, ^{i}Bu, ^{t}Bu, and pentyl, the selectivity for the corresponding alcohol surpassed 95% up to 523 K [33].

Cyclopentanone and cyclohexanone were similarly converted to their respective alcohols through reduction.

The MPV reactions involving different α, β-unsaturated aldehydes and 2-propanol were conducted under reflux conditions (355 K) with the aid of $MgO\text{-}Al_2O_3$ mixed oxide catalysts prepared *via* the calcination of hydrotalcite precursors. The outcomes are presented below (Table **3**).

The mixed oxides are suitable for vapor-phase MPV reactions. In the MPV reaction of propiophenone at 523 K, Mg-Al mixed oxide exhibits the highest activity compared to Mg-Al, Co-Al, Ni-Al, and Cu-Al mixed oxides. The resulting products comprise a mixture of 1-phenylpropyl alcohol and 2-methylstyrene, with the ratio varying based on the reaction temperature. When utilizing $MgO\text{-}Al_2O_3$, propiophenone undergoes complete conversion into 2-

methylstyrene. Conversely, on MgO, the predominant formation of the alcohol occurs within the temperature range of 523–548 K [34, 35].

The reduction of carbonyl compounds can typically be accomplished with hydrogen using supported metal catalysts like Pt/C and Pd/C. However, carbonyl compounds incorporating sulfur or nitrogen moieties are unsuitable for reduction with metals due to metal poisoning by sulfur. Consequently, transfer hydrogenation serves as a valuable method for reducing these compounds. Heterocyclic carbaldehydes can undergo hydrogenation using alcohols in the presence of $MgO\text{-}Al_2O_3$ mixed oxides.

Table 3. Catalytic activity of $MgO\text{-}Al_2O_3$ for the transfer hydrogenation.

S. No.	Reductant	Product	Yield %
1	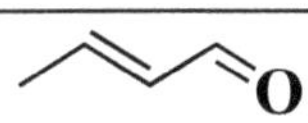	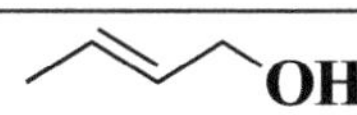	100
2	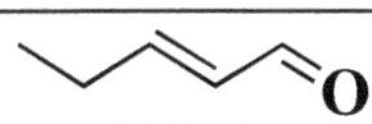	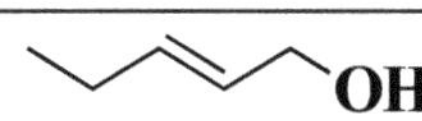	100
3	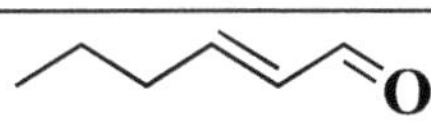	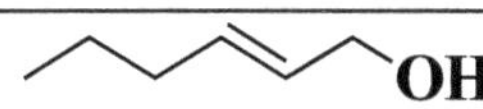	98
4	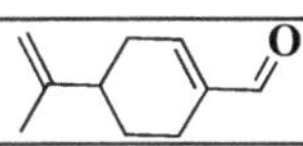	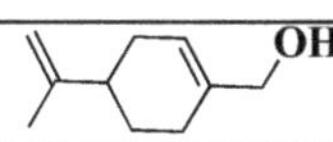	41
5	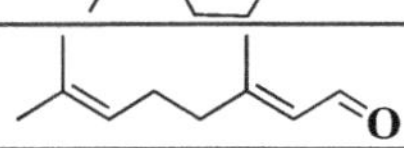	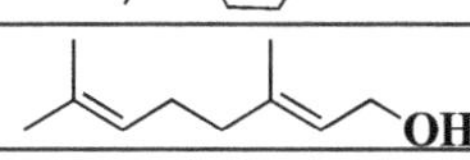	76
6	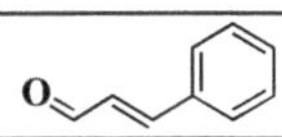	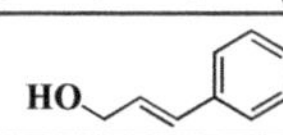	71

Vu *et al.* utilized a $MgO\text{-}Al_2O_3$ catalyst with a ratio of Mg to Al of 2 for the hydrogenation of sulfur compounds, namely thiophene carbaldehyde, 5-bromothiophenecarbaldehyde, and 4-methylthiobenzaldehyde [36].

The three aldehydes underwent complete reduction with 2-butanol to form the corresponding alcohols within 3 hours at 371 K [37]. Various heterocyclic carbaldehydes were hydrogenated with 2-propanol in the presence of $MgO\text{-}Al_2O_3$ (with a Mg/Al ratio of 2/0.84) and $MgO\text{-}Al_2O_3\text{-}ZrO_2$ (with Mg/Al/Zr ratios of 2.0/0.81/0.11). According to Table **4**, these reactions achieved conversions close to 90% with 100% selectivity within 24 hours at 355 K in all cases. In terms of the heteroatom present in the carbaldehyde, the reaction rate decreased in the following order: thiophene > furan > pyrrole. Moreover, the reaction proceeded faster with carbonyl substituents at position 2 than at position 3.

Shabtai *et al.* examined zeolites' efficacy in hydrogen transfer reactions using 2-propanol in a continuous flow reactor. They observed that *n*-Alkanals (C5-C12) could be transformed into alcohols at 453 K with high conversions and selectivity

(>95%) over NaX zeolite. Additionally, they investigated non-substituted and alkyl-substituted cycloalkanones of 4–15 membered rings. Conversion rates were influenced by the alkyl substituent's position and size. Regarding citronellal **1**, two competing reactions were observed: cyclization to isoplegol **2** and normal MPV reduction to citronellol **3**. The reaction direction was markedly affected by the exchangeable cations. CsX zeolite primarily produced the MPV reduction product **3** with 92.3% selectivity at 423 K, while LiX, NaX, and CaX zeolites favored the production of cyclic compound **2** (Scheme **2**).

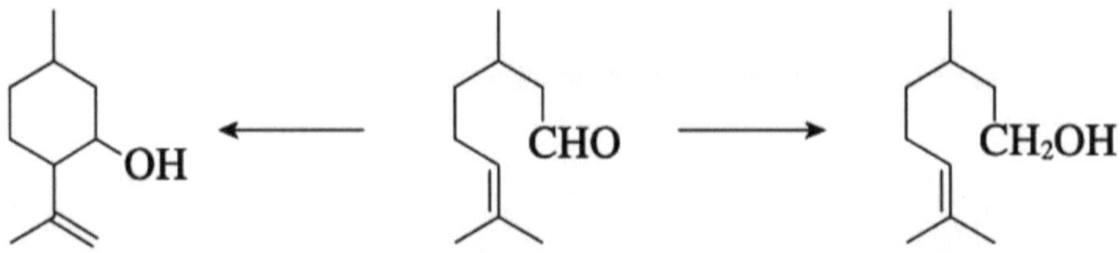

2 **1** **3**

Scheme (2). Cyclization and MPV reaction of citronellal.

Table 4. Hydrogenation of heterocyclic carbaldehyde

S. No.	Substrate	Product	Conversion %	
			$MgO\text{-}Al_2O_3$	$MgO\text{-}Al_2O_3.ZrO_2$
1	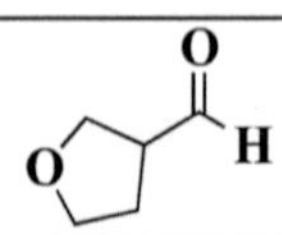	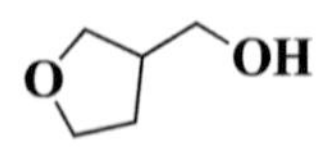	69	90
2	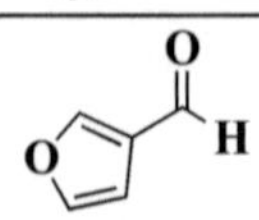	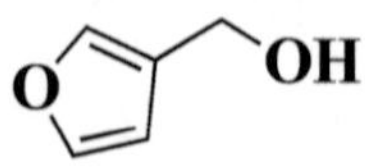	97	87
3	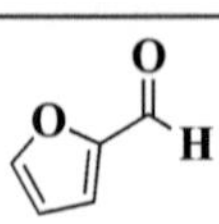	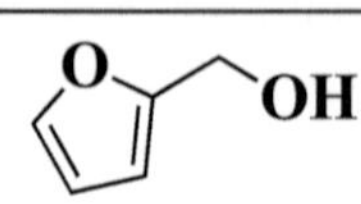	99	95
4	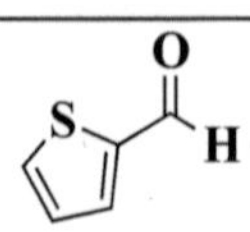	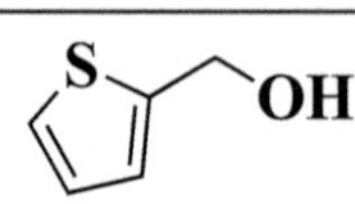	91	98
5	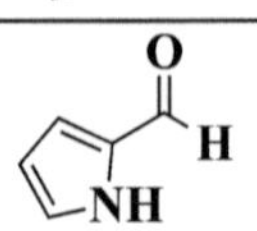	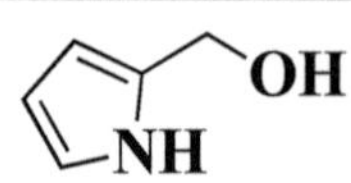	89	100

Hydrous zirconia demonstrates remarkable efficacy as a hydrogen transfer catalyst. In the case of ketones, the presence of an α-substituted methyl group impedes the reaction. Consequently, cyclohexanone and 4-methylcyclohexanone exhibit faster reduction rates compared to 2-methylcyclohexanone. Furthermore, the presence of a double bond conjugated to a carbonyl group tends to decelerate certain reactions. Notably, cinnamaldehyde is selectively converted into cinnamyl alcohol [38]. Aldehydes and ketones can also be efficiently hydrogenated with 2-propanol in the vapor phase [39].

Hydrated zirconia demonstrates effectiveness in reducing carboxylic acids and esters using 2-propanol [40]. These reactions were conducted in a flow reactor at 573 K, resulting in high yields of the corresponding alcohols, as illustrated in Table **5**. Additionally, the ester of the carboxylic acid was obtained as a by-product.

Table 5. Hydrogenation of carboxylic acid over zirconia.

S. No.	Reactant	Conversion %	Selectivity %	
			Alcohol	Ester
1	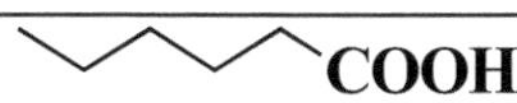	100	81	tr.
2		100	64	10
3		-	34	19
4	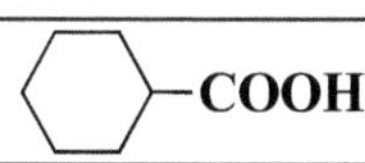	100	95	4
5	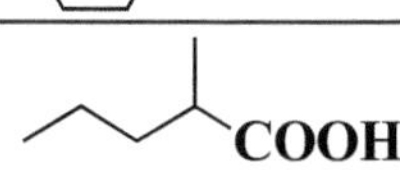	100	84	tr.
6	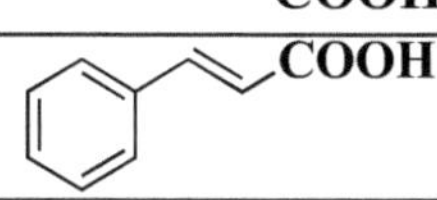	100	78	tr.
7	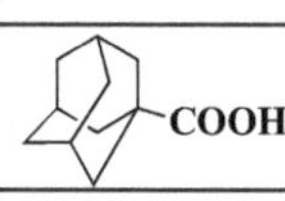	89	53	2
8	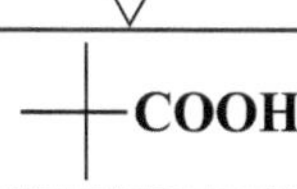	100	98	tr.
9	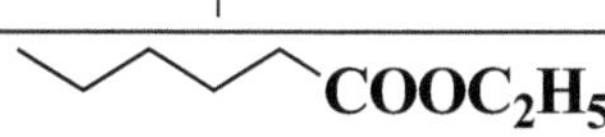	100	80	--
10	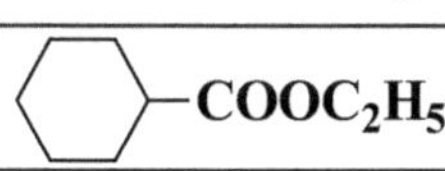	100	92	--
11	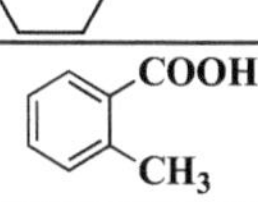	43	0	67

(Table 5) cont.....

S. No.	Reactant	Conversion %	Selectivity %	
			Alcohol	Ester
12	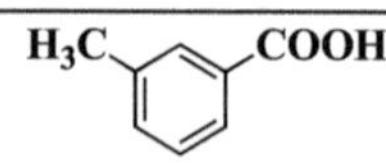	98	25	59
13	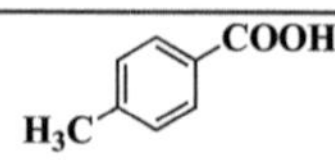	100	31	33

Reduction of Nitro-compounds and Nitriles with Alcohols

Nitro-compounds can be hydrogenated into the corresponding amines using alcohols over MgO [41]. In the nitrobenzene reaction, the order of reactivity for alcohol hydrogen donation was observed as follows: methanol ≈ 2-propanol > 2-butanol > 1-propanol > ethanol > 1-butanol > isobutyl alcohol.

Nitriles undergo reduction into the corresponding alcohols with 2-propanol in the vapor phase at 573 K using hydrous zirconia [42]. The reduction of aliphatic nitriles proceeds efficiently to yield the corresponding alcohols. Similarly, aromatic nitriles are converted into the corresponding alcohols as primary products. Furthermore, the reduction of cyanopyridines predominantly yields the corresponding picolines.

Reduction with Hydrocarbons

Hydrogen transfer reductions can be achieved using hydrocarbons like cyclohexene or *n*-alkanes over MgO [43]. Additionally, the hydrogenation of 1,3-butadiene with 1,3-cyclohexadiene over metal oxides such as MgO, CeO_2, and ZrO_2 has been investigated [44, 45]. The rates of transfer hydrogenation with 1,3-cyclohexadiene are notably higher by one or two orders of magnitude compared to hydrogenation with H_2. Specifically, over ZrO_2, the primary product in transfer hydrogenation is 1-butene, whereas hydrogenation with H_2 results in 2-butenes as the main products.

ADVANCEMENT OF SOLID BASE IN REDUCTION REACTION

Recent Breakthroughs in Catalyst Development for the Conversion of CO_2 to Methanol through Hydrogenation

A catalyst suited for CO_2 hydrogenation into methanol must maintain stability under high water partial pressures and outperform the Reverse Water-Gas Shift (RWGS) reaction [46]. Although initially perceived as a significant influencer of methanol selectivity [47], water vapor underscores the critical importance of the catalyst's resilience in such environments.

Cu-Based Catalysts

Originally, Cu-based catalysts, particularly $Cu/ZnO/Al_2O_3$, were utilized in the industrial production of methanol from syngas (CO, CO_2, and H_2) at temperatures around 250 °C and pressures ranging from 50 to 100 bar [48, 49]

The predominant view considered CO_2 as the primary carbon source for methanol [50]. This suggests the potential activity of Cu/ZnO catalysts in methanol synthesis through the hydrogenation of pure CO_2. Currently, Cu-based catalysts remain at the forefront of research in the field of CO_2 hydrogenation for methanol production [51]. Despite ongoing debates surrounding the reaction mechanisms of commercial $Cu/ZnO/Al_2O_3$ catalysts, notable advancements in catalyst development have been evident in recent years.

Cu/ZnO catalysts have been found to exhibit high activity and stability for methanol synthesis from both CO and CO_2. However, their combination demonstrates superior performance [52]. This enhanced performance can be attributed to various factors, such as stabilized Cu on ZnO, partially oxidized Cu^{2+} species and the creation of oxygen vacancies [53].

In_2O_3-Based Catalysts

Initially, In_2O_3 was found to demonstrate remarkable catalytic activity in methanol steam reforming, boasting significantly higher selectivity towards CO_2 compared to Cu/ZnO catalysts.

In 2012, Ge's team released a theoretical paper projecting In_2O_3's methanol synthesis potential *via* CO_2 hydrogenation [54]. Subsequent research identified oxygen vacancies (D4) as pivotal in CO_2 activation and stabilizing crucial intermediates. Since then, numerous studies have highlighted the superior activities and selectivity of either pure In_2O_3 or In_2O_3-based catalysts [55, 56].

Liu's team recently reported on an Ir catalyst supported by In_2O_3 (Ir/In_2O_3), demonstrating improved activity, methanol selectivity, and stability under reaction conditions of 300 °C, 5 MPa, and 2100 h^{-1}, with a maintained H_2/CO_2 ratio of 4 [57].

Rhodium supported on In_2O_3, synthesized using the deposition-precipitation method, also demonstrated notable activity and selectivity in the conversion of CO_2 to methanol [58].

Enhanced H_2 dissociation efficiency is induced by the introduction of Rh, with subsequent spillover to In_2O_3 promoting the formation of surface oxygen vacancies. Comparable effects were also observed for Ni, Pt, Rh, Au, and Ru when supported on In_2O_3.

Nanoalloy Catalysts

Shi and colleagues [59] synthesized a Cu-In intermetallic catalyst and explored the impact of reduction temperature on its reaction activity. They discovered that the reduction temperature significantly affects alloy formation, crystallite size, and gas adsorption. Specifically, CuO underwent reduction to metallic Cu at 250 °C, while the $Cu_{11}In_9$ alloy formed when the reduction temperature surpassed 300 °C. Furthermore, higher reduction temperatures led to a slight increase in the crystallite size of $Cu_{11}In_9$, accompanied by reduced H_2 adsorption capability.

Elevating the reduction pressure gradually shifts the maximum reduction rate (resulting in the simultaneous formation of copper (0) to lower temperatures. Methanol synthesis activity of the catalysts commenced when the total pressure exceeded 1 bar, accompanied by heightened formation of oxygen vacancies and other structural distortions in the ZnO phase.

Selective hydrogenation of nitro styrene to amino styrene over Pt/TiO_2 catalysts

Anilines, essential intermediates in the production of polyurethanes, dyes, pharmaceuticals, explosives, and agricultural products, are commonly manufactured through the catalytic hydrogenation of nitrobenzenes. In industrial applications, the reaction is commonly conducted in the gas phase with Cu or Ni catalysts, utilizing Hydrogen gas at pressures near or slightly above atmospheric and temperatures around 250 °C. Another approach involves performing the reaction in the liquid phase with supported metal catalysts such as Platinum, Palladium, or Nickel in organic solvents, using Hydrogen at higher pressures and temperatures around 50 °C [60].

For the green production of anilines, it is crucial to avoid the formation and accumulation of harmful reaction intermediates such as azoxybenzenes, nitrosobenzenes, azobenzene and phenylhydroxylamines which can occur with these catalysts, leading to low selectivity for anilines.

In 2011, Corma *et al.* investigated the selective hydrogenation of nitrostyrene to amino styrene using TiO_2-supported metal catalysts in toluene. These catalysts required high-temperature reduction to achieve high selectivity for amino styrene. However, this procedure occasionally resulted in lower hydrogenation activity due to strong metal-support interactions.

Non-chemo-selective hydrogenation metals like Pt, Ru, and Ni can be converted into chemo-selective catalysts by supporting nanosized crystals of these metals on TiO_2 through a straightforward catalyst activation process (Scheme **3**).

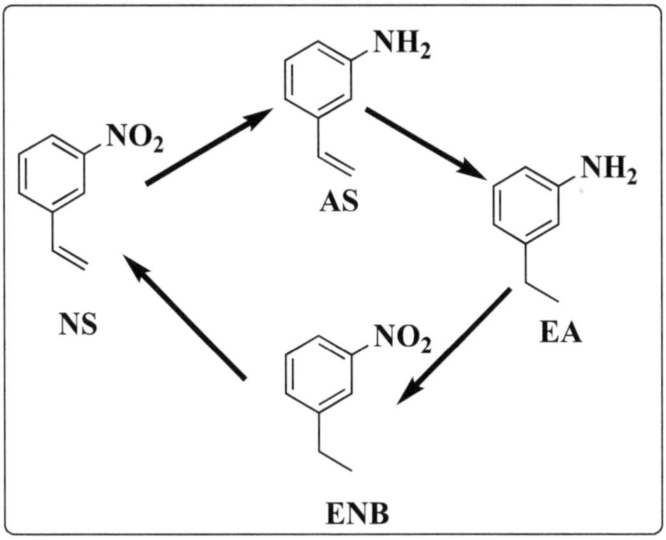

Scheme (3). Hydrogenation of nitrostyrene.

Recent Progress on Upgrading of Bio-oil to Hydrocarbons

Enhancing bio-oil is crucial for converting biomass into high-quality hydrocarbons, such as transportation fuels and petrochemicals, thereby reducing CO_2 emissions from fossil-based materials (Fig. **4**).

There have been significant recent advances in upgrading bio-oil using metal/zeolite bifunctional catalysts, involving studies with both model compounds and real bio-oil. This section will highlight the deoxygenation mechanisms at various metal active centers and the acid-catalyzed reactions, including isomerization and cracking, at zeolitic acid sites, which play a crucial role in the upgrading of bio-oil (Fig. **5**).

Phenolics upgrading

Lignocellulosic biomass, which includes cellulose, hemicellulose, and lignin, is attracting significant attention due to its sustainable and environmentally friendly properties. Through thermochemical processing, it can be converted into bio-oil, suitable for use as gasoline. This bio-oil contains up to 30 wt% of phenolic compounds, including phenols, guaiacols, anisole and their derivative (Fig. **6**).

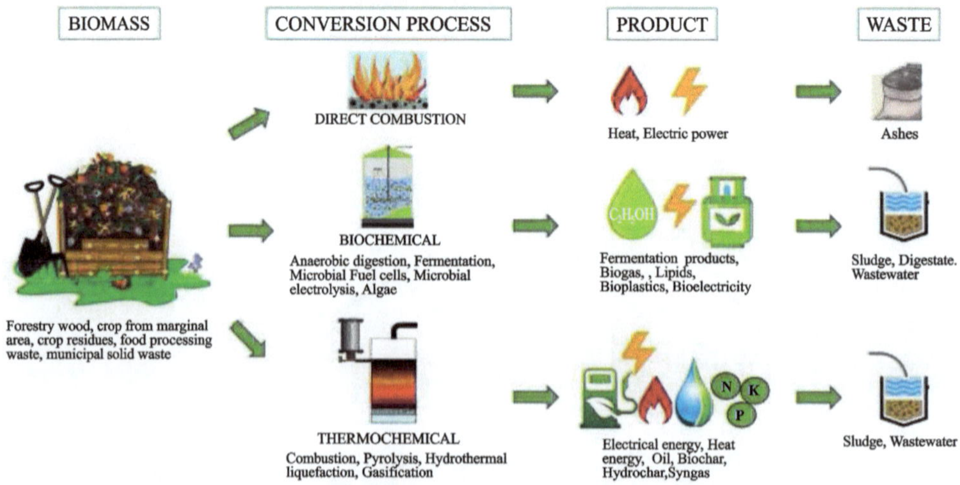

Fig. (4). Biomass conversion routes and product application.

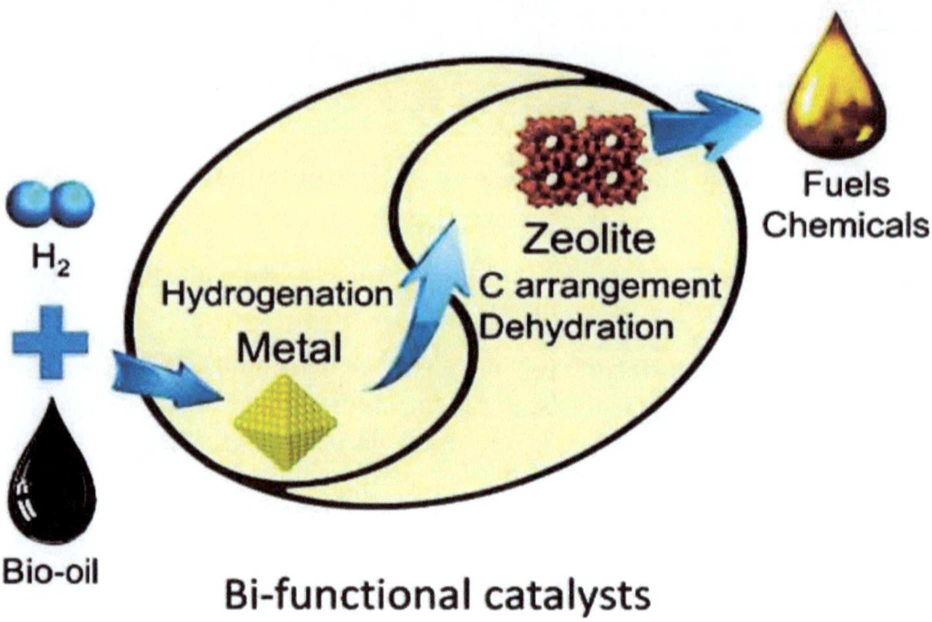

Fig. (5). Bifunctional catalyst for bio-oil upgrading.

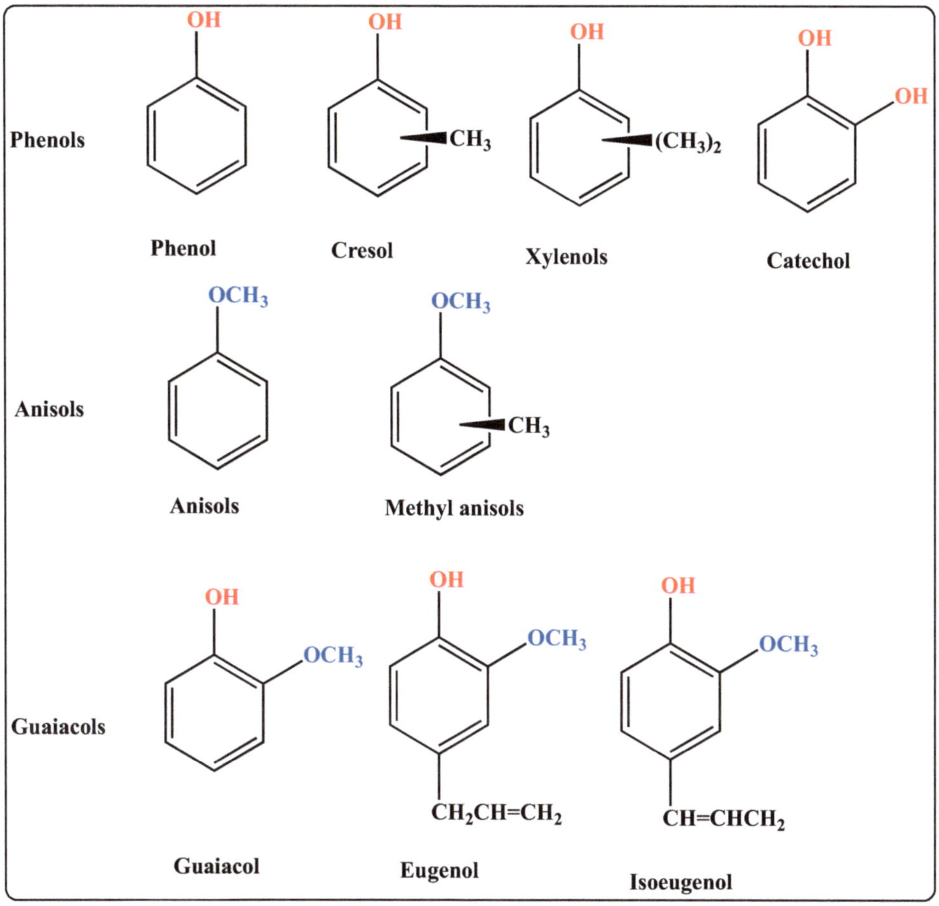

Fig. (6). Typical phenolic compounds in bio-oil.

This section examines and summarizes a range of bifunctional metal/zeolite catalysts for the upgrading of phenolic bio-oil, focusing on their integrated hydrogenation, hydrolysis, and acid-catalyzed reactions.

Resasco and colleagues [61] conducted a detailed study on the conversion of *m*-cresol over HY and H-ZSM-5 zeolites (with a Si/Al ratio of 40) within the temperature range of 400 to 600 °C. They proposed the reaction pathway depicted in (Scheme **4**).

m-Cresol underwent a series of acid-catalyzed reactions on acid zeolites. At lower temperatures, toluene was the primary product of pool cracking, while higher temperatures increased the selectivity towards benzene. This underscores the crucial role of reaction temperature in influencing product selectivity during phenolic conversion over acid sites on zeolites, irrespective of the degree of conversion achieved (Scheme **5**).

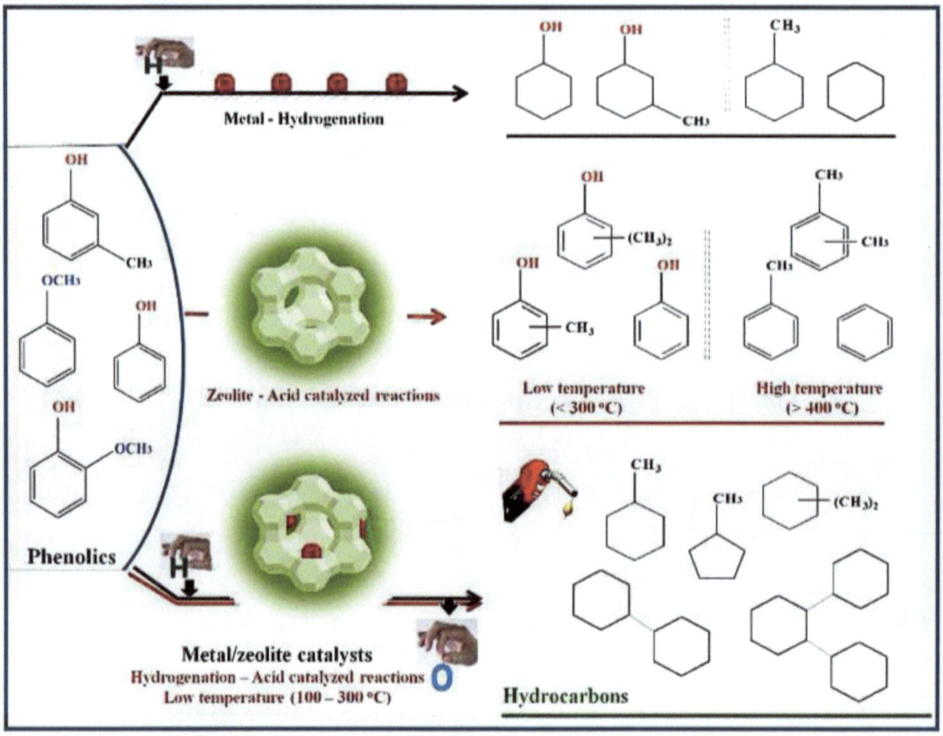

Scheme (4). Metal/zeolites during upgrading of phenolics based.

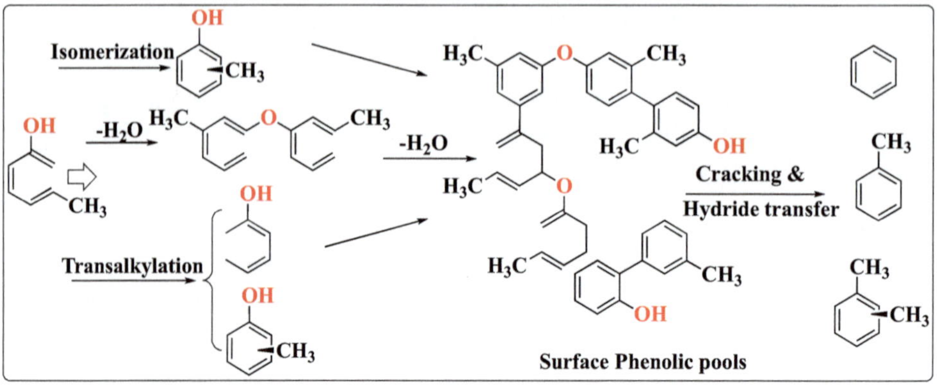

Scheme (5). Reaction pathway of *m*-cresol conversion using zeolite catalyst.

Typical bifunctional catalysts combine metal centers and acid sites in a single material, exerting significant influence over reaction pathways to enhance phenolic conversion and control product distribution through reaction route manipulation.

The synergy between metals and zeolites in bifunctional catalysts can customize the chemical reaction pathway, resulting in improved phenolic conversion and enhanced selectivity of final products.

Phenols undergo hydrodeoxygenation to form corresponding alcohols on metal centers. Then, on acid sites of zeolitic supports, these alcohols are dehydrated to yield corresponding olefins, which are subsequently hydrogenated back to alkanes on metal centers (Scheme **6**).

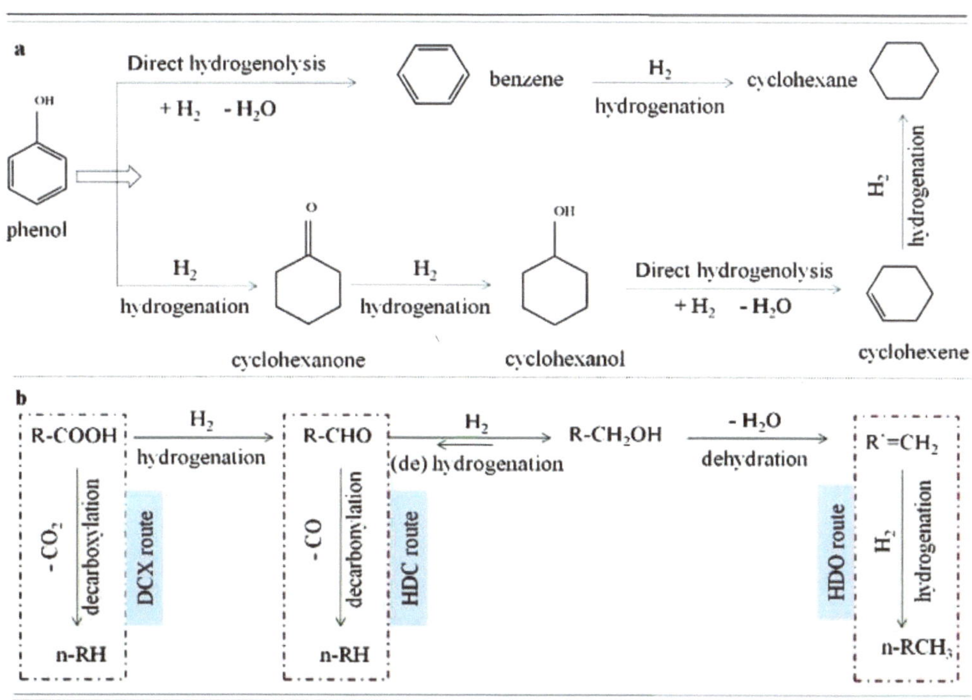

Scheme (6). Deoxygenation of phenol and carboxylic acids over catalysts.

Carboxylic acids upgrading

Stearic acid (C18) and palmitic acid (C16), originating primarily from lipids and microalgae in bio-oil, are utilized as model compounds for exploring the transformation of carboxylic acids into transportation fuels [62].

There are three pathways for catalytic deoxygenation of carboxylic acids, discussed in (Scheme 7).

Hydrodeoxygenation (HDO: R-COOH + $3H_2$ → R-CH_3 + $2H_2O$),

Hydro-de carbonylation (HDC: R-COOH + H_2 → R-H + CO + H_2O), and

Decarboxylation (DCX: R-COOH → R-H + CO_2)

Scheme (7). Catalytic deoxygenation of carboxylic acids.

Among these, hydrodeoxygenation (HDO) stands out as the most promising pathway for upgrading bio-oil, ensuring minimal carbon loss by selectively removing oxygen in the form of water.

During the upgrading of carboxylic acids (Fig. 7), both the chemical removal of oxygen atoms and the appropriate rearrangement of carbon chains must be achieved.

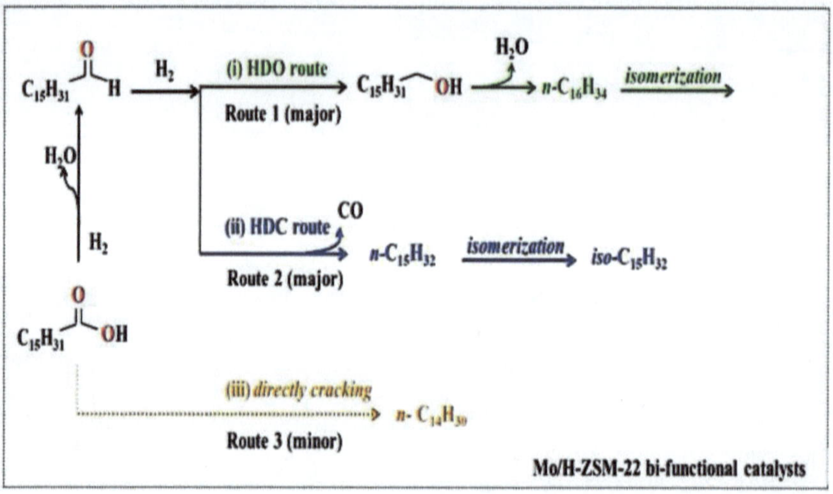

Fig. (7). Reaction pathway for upgrading of palmitic acid using H_2 and isomerization of products.

Researchers have incorporated zeolites into the refining industry to enhance deoxygenation catalysts, facilitating a direct conversion of carboxylic acids into premium transportation fuels like high-octane gasoline, diesel, and jet fuels.

Other oxygenated compounds Upgrading

In addition to phenolics and carboxylic acids, bio-oil contains various other components, such as aldehydes, ketones, furans/furfurals, polyhydric alcohols, and more, which are closely linked to the feedstock used.

Aldehyde

Prochazkova *et al.* [63] investigated the hydrodeoxygenation of aldehydes (including benzaldehyde, and its derivatives) using a bifunctional Pd/H-ZSM-5 catalyst.

Perez-Ramirez and colleagues [64, 65] investigated the application of alkali-metal-modified zeolites like USY, beta, and ZSM-5 in both liquid and gas phases for aldol condensation of propanal. Their findings underscored these modified zeolites as effective catalysts with strong activity and selective performance in enhancing short-chain aldehyde conversions.

However, aldol condensation alone does not fully eliminate oxygen atoms from bio-oil, requiring additional upgrading to transform it into hydrocarbons using metals supported by acidic catalysts.

Ketones

Witsuthammakul *et al.* [66] reported the use of bifunctional metal/zeolite catalysts for hydrogenation and dehydration offers a promising method to deoxygenate ketones into alkanes.

Bejblová *et al.* [67] investigated the hydrodeoxygenation of benzophenone using Pd supported on various materials, including activated carbon, alumina, ZSM-5 zeolites with different Si/Al ratios, and Al-MCM-41.

Furan/furfurals and derivatives

During the pyrolysis of biomass, anhydro-sugars are generated from cellulose and hemicellulose. These compounds then undergo dehydration to produce furans/furfurals or their derivatives, which are frequently present in bio-oil.

In recent years, there has been considerable interest in the upgrading of compounds based on furans and furfurals due to their environmental, economic, and strategic advantages [68, 69]. Abu-Omar *et al.* [70] studied the catalytic upgrading of furfural derivatives to alkanes using a Pd/H-beta catalyst. Cheng *et al.* [71] investigated furan as a model compound for catalytic fast pyrolysis of wood. Their work demonstrated that incorporating gallium into ZSM-5 zeolite increased aromatics yield by 40% compared to standard ZSM-5 (Fig. **8**).

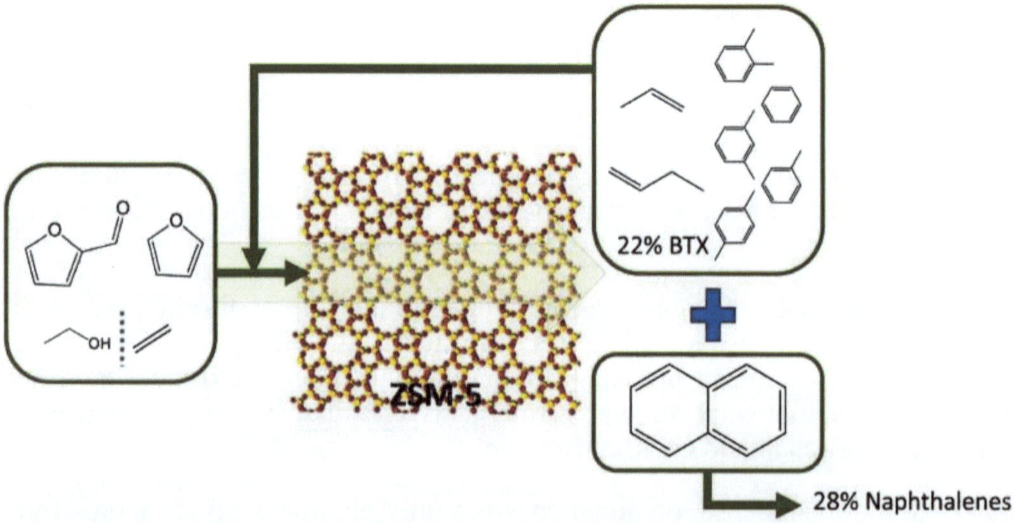

Fig. (8). Schematic diagram of furan derivatives into aromatics over ZSM catalysts.

Polyhydric alcohols

Zhang *et al.* [72, 73] designed a bifunctional Ni/H-ZSM-5 catalyst to generate iso-paraffins *via* aqueous-phase reforming of sorbitol. Their research highlighted the critical influence of both reaction temperatures 210-250 °C and catalyst calcination temperature 400 – 700 °C on catalytic performance (Scheme **8**).

Real bio-oil upgrading

Bio-oil is a complex mixture influenced by various factors such as the chemical composition of biomass feedstocks, reaction temperature, residence time, heating rate, efficiency of product removal systems, and more [74, 75].

Bio-oil, sourced from diverse feedstocks, contains key components such as phenolics, acids, aldehydes, ketones, sugars, furans, esters, and more, typically with an oxygen content ranging from 10% -40%. As a result, deoxygenation has become the primary method for upgrading bio-oil to produce high-quality transportation fuels or chemicals, highlighting its significant and ongoing importance.

Overall, the conversion of bio-oil into transportation fuels and petrochemicals requires two fundamental processes: the removal of oxygen atoms through metal catalysts and the rearrangement of C–C bonds facilitated by zeolitic catalysts.

Zeolites are extensively used in the refinery and petrochemical industries for their effectiveness in catalyzing C–C bond rearrangements. The overarching aim is to develop an integrated, single-step process that combines both functions, prompting the design of specialized bifunctional metal/zeolite catalysts for this application.

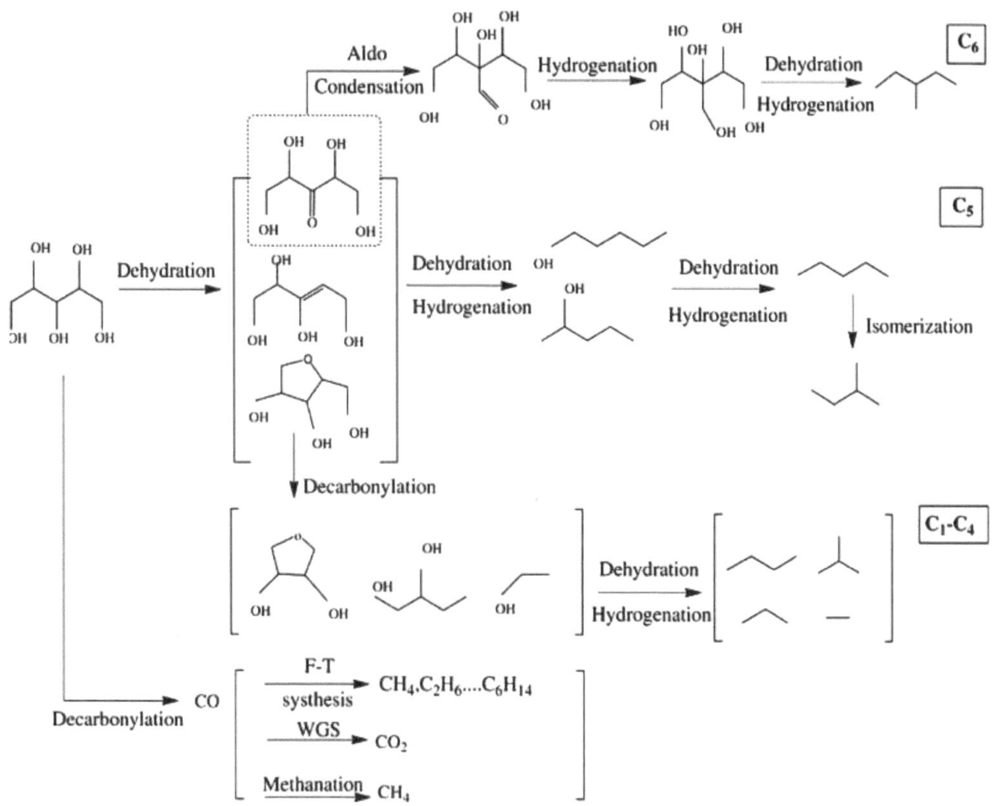

Scheme (8). Reaction pathway of xylitol conversion to alkene.

Hydrodeoxygenation of Phenol Over Sulfided CoMo Catalysts Supported on a Mixed Aluminium and titanium oxide

J. A. de los Reyes investigated the influence of mixed Al_2O_3–TiO_2 (Al/Ti = 2, denoted as AT2) supports on support–metal interactions in the oxide state, MoS_2 dispersion, and their effects on the hydrodeoxygenation of phenol [76]. The catalytic activity in the hydrodeoxygenation (HDO) of phenol was significantly enhanced by using mixed Al_2O_3-TiO_2 oxide (Al/Ti = 2, AT2) as a support (Table 6).

Table 6. Reaction rates & products conc. in the HDO of phenol over mixed oxide catalysts

Catalyst	Initial reaction rate [x 10^{-7} mol $(gmoS)^{-1}$]	Product Concentration (at 20% conversion) [x 10^{-4} mol L^{-1}]		
	-	Benzene	Cyclohexene	Cyclohexane
CoMo/Al_2O_3	27±4	27±2	9±1	10±1
CoMo/AT2	50±4	34±3	8±1	6±1

Recent advances in catalysts for photocatalytic selective hydrogenation of nitrobenzene to aniline

Selective Hydrogenation of Nitrobenzene (SHN) is essential for producing aniline, a valuable intermediate in textiles, pharmaceuticals, and dyes. Typically, the SHN reaction requires high temperatures and high hydrogen pressures in conventional thermal-driven catalytic processes. Tailoring effective photocatalysts is essential for SHN. So far, several photocatalysts, including TiO_2, Cu/graphene, CdS, and Eosin Y, have been investigated for their potential in photocatalytic SHN [77] (Fig. **9**).

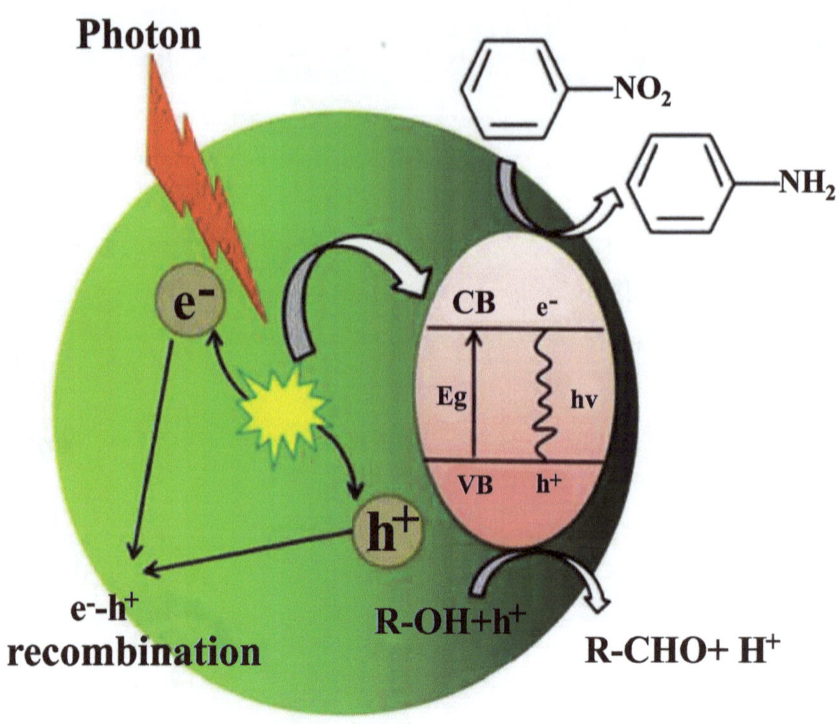

Fig. (9). Mechanism of selective hydrogenation of nitrobenzene by semiconductor.

CO_2 Hydrogenation into Methane with mixed oxide

Cerium-promoted NiMgAl mixed-oxide catalysts (NiCex-C, with x = 0, 1, 5, 10) were synthesized from quaternary hydrotalcite precursors for CO_2 hydrogenation to methane.

Among all, $NiCe_5$-C exhibited significantly better catalytic performance in low-temperature CO_2 hydrogenation, with activity three times higher than that of the Ce-free NiCe0-C catalyst.

The findings suggest that the presence of high medium-strength and Cerium-assisted surface Ni (0) species contributes to improved catalytic activity at low temperatures in the CO_2 hydrogenation to methane process.

This study demonstrated a straightforward method to prepare Ce-promoted NiMgAl catalysts derived from hydrotalcite *via* co-precipitation. These catalysts were designed to enhance CO_2 methanation at low temperatures, with the NiCe5-C sample showing superior activity in this regard [78].

The characterization results indicated that adding the right amount of Cerium to NiMgAl weakened the strong interaction between the metal and the support, leading to enhanced reduction of the NiO species.

Zeolites as Acid/Basic Solid Catalysts

Deacetylation of carbohydrate derivatives

Zeolites, which can be either naturally occurring or synthetic, are crystalline aluminosilicate solids characterized by microporous structures with pore sizes under 10 Å. Their framework is made up of SiO_4 and AlO_4 tetrahedrons linked together at their corners by shared oxygen ions.

Zeolite, being a solid catalyst, can be tunable and separated from the reaction mixture by simple filtration and can be recycled several times.

Ravindranathan Kartha and coauthors demonstrated that activated 4 Å molecular sieves can promote the selective deacetylation of various *O*-acetylated and *O*-benzylated carbohydrate derivatives (Scheme **9**). The procedure involves using methanol as a solvent over activated 4 Å molecular sieves, thereby generating methanolic sodium methoxide [79].

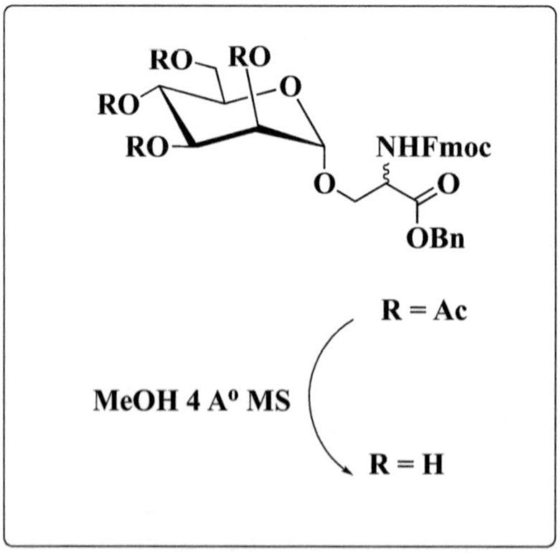

Scheme (9). De-*O*-acetylation of saccharide by activated 4 A° MS.

Microwave Irradiation in the Presence of a Basic Zeolite as Catalyst

Costarrosa and coauthors conducted an alkylation reaction on imidazole using various alkylating agents, with an alkaline-exchanged X-type zeolite as the catalyst, and promoted the process with microwave irradiation [80].

In his work, his team has taken 9 equiv. of alkylating agents (1-Br-butane, 2-B- -butane, and 1-Br-hexane) and was added to a Teflon vessel. The reactor vessel was then placed in a microwave oven and irradiated at 300 °C and 700 W for varying time intervals (1, 3, and 5 minutes). The reaction products were subsequently extracted using acetone [81] (Scheme **10**).

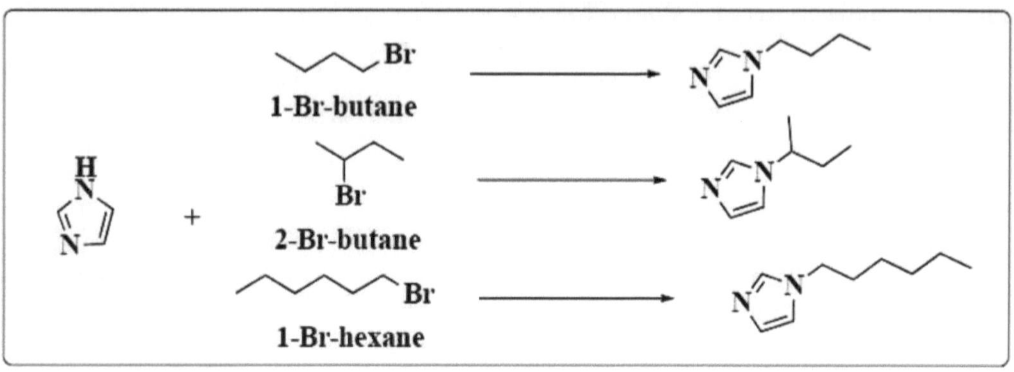

Scheme (10). *N*-alkylation of imidazole with different alkylating agents with basic zeolites.

Overall, after analyzing the experiments comprehensively, it was found that the catalytic activity depends on the basicity of the zeolite used, the polarizing ability of the exchanged cation, and the intensity of the microwave irradiation employed. The optimal catalyst for 1-Br-butane and 2-Br-butane was determined considering these parameters.

Different Base-Catalyzed Reactions Promoted by Methylammonium- Faujasite

As the electronegativity of the compensating cations decreases, the basicity of zeolites increases, leading to enhanced electron density on the framework oxygens. This results in a cationic zeolite basic strength order of $Li^+ < Na^+ < K^+ < Cs^+$.

With this in consideration, Martins and colleagues prepared Cs-exchanged FAU and methylammonium-exchanged FAU, the latter exhibiting the highest ion exchange capacity.

The catalytic efficacy of $(CH_3)\ NH_3^+$-FAU was particularly notable in various reactions, such as Knoevenagel and Claisen condensations, Schmidt condensation, and alcoholysis of propylene oxide (Scheme **11**), showing superior performance.

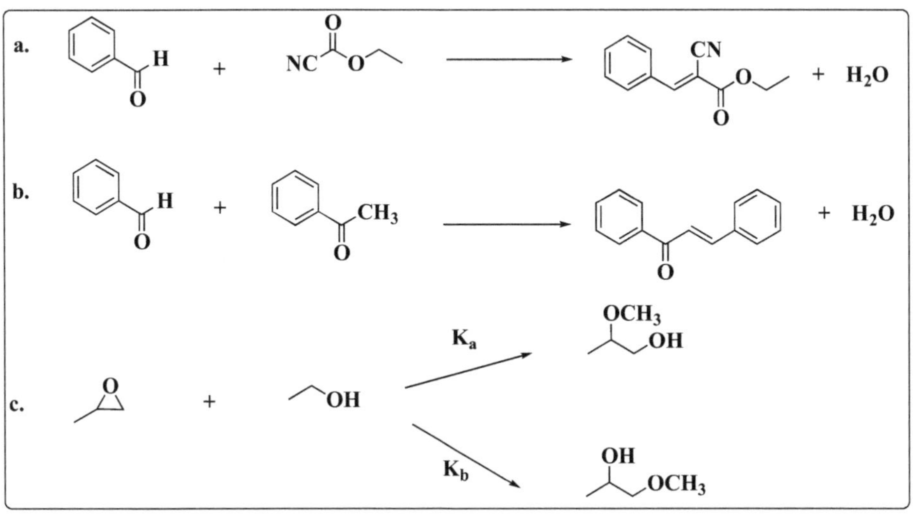

Scheme (11). Different base-catalyzed reactions promoted by FAU zeolites.

Basic Mesoporous Zeolite: An Efficient Catalyst for Several Reactions

Sarmah and coauthors examined the catalytic activity of this basic zeolite prepared under basic conditions and then treated with NH_4OH in its calcined form named ZSM-5 zeolite and several base-catalyzed reactions, including the synthesis of derivatives of styrenes, naphtho-pyrans, carbinol-amides and cyclic carbonates *via* Knoevenagel reactions, one-pot multicomponent condensation, amide hydroxylation, and cycloaddition reactions [82] (Scheme **12**).

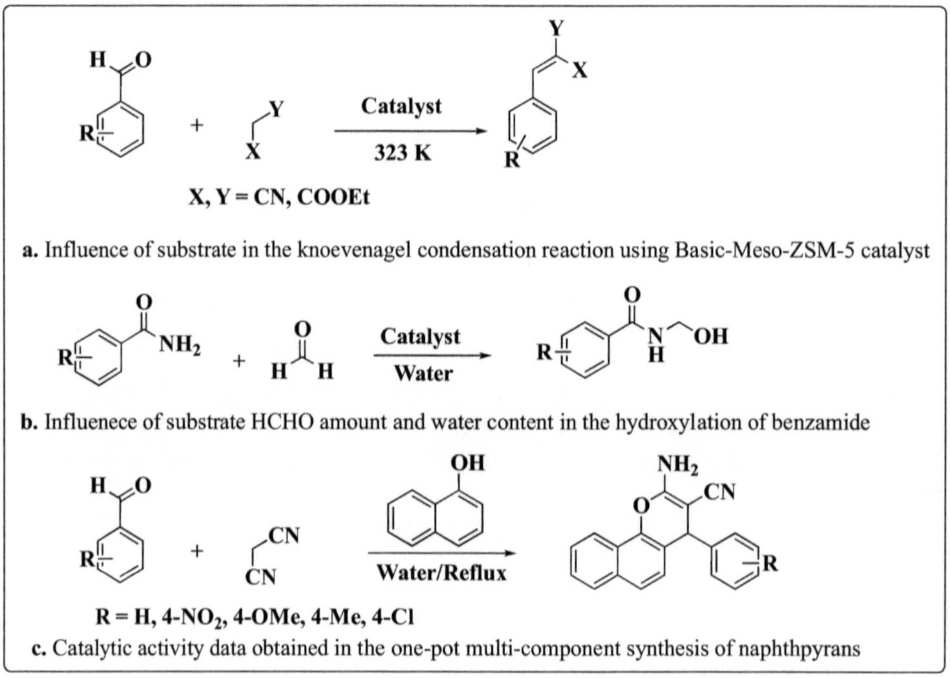

Scheme (12). Synthesis of substituted styrene, carbinol amides and others by using basic zeolites.

Poly substituted Cyclopropanes Synthesis Catalyzed by Basic Zeolite

Rama and his team described an efficient procedure for the one-pot, two-step synthesis of substituted cyclopropane. The reaction, which employs aromatic aldehydes, substituted benzyl halides, pyridine, derivatives of acetonitrile, and potassium-exchanged Y zeolites as catalysts (Scheme **13**), failed to produce any product without the basic zeolite. This was because the basicity of pyridine alone was insufficient to catalyze the final addition needed for cyclopropane formation, though it could facilitate the Knoevenagel reaction [83].

The yield of the catalyzed reactions correlated with the basicity degree of the zeolites, following the order: CsY > KY > NaY > LiY. Undeniably, KY was preferred over CsY for its cost-effectiveness and ease of handling.

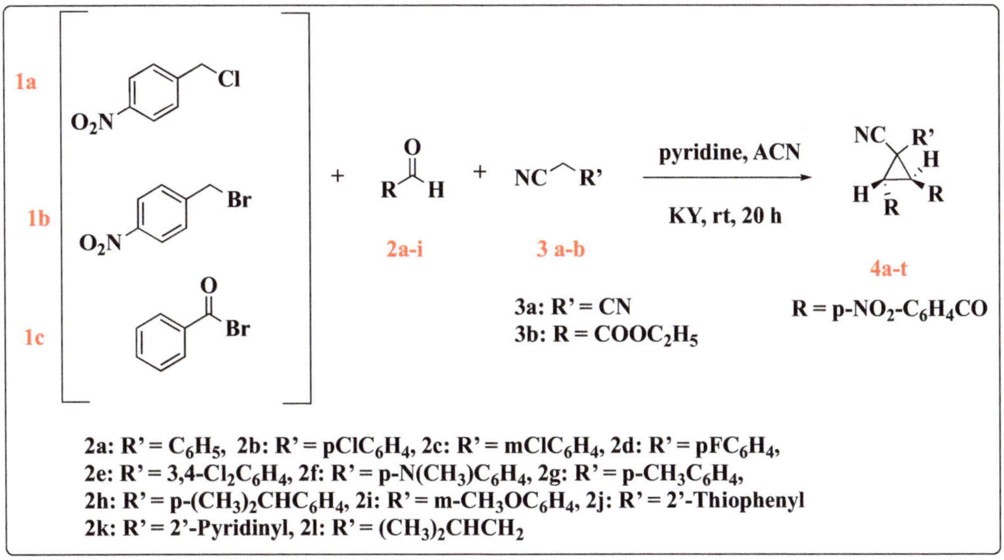

Scheme (13). Synthesis of poly-substituted cyclopropanes catalyzed by basic zeolites.

A plausible reaction pathway is proposed in (Scheme **14**). Initially, the reaction between *p*-nitrobenzyl bromide and pyridine, catalyzed by KY zeolite, leads to the formation of the intermediate *N-p*-nitrobenzylpyridinium salt. Simultaneously, the Knoevenagel condensation of the corresponding aromatic aldehyde with malononitrile, also catalyzed by KY zeolite, yields the arylidenemalononitrile.

In the second step, the pyridinium ylide, generated by the deprotonation of the pyridinium salt by KY zeolite, undergoes a Michael addition to arylidenemalononitrile, forming a carbanion intermediate. Due to steric repulsion, the bulky aryl and *p*-nitrophenyl groups preferentially adopt anti-positions. In the final step, an intramolecular cyclization occurs, where the carbanion displaces pyridine, leading to the formation of a cyclopropane ring with a trans configuration. This cyclization ensures that the substituents occupy the thermodynamically more stable transpositions.

Notably, the *in situ* formation of the pyridinium ylide occurs within the zeolite pore channels, as their small size favors the reaction between *p*-nitrobenzyl bromide and pyridine, along with the KY zeolite-catalyzed Knoevenagel condensation of the aromatic aldehyde with malononitrile. However, the subsequent Michael addition of the pyridinium ylide and the final cyclopropanation occur on the zeolite surface, as the larger size of polysubstituted cyclopropanes prevents them from forming within the confined pore spaces.

In conclusion, the developed protocol operates under mild conditions, delivering good yields with a straightforward work-up and allowing for catalyst reusability. Throughout three consecutive reactions, the catalyst maintained its full activity (Scheme **14**).

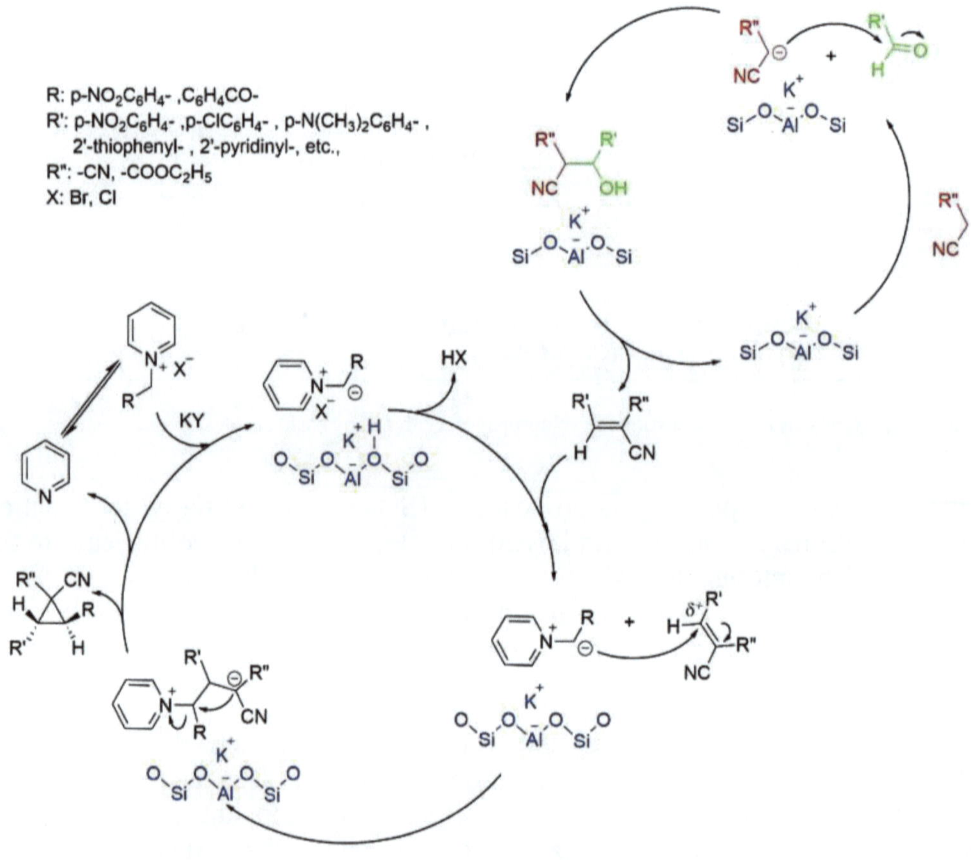

Scheme (14). Plausible mechanism for poly-substituted cyclopropanes catalyzed by basic zeolites.

Layered double hydroxides-based solid acid-base catalysts for eco-friendly catalytic transformations

Layered Double Hydroxides (LDHs) are gaining prominence as catalysts for acid-base reactions due to their unique structure and adjustable dual acid-base properties. They provide environmentally friendly and sustainable discretions to conventional liquid catalysts.

Layered Double Hydroxides (LDHs) feature a unique structure comprising positively charged metal hydroxide nanosheets, typically containing divalent metal cations intermingled with anions and water molecules. Adjusting their composition, structure, and interlayer anions can enhance the catalytic performance and stability of LDHs, and selectivity, making them a promising option for environmentally welcoming acid-base catalytic reactions [84, 85] (Fig. **10**).

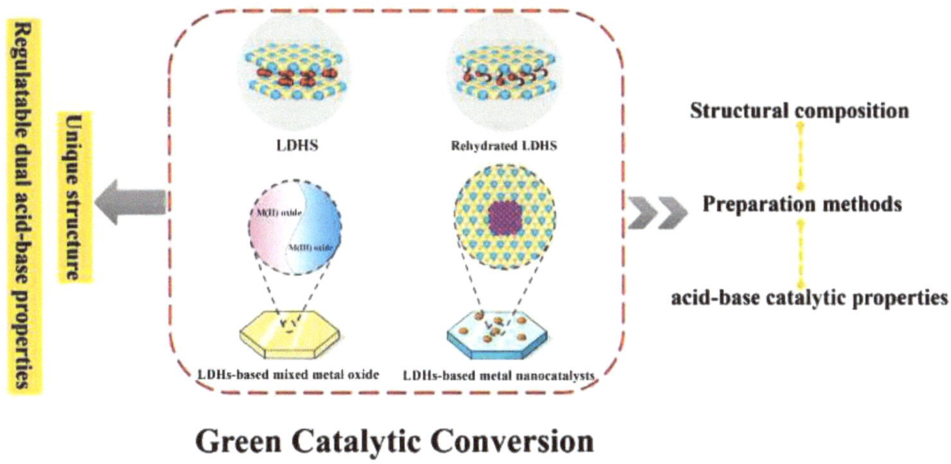

Fig. (10). Schematic illustration of LDHs as solid base catalyst for green transformations.

Application of LDH-based catalysts

Zhang *et al.* developed trimetallic NiCoFe LDHs through a one-step controllable synthesis. These LDHs exhibit efficient catalytic activity for both the oxygen Evolution Reaction (OER) and the selective oxidation of biomass-derived 5-hydroxymethylfurfural (HMF) to 2,5-furan-dicarboxylic acid (FDCA), as depicted in Fig. (**11**) [86].

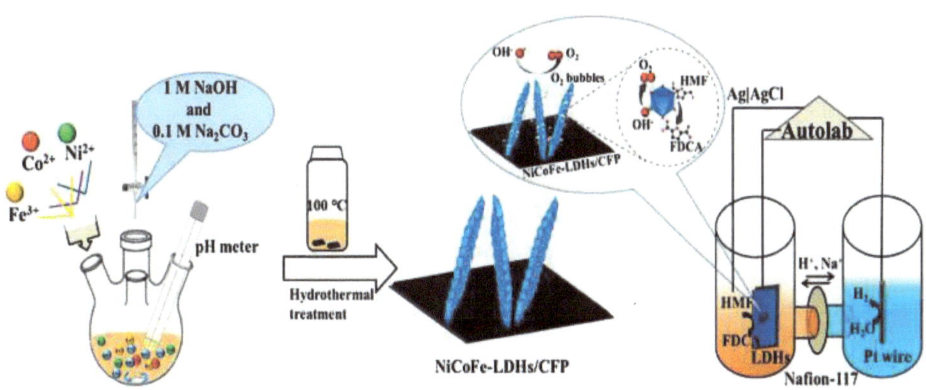

Fig. (11). Selective oxidation of biomass-based 5-hydroxymethylfurfural to 2,5-furan-dicarboxylic acid.

Wang *et al.* studied the catalytic transfer hydrogenation of furfural to furfuryl alcohol using NiFe LDHs, with 2-propanol serving as the hydrogen donor [87]. The catalyst achieved a high conversion rate, transforming 97.0% of furfural into furfuryl alcohol with a yield of 90.2% under conditions of 140°C for 5 hours.

Yabushita *et al.* explored the isomerization of glucose to fructose in ethanol solvent catalyzed by rehydrated MgAl LDHs, achieving a maximum yield of 56% with high selectivity reaching 80% [88].

LDHs-based metal nanocatalysts

5-hydroxymethylfurfural (HMF) holds promise as a biomass-derived intermediate sourced from fructose, glucose, and cellulose.

Wang *et al.* created diverse Cu-based catalysts featuring multiple interfaces derived from LDHs to control the reaction pathway and product selectivity during the hydrogenation of HMF [89]. Their research highlighted that the Cu/Co ratio at the metal nanoparticle interface played a crucial role in determining the hydrogenation of C=O bonds in HMF. (Scheme **15**) illustrates the varying hydrogenation pathways of HMF on these catalysts.

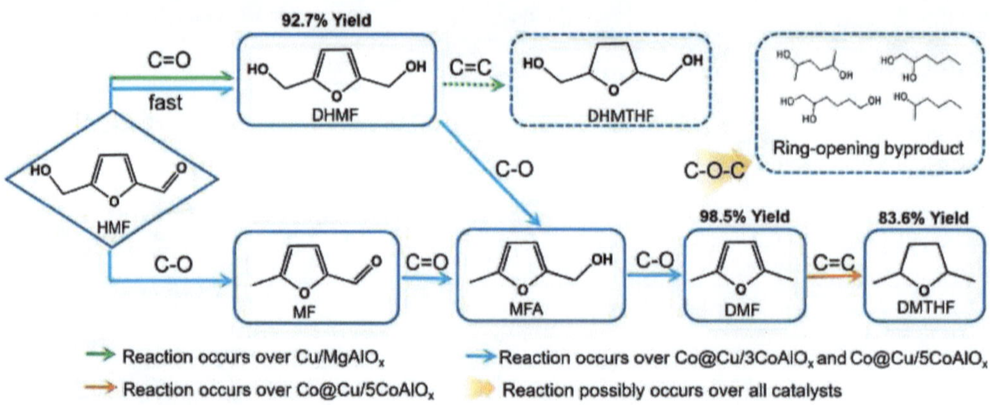

Scheme (15). Reaction pathway for hydrogenation of HMF on different catalysts.

Zhe *et al.* developed uniformly distributed Co-based catalysts using *in situ* H_2 reduction of LDHs, applying them for the efficient selective hydrogenation of HMF to DMF [90]. The Co-based catalysts demonstrate significant activity in hydrogenating C−O and C=O bonds, highlighting their potential for biomass utilization.

CONCLUSION AND FUTURE PERSPECTIVES

Recent advancements in solid base catalysis for reduction reactions have underscored their efficacy in promoting selective and efficient transformations. These catalysts have demonstrated superior performance in facilitating the reduction of various functional groups, such as C=O and C−O bonds, crucial for biomass conversion and fine chemical synthesis. Compared to conventional liquid bases, solid base catalysts offer enhanced stability, recyclability, and reduced environmental footprint, making them highly attractive for sustainable chemical processes.

Looking ahead, future research could focus on several key areas to further enhance solid base catalysis. Firstly, there is a need to explore novel materials and modify existing ones to improve catalytic efficiency and selectivity under diverse reaction conditions. Secondly, advancing understanding of the structure-function relationships in solid base catalysts will aid in designing tailored catalysts with optimized performance. Moreover, integrating these catalysts into tandem or cascade reactions could unlock new pathways for complex molecule synthesis with minimal waste generation.

In conclusion, the ongoing development of solid base catalysis represents a pivotal advancement in the field of reduction reactions. Continued research efforts aimed at refining catalyst design, elucidating reaction mechanisms, and expanding application scopes will be essential for realizing the full potential of solid base catalysis in sustainable chemistry and industrial processes.

ABBREVIATIONS

AlO_4	Aluminate
D	Deuterium
DMSO	Dimethyl sulphoxide
ESR	Electron Spin Resonance
FDCA	2,5-Furandicarboxylic Acid
FUS	Methylammonium-Faujasite Zeolites
H-ZSM-5	Zeolite Socony Mobil-5
HDO	Hydrodeoxygenation
HMF	5-Hydroxymethylfurfural
IR	Infrared
LDH	Layered Double Hydroxide
MEK	Methyl Ethyl Ketone
MPV	Meerwein–Ponndorf–Verley

NH_4OH	Ammonium Hydroxide
OER	Oxygen Evolution Reaction
RWGS	Reverse Water Gas Shift
SHN	Hydrogenation of Nitrobenzene
SiO_4	Silicate
UV	Ultraviolet
XPS	X-ray Photoelectron Spectroscopy

REFERENCES

[1] Pines, H.; Haag, W. Communications - stereoselectivity in the carbanion-catalyzed isomerization of 1-butene. *J. Org. Chem.,* **1958**, *23*(2), 328-329. [http://dx.doi.org/10.1021/jo01096a627]

[2] Tanabe, K.; Yoshii, N.; Hattori, H. Catalytic activity and selectivity of evacuated calcium oxide for isomerization of but-1-ene. *J. Chem. Soc. D,* **1971**, (9), 464. [http://dx.doi.org/10.1039/c29710000464]

[3] Hattori, H.; Yoshii, N.; Tanabe, K. *Proceedings of the 5th international congress on catalysis,* **1972**Miami Beach, p. 233.

[4] Tanabe, K.; Misono, M.; Ono, Y.; Hattori, H. *New Solid Acids and Bases*; Kodansha/Elsevier, **1989**.

[5] Laszlo, P. Catalysis of organic reactions by inorganic solids. *Acc. Chem. Res.,* **1986**, *19*(4), 121-127. [http://dx.doi.org/10.1021/ar00124a004]

[6] Baba, T.; Kato, A.; Yuasa, H.; Toriyama, F.; Handa, H.; Ono, Y. New Si–C bond forming reactions over Solid-Base Catalysts. *Catal.* Today, **1998**, 44 (1–4), 271–276. [http://dx.doi.org/10.1016/S0920-5861(98)00199-0]

[7] Baba, T.; Yuasa, H.; Handa, H.; Ono, Y. Catalytic Si-C bond formation by nucleophilic substitution at silicon by benzyl anions generated over KNH2 loaded on alumina. *Catal. Lett.,* **1998**, *50*(1/2), 83-85. [http://dx.doi.org/10.1023/A:1019006816637]

[8] Dwyer, J.; Schofield, H. *in Acidity and Basicity of Solids, Theory, Assessment and Utility*; (ed. by J. Fraissard, L. Petrakis), NATO ASI Ser., C444, Kluwer Academic Publishers, **1994**, p. 13.

[9] Hattori, H. Solid Base Catalysts: Generation of basic sites and application to organic synthesis. *Appl. Catal.*, A **2001**, 222 (1–2), 247–259. [http://dx.doi.org/10.1016/S0926-860X(01)00839-0]

[10] Ono, Y.; Baba, T. Selective reactions over solid base catalysts. *Catal. Today,* **1997**, *38*(3), 321-337. [http://dx.doi.org/10.1016/S0920-5861(97)81502-5]

[11] Tanabe, K.; Hoelderich, W.F. Industrial application of solid acid–base catalysts. *Appl. Catal. A Gen.,* **1999**, *181*(2), 399-434. [http://dx.doi.org/10.1016/S0926-860X(98)00397-4]

[12] Hattori, H.; Tanaka, Y.; Tanabe, K. A novel catalytic property of magnesium oxide for hydrogenation of 1,3-butadiene. *J. Am. Chem. Soc.,* **1976**, *98*(15), 4652-4653. [http://dx.doi.org/10.1021/ja00431a056]

[13] Tanaka, Y.; Imizu, Y.; Hattori, H.; Taanabe, K. *Proceedings of the 7th International Congress on Catalysts,* Tokyo**1980**, p. 1254.

[14] Voltz, S.E. The catalytic properties of supported sodium and lithium catalysts. *J. Phys. Chem.,* **1957**, *61*(6), 756-758. [http://dx.doi.org/10.1021/j150552a012]

[15] Hattori, H.; Tanaka, Y.; Tanabe, K. hydrogenation of olefins catalyzed by alkaline earth metal oxides. *Chem. Lett.,* **1975**, *4*(7), 659-660.
[http://dx.doi.org/10.1246/cl.1975.659]

[16] Minachev, K.; Khodakov, Yu.S.; Nakhshunov, V.S. Activity and mechanism of the catalytic action of rare-earth oxides in low-temperature ethylene hydrogenation. *J. Catal.,* **1977**, *49*(2), 207-215.
[http://dx.doi.org/10.1016/0021-9517(77)90256-1]

[17] Conner, W.C.; Innes, R.A.; Kokes, R.J. Catalysis of addition of deuterium to ethylene over zinc oxide. *J. Am. Chem. Soc.,* **1968**, *90*(24), 6858.
[http://dx.doi.org/10.1021/ja01026a061]

[18] Conner, W.C.; Kokes, R.J. Addition of hydrogen-deuterium mixtures to ethylene over chromia and zinc oxide. *J. Phys. Chem.,* **1969**, *73*(7), 2436-2438.
[http://dx.doi.org/10.1021/j100727a060]

[19] Dent, A.L.; Kokes, R.J. Hydrogenation of ethylene by zinc oxide. I. Role of slow hydrogen chemisorption. *J. Phys. Chem.,* **1969**, *73*(11), 3772-3780.
[http://dx.doi.org/10.1021/j100845a035]

[20] Tanaka, Y.; Hattori, H.; Tanabe, K. Selective hydrogenation of 1,3-butadiene over alkaline earth metal oxides. *Chem. Lett.,* **1976**, *5*(1), 37-38.
[http://dx.doi.org/10.1246/cl.1976.37]

[21] Hattori, H.; Tanaka, Y.; Tanabe, K. A novel catalytic property of magnesium oxide for hydrogenation of 1,3-butadiene. *J. Am. Chem. Soc.,* **1976**, *98*(15), 4652-4653.
[http://dx.doi.org/10.1021/ja00431a056]

[22] Imizu, Y.; Sato, K.; Hattori, H. A mechanistic study of hydrogenation of conjugated dienes over a lanthanum oxide catalyst. *J. Catal.,* **1982**, *76*(1), 65-74.
[http://dx.doi.org/10.1016/0021-9517(82)90237-8]

[23] Nakano, Y.; Yamaguchi, T.; Tanabe, K. Hydrogenation of conjugated dienes over ZrO_2 by H_2 and cyclohexadiene. *J. Catal.,* **1983**, *80*(2), 307-314.
[http://dx.doi.org/10.1016/0021-9517(83)90256-7]

[24] Nakano, Y.; Yamaguchi, T.; Tanabe, K. Hydrogenation of conjugated dienes over ZrO_2 by H_2 and cyclohexadiene. *J. Catal.,* **1983**, *80*(2), 307-314.
[http://dx.doi.org/10.1016/0021-9517(83)90256-7]

[25] Yokoyama, T.; Setoyama, T.; Fujita, N.; Nakajima, M.; Maki, T. *Appl. Catal., A,* **1992**, *88*, 149.
https://www.sciencedirect.com/science/article/abs/pii/0926860X9280212U

[26] Kondo, J. N, Ding, K. Maruya, K. Domen, T. Yokoyama, N. Fujita, T. Maki. Infrared study of hydrogenation of benzoic acid to benzaldehyde on ZrO_2 catalysts *Bull. Chem. Soc. Jpn.,* **1993**, *66*, 3085.
[http://dx.doi.org/10.1246/bcsj.66.3085]

[27] Sakata, Y.; van Tol-Koutstaal, C.A.; Ponec, V. Selectivity problems in the catalytic deoxygenation of benzoic acid. *J. Catal.,* **1997**, *169*(1), 13-21.
[http://dx.doi.org/10.1006/jcat.1997.1665]

[28] Niiyama, H.; Echigoya, E. *Bull. Chem. Soc. Jpn.,* **1972**, *45*, 939.

[29] Ivanov, V.A.; Bachelier, J.; Audry, F.; Lavalley, J.C. Study of the Meerwein—Pondorff—Verley reaction between ethanol and acetone on various metal oxides. *J. Mol. Catal.,* **1994**, *91*(1), 45-59.
[http://dx.doi.org/10.1016/0304-5102(94)00017-4]

[30] Okamoto, Y.; Imanaka, T.; Teranishi, S. Hydrogen-transfer reaction between ketones and alcohols over metal oxide catalysts. *Bull. Chem. Soc. Jpn.,* **1972**, *45*(10), 3207-3208.
[http://dx.doi.org/10.1246/bcsj.45.3207]

[31] Posner, G.H.; Runquist, A.W.; Chapdelaine, M.J. Organic reactions at alumina surfaces. 6. Isopropyl

alcohol and diisopropylcarbinol on dehydrated alumina as reagents for very selective carbonyl reductions. *J. Org. Chem.*, **1977**, *42*(7), 1202-1208. [http://dx.doi.org/10.1021/jo00427a022]

[32] Ramana, D.V.; Pillai, C.N. Hydrogen transfer reactions. I. Reduction of carbonyl compounds by alcohols catalyzed by alumina. *Can. J. Chem.*, **1969**, *47*(19), 3705-3707. [http://dx.doi.org/10.1139/v69-616]

[33] Kijeski, J.; Gliski, M.; Czarecki, J. *J. Chem. Soc., Perkin Trans.*, **1991**.

[34] Joshi, T. M.; Raja, T.; Sreekumar, K.; Rao, B. S. *Appl. Catal., A,* **2001**, *157*, 193.

[35] Joshi, T.M.; Raja, T.; Rao, B.S. Catalytic transfer reduction (CTR) of alkyl alkyl, alkyl aryl, cyclic and unsaturated ketones over calcined Mg–Al hydrotalcites. *J. Mol. Catal.*, **2001**, *168*, 187. [http://dx.doi.org/10.1016/S1381-1169(00)00527-6]

[36] Vu, T.T.H.; Kumbhar, P.S.; Figueras, F. Base-catalysed hydrogenation of sulphur-containing aldehydes. *Adv. Synth. Catal.*, **2003**, *345*(4), 493-496. [http://dx.doi.org/10.1002/adsc.200390056]

[37] Jiménez-Sanchidrián, C.; Hidargo, J. M.; Ruiz, J. R. *Appl. Catal., A,* **2006**, *303*, 23.

[38] Liu, S.H.; Jaenicke, S.; Chuah, G.K. Hydrous zirconia as a selective catalyst for the meerwein–ponndorf–verley reduction of cinnamaldehyde. *J. Catal.*, **2002**, *206*(2), 321-330. [http://dx.doi.org/10.1006/jcat.2001.3480]

[39] Shibagaki, M.; Takahashi, K.; Kuno, H.; Kawakami, H.; Matsushita, H. Vapor-phase reduction of aldehydes and ketones with 2-propanol over hydrous zirconium oxide. *Chem. Lett.*, **1988**, *17*(10), 1633-1636. [http://dx.doi.org/10.1246/cl.1988.1633]

[40] Takahashi, K.; Shibagaki, M.; Kuno, H.; Matsushita, H. Reduction of carboxylic acid with 2-propanol over hydrous zirconium oxide. *Chem. Lett.*, **1989**, *18*(7), 1141-1144. [http://dx.doi.org/10.1246/cl.1989.1141]

[41] Kijeski, J.; Gliski, M.; Winiewski, R.; Murghani, S. Stud. Surf. Sci. Catal., 1991, 59, 169. $_4$**2.** K. Takahashi, M. Shibagaki, H. Matsushita. *Chem. Lett.*, **1990**, *19*, 311.

[43] Gliski, M.; Wolski, R. Catalytic hydrogen transfer over magnesia, xvii. a hydrocarbons as hydrogen donors. *React. Kinet. Catal. Lett.*, **2001**, *73*, 21. [http://dx.doi.org/10.1023/A:1013900232129]

[44] Nakano, Y.; Yamaguchi, T.; Tanabe, K. Hydrogenation of conjugated dienes over ZrO_2 by H_2 and cyclohexadiene. *J. Catal.*, **1983**, *80*(2), 307-314. [http://dx.doi.org/10.1016/0021-9517(83)90256-7]

[45] Shima, H.; Yamaguchi, T. Transfer hydrogenation of conjugated dienes over oxide catalysts. *J. Catal.*, **1984**, *90*(1), 160-164. [http://dx.doi.org/10.1016/0021-9517(84)90098-8]

[46] Yang, H.; Zhang, C.; Gao, P.; Wang, H.; Li, X.; Zhong, L.; Wei, W.; Sun, Y. A review of the catalytic hydrogenation of carbon dioxide into value-added hydrocarbons. *Catal. Sci. Technol.*, **2017**, *7*(20), 4580-4598. [http://dx.doi.org/10.1039/C7CY01403A]

[47] Khayoon. M.S.; Hameed. B.H. Mg1+xCa1−xO_2 as reusable and efficient heterogeneous catalyst for the synthesis of glycerol carbonate via the transesterification of glycerol with dimethyl carbonate. *Appl. Catal. A Gen.* **2013**, *466*, 272–281.

[49] Klier, K. Methanol synthesis. *Adv. Catal.*, **1982**, *31*, 243-313. [http://dx.doi.org/10.1016/S0360-0564(08)60455-1]

[50] Chinchen, G.C.; Denny, P.J.; Parker, D.G.; Spencer, M.S.; Whan, D.A. Mechanism of methanol synthesis from $CO_2/CO/H_2$ mixtures over copper/zinc oxide/alumina catalysts: use of 14C-labelled

reactants. *Appl. Catal.,* **1987**, *30*(2), 333-338.
[http://dx.doi.org/10.1016/S0166-9834(00)84123-8]

[51] Zhang, L.; Zhang, Y.; Chen, S. Effect of promoter SiO_2, TiO_2 or SiO_2-TiO_2 on the performance of CuO-ZnO-Al_2O_3 catalyst for methanol synthesis from CO_2 hydrogenation. *Appl. Catal. A Gen.,* **2012**, *415-416*, 118-123.
[http://dx.doi.org/10.1016/j.apcata.2011.12.013]

[52] Perez, S.; Ramirez, P.J.; Gutierrez, R.A.; Stacchiola, D.J.; Vines, F.; Liu, P.; Illas, F.; Rodriguez, J.A. The conversion of CO_2 to methanol on orthorhombic β-Mo2C and Cu/β-Mo2C catalysts: mechanism for admetal induced change in the selectivity and activity. *Catal. Sci. Technol.,* **2016**, *6*, 6766.
[http://dx.doi.org/10.1039/C5CY02143J]

[53] Elnabawy, A.O.; Schimmenti, R.; Cao, A.; Nørskov, J.K. Why ZnO is the support for Cu in methanol synthesis? A systematic study of the strong metal support interactions. *ACS Sustain. Chem.& Eng.,* **2022**, *10*(4), 1722-1730.
[http://dx.doi.org/10.1021/acssuschemeng.1c07980]

[54] Ye, J.; Liu, C.; Ge, Q. DFT study of CO $_2$ adsorption and hydrogenation on the in $_2$ O $_3$ surface. *J. Phys. Chem. C,* **2012**, *116*(14), 7817-7825.
[http://dx.doi.org/10.1021/jp3004773]

[55] Frei, M.S.; Mondelli, C.; García-Muelas, R.; Kley, K.S.; Puértolas, B.; López, N.; Stewart, J.A.; Ferré, D.C. Pérez-Ramírez. Atomic-scale engineering of indium oxide promotion by palladium for methanol production *via* CO_2 hydrogenation. *J. Nat. Commun.,* **2019**, *10*, 3377.
[http://dx.doi.org/10.1038/s41467-019-11349-9]

[56] Frei, M.S.; Mondelli, C.; Garcia-Muelas, R. Morales- Vidal, J.; Philipp, M.;Safonova,O.V.; Lopez,N.; Stewart, J. A.; Ferre, D. C.; Perez-Ramirez. Nanostructure of nickel-promoted indium oxide catalysts drives selectivity in CO_2 hydrogenation. *J. Nat. Commun.,* **2021**, *12*, 1960.
[http://dx.doi.org/10.1038/s41467-021-22224-x]

[57] Shen, C.; Sun, K.; Zhang, Z.; Rui, N.; Jia, X.; Mei, D.; Liu, C. Highly active Ir/In $_2$ O $_3$ catalysts for selective hydrogenation of CO $_2$ to methanol: Experimental and theoretical studies. *ACS Catal.,* **2021**, *11*(7), 4036-4046.
[http://dx.doi.org/10.1021/acscatal.0c05628]

[58] Wang, J.; Sun, K.; Jia, X.; Liu, C. CO_2 hydrogenation to methanol over Rh/In_2O_3 catalyst. *Catal. Today,* **2021**, *365*, 341-347.
[http://dx.doi.org/10.1016/j.cattod.2020.05.020]

[59] Shi, Z.; Tan, Q.; Tian, C.; Pan, Y.; Sun, X.; Zhang, J.; Wu, D. CO_2 hydrogenation to methanol over Cu-In intermetallic catalysts: Effect of reduction temperature. *J. Catal.,* **2019**, *379*, 78-89.
[http://dx.doi.org/10.1016/j.jcat.2019.09.024]

[60] Corma, A.; Serna, P.; Concepción, P.; Calvino, J.J. Transforming nonselective into chemoselective metal catalysts for the hydrogenation of substituted nitroaromatics. *J. Am. Chem. Soc.,* **2008**, *130*(27), 8748-8753.
[http://dx.doi.org/10.1021/ja800959g] [PMID: 18597431]

[61] Ausavasukhi, A.; Huang, Y.; To, A.T.; Sooknoi, T.; Resasco, D.E. Hydrodeoxygenation of m-cresol over gallium-modified beta zeolite catalysts. *J. Catal.,* **2012**, *290*, 90-100.
[http://dx.doi.org/10.1016/j.jcat.2012.03.003]

[62] Shi, Y.; Cao, Y.; Duan, Y.; Chen, H.; Chen, Y.; Yang, M.; Wu, Y. Upgrading of palmitic acid to iso-alkanes over bi-functional Mo/ZSM-22 catalysts. *Green Chem.,* **2016**, *18*(17), 4633-4648.
[http://dx.doi.org/10.1039/C6GC01367H]

[63] Procházková, D.; Zámostný, P.; Bejblová, M.; Červený, L.; Čejka, J. Hydrodeoxygenation of aldehydes catalyzed by supported palladium catalysts. *Appl. Catal. A Gen.,* **2007**, *332*(1), 56-64.
[http://dx.doi.org/10.1016/j.apcata.2007.08.009]

[64] Keller, T.C.; Isabettini, S.; Verboekend, D.; Rodrigues, e.g.; Pérez-Ramírez, J. Hierarchical high-silica zeolites as superior base catalysts. *Chem. Sci. (Camb.),* **2014**, *5*(2), 677-684. [http://dx.doi.org/10.1039/C3SC51937F]

[65] Keller, T.C.; Desai, K.; Mitchell, S.; Pérez-Ramírez, J. Design of base zeolite catalysts by alkali-metal grafting in alcoholic media. *ACS Catal.,* **2015**, *5*(9), 5388-5396. [http://dx.doi.org/10.1021/acscatal.5b00761]

[66] Witsuthammakul, A.; Sooknoi, T. Selective hydrodeoxygenation of bio-oil derived products: ketones to olefins. *Catal. Sci. Technol.,* **2015**, *5*(7), 3639-3648. [http://dx.doi.org/10.1039/C5CY00367A]

[67] Bejblová, M.; Zámostný, P.; Červený, L.; Čejka, J. Hydrodeoxygenation of benzophenone on Pd catalysts. *Appl. Catal. A Gen.,* **2005**, *296*(2), 169-175. [http://dx.doi.org/10.1016/j.apcata.2005.07.061]

[68] Bohre, A.; Dutta, S.; Saha, B.; Abu-Omar, M.M. Upgrading furfurals to drop-in biofuels: An overview. *ACS Sustain. Chem.& Eng.,* **2015**, *3*(7), 1263-1277. [http://dx.doi.org/10.1021/acssuschemeng.5b00271]

[69] Hu, X.; Westerhof, R.J.M.; Dong, D.; Wu, L.; Li, C.Z. Acid-catalyzed conversion of xylose in 20 solvents: insight into interactions of the solvents with xylose, furfural, and the acid catalyst. *ACS Sustain. Chem.& Eng.,* **2014**, *2*(11), 2562-2575. [http://dx.doi.org/10.1021/sc5004659]

[70] Bohre, A.; Saha, B.; Abu-Omar, M.M. Catalytic upgrading of 5-Hydroxymethylfurfural to Drop-in Biofuels by Solid Base and Bifunctional Metal–Acid Catalysts. *ChemSusChem,* **2015**, *8*(23), 4022-4029. [http://dx.doi.org/10.1002/cssc.201501136] [PMID: 26549016]

[71] Y. T. Cheng, J. Jae, J. Shi, W. Fan and G. W. Huber. *Angew. Chem. Int. Ed.,* **2012**, *51*, 1387.

[72] Q. Zhang, K. Qiu, B. Li, T. Jiang, X. H. Zhang, L. L. Ma and T. J. Wang. *Fuel,* **2011**, *90*, 3468.

[73] Q. Zhang, T. J. Wang, B. Li, T. Jiang, L. L. Ma, X. H. Zhang and Q. Y. Liu. *Appl. Energy,* **2012**, *97*, 509.

[74] G. W. Huber, S. Iborra and A. Corma. *Chem. Rev.,* **2006**, *106*, 4044.

[75] P. M. Mortensen, J. D. Knudsen and A. D. Jensen. *Appl. Catal. A Gen.,* **2011**, *407*, 1.

[76] J. A. Tavizón-Pozos, V. A. Suárez-Toriello, P. del Ángel and J. A. de los Reyes. *Int. J. Chem. React. Eng.,* **2016**, *14*, 1211.

[77] Guo J, Liu H, Li Y, Li D and He D. *Front Chem.,* **2023**, *11*, 1162183.

[78] Chu, W.; Xiao, X.; Chu, W.; Song, L.; Wang, N.; Chu, W. Engineering the quaternary hydrotalcite-derived ce-promoted ni-based catalysts for enhanced low-temperature co2 hydrogenation into methane. *Materials (Basel),* **2023**, *16*, 4642. [http://dx.doi.org/10.3390/ma16134642] [PMID: 37444955]

[79] Kartha, R.K.P.; Mukhopadhyay, B.; Field, R.A. Practical de-O-acylation reactions promoted by molecular sieves. *Carbohydr. Res.,* **2004**, *339*, 729. [http://dx.doi.org/10.1016/j.carres.2003.11.021] [PMID: 15013413]

[80] Costarrosa, L.; Ruiz-Martínez, J.; Rios, R.V.R.A.; Silvestre-Albero, J. Sepúlveda- Escribano, A. Basic zeolites as catalysts in the N-alkylation of imidazole: Activation by microwave irradiation. *Microporous Mesoporous Mater.,* **2009**, *120*, 115. [http://dx.doi.org/10.1016/j.micromeso.2008.07.046]

[81] Martins, L; Hölderich, W; Cardoso, DJ *Catal,* **2008**, *258*

[82] Sarmah, B; Satpati, B; Srivastava, R *J Colloid Interface Sci,* **2017**, *493*, 307.

[83] Rama, V; Kanagaraj, K; Subramanian, T; Pitchumani, K *Catal Commun,* **2012**, *26*, 39.

[84] Yan, Q; Hou, X; Liu, G; Li, Y; Zhu, T; Xin, Y *Hazard Mater,* **2020**, *400*, 123260.

[85] Hu, S-L; Liang, S; Mo, L-Z; Su, H-H; Zhang, P-J; Qin, J-N , **2023**, *32*, 119254.

[86] Zhang, M; Liu, Y; Liu, B; Chen, Z; Yan, K *ACS Catal,* **2020**, *10*, 5179-5189.

[87] Wang, T; Hu, A; Wang, H; Xia, Y *J Chin Chem Soc-TAIP,* **2019**, *66*, 1610.

[88] Yabushita, M; Shibayama, N; Fukuoka, A *ACS Catal,* **2019**, *9*, 2101.

[89] Wang, Q; Feng, J; Zheng, L; Bi, R; He, Y; Liu, H; Li, D *ACS Catal,* **2019**, *10*, 1353.

[90] An, Z.; Wang, W.; Dong, S.; He, J. Well-distributed cobalt-based catalysts derived from layered double hydroxides for efficient selective hydrogenation of 5-hydroxymethyfurfural to 2,5-methylfuran. *Catal. Today,* **2019**, *319*, 128-138. [http://dx.doi.org/10.1016/j.cattod.2018.03.052]

SUBJECT INDEX

A

B

Ravi Tomar, Vivek Mishra, Kapil Arya & Garima Pandey (Eds.)

E

F

G

H

U

W

X

www.ingramcontent.com/pod-product-compliance
Lightning Source LLC
LaVergne TN
LVHW070947210926
848482LV00002B/127